CONSERVATION'S ROOTS

The Environment in History: International Perspectives

Series Editors: Dolly Jørgensen, *University of Stavanger*; Kieko Matteson, *University of Hawai'i at Mānoa*; Christof Mauch, *LMU Munich*; Helmuth Trischler, *Deutsches Museum, Munich*

eseh european society for environmental history

Rachel Carson Center
ENVIRONMENT AND SOCIETY

Recent volumes:

Volume 19
Conservation's Roots: Managing for Sustainability in Preindustrial Europe, 1100–1800
Edited by Abigail P. Dowling and Richard Keyser

Volume 18
Colonial Seeds in African Soil: A Critical History of Forest Conservation in Sierra Leone
Paul Munro

Volume 17
Hazardous Chemicals: Agents of Risk and Change, 1800–2000
Edited by Ernst Homburg and Elisabeth Vaupel

Volume 16
Planning for the Planet: Environmental Expertise and the International Union for Conservation of Nature and Natural Resources, 1960–1980
Simone Schleper

Volume 15
Changes in the Air: Hurricanes in New Orleans from 1718 to the Present
Eleonora Rohland

Volume 14
Ice and Snow in the Cold War: Histories of Extreme Climatic Environments
Edited by Julia Herzberg, Christian Kehrt, and Franziska Torma

Volume 13
A Living Past: Environmental Histories of Modern Latin America
Edited by John Soluri, Claudia Leal, and José Augusto Pádua

Volume 12
Managing Northern Europe's Forests: Histories from the Age of Improvement to the Age of Ecology
Edited by K. Jan Oosthoek and Richard Hölzl

Volume 11
International Organizations and Environmental Protection: Conservation and Globalization in the Twentieth Century
Edited by Wolfram Kaiser and Jan-Henrik Meyer

Volume 10
In the Name of the Great Work: Stalin's Plan for the Transformation of Nature and Its Impact in Eastern Europe
Edited by Doubravka Olšáková

For a full volume listing, please see the series page on our website: http://berghahnbooks.com/series/environment-in-history

Conservation's Roots

Managing for Sustainability in Preindustrial Europe, 1100–1800

Edited by
Abigail P. Dowling and Richard Keyser

berghahn
NEW YORK • OXFORD
www.berghahnbooks.com

First published in 2020 by
Berghahn Books
www.berghahnbooks.com

Library of Congress Cataloging-in-Publication Data

Names: Dowling, Abigail Paige, editor.
Title: Conservation's roots : managing for sustainability in preindustrial Europe, 1100-1800 / edited by Abigail P. Dowling and Richard Keyser.
Description: First edition. | New York : Berghahn Books, 2020. | Series: Environment in history : international perspectives ; volume 19 | Includes bibliographical references and index.
Identifiers: LCCN 2020009806 (print) | LCCN 2020009807 (ebook) | ISBN 9781789206920 (hardback) | ISBN 9781789206937 (ebook)
Subjects: LCSH: Conservation of natural resources--Europe--History. | Human ecology--History. | Environmental management--Europe--History. | Sustainability--Europe--History.
Classification: LCC GF13.3.E85 C64 2020 (print) | LCC GF13.3.E85 (ebook) | DDC 333.72094--dc23
LC record available at https://lccn.loc.gov/2020009806
LC ebook record available at https://lccn.loc.gov/2020009807

British Library Cataloguing in Publication Data

A catalogue record for this book is available from the British Library

ISBN 978-1-78920-692-0 hardback
ISBN 978-1-78920-693-7 ebook

Contents

Illustrations

Figures

Maps

Tables

Acknowledgments

This project was conceived when the editors exchanged views about what they both saw as the relative lack of awareness about premodern or preindustrial environmental history in popular and academic circles. In fact, a similar lack of awareness is still often prevalent among those who specialize in the cultural, intellectual, and political history of early periods, even if such specialists may be assumed to have at least some interest in the environmental history of the times and places they study. We hope the present volume serves to make some aspects of preindustrial European environmental history more well-known across these and other scholarly communities and perhaps, too, among the wider public.

We wish to thank Matthew Irvin and the other organizers of the University of the South's Medieval Colloquium on "Medieval Natures" (Sewanee, TN, 1–2 April 2016), where the editors first met and launched the idea for this book. We also thank the organizers and panelists for our respective panels at Sewanee, and in particular Kellie Robertson, the convener of the seminar on "Defining Nature," where an animated discussion about Richard Keyser's paper on "Medieval Conservation" prompted Abigail P. Dowling to suggest that the topic merited not just an article, but a book. We also thank Scott Bruce, Anne Lester, and the other organizers of the University of Colorado at Boulder's Center for Medieval and Early Modern Studies Symposium on "Premodern Ecologies" (21–22 October 2016), where we further developed our ideas. At Boulder, we did so in collaboration with Richard C. Hoffmann, Eva Jakobsson, and Sara Morrison, all of whom helped us to begin sketching the outline and contents for *Conservation's Roots*.

We also owe thanks to all the contributors to the volume who signed on when the project was still just in a formative stage. We have learned an enormous amount not only from their essays, but also from their responses to our queries, as we became more aware of the variety of sources, historiographical traditions, and approaches to conserving or sustaining resources across preindustrial Europe. As we took on new ideas, we expanded the project beyond its initial medieval focus to embrace also the early modern period (ca. 1500–1800 CE), when scholars have traditionally identified important changes in how people conserved or sustained natural resources.

We owe more specific debts to three of our contributors. We would like to laud Michael Zeheter, whose careful comments on an early draft of the introduction have led, we hope, to a volume that represents a less U.S.-centric perspective on the history of conservation and sustainability. We would also like to thank Richard C. Hoffmann, a long-time leader of the field and mentor to those of us in North America working on premodern environmental history, for stepping in to write a chapter after the tragic loss of Alasdair Ross in 2017. We are also grateful to our afterword author, Paul Warde, for coming in at the eleventh hour to compose an excellent afterword.

In addition, we are very grateful to Berghahn editor and senior editor, Chris Chappell and Christof Mauch, for their early belief in the idea of a book on "conservation's roots." Further, we thank Chappell and editorial assistant Mykelin Higham for their patience and professionalism throughout the editing process. We are also indebted to our two anonymous peer reviewers for their careful and thoughtful comments on the first draft of the volume; their insights were invaluable and helped shape the final, coherent form of the volume.

Additionally, we would both like to thank our respective institutions for research support. In particular, Dowling would like to thank the library faculty at Tarver Library for their invaluable help with research and locating seemingly innumerable Interlibrary Loans, and Keyser would like to thank the University of Wisconsin-Madison's Cartography laboratory for making three of the book's maps.

Finally, none of this would be possible without the extensive support of our families. We thank our spouses, children, and extended families for their patience, love, and assistance as we shepherded this volume to fruition.

Europe. Map created by University of Wisconsin–Madison, Cartography Laboratory. Published with permission.

Introduction

Richard Keyser and Abigail P. Dowling

Conservation's Roots grew—pardon the pun—from the frustration that we, the editors, shared concerning the unawareness and simplistic narratives about preindustrial environmental history that circulate widely in popular and academic circles, despite the growing body of work in this field. To address at least some of this lack of awareness and oversimplification, *Conservation's Roots* presents twelve studies that explore how Europeans of premodern times, before about 1800 CE, managed the natural resources on which they depended.

Whereas earlier scholarly debates asked questions such as whether preindustrial peoples lived more in "harmony" with nature than we do, or whether they already exhibited domineering and destructive environmental tendencies, we show that while approaches varied, overall preindustrial management practices and polices reflect practicality and a desire to ensure the "sustainability" of natural resources. Further, we argue that using modern environmental concepts, such as conservation and sustainability, to write resource-focused history deepens our understanding of the interactions between humans and the natural world before the rise of modern industry. We contend that these concepts force us to reconsider the broader narratives of medieval and early modern European history and environmental history.

In this Introduction, we address some of the challenges that works on preindustrial environmental history must confront. We then explain the deeply rooted efforts to conserve and sustain such "uncultivated" or naturally regenerating resources as pastures, woodlands, wild game, fish, and fresh water itself. The Introduction also suggests that, while recent scholarship into sustainability and conservation within preindustrial Europe has focused mostly on ideas and cultural history, *Conservation's Roots* enables a shift in focus to the various ways in which preindustrial peoples interacted with their environments to try to guarantee themselves the resources they sought.

Impediments to Investigating Preindustrial Sustainability

There are at least three major challenges that any historical research, especially in environmental topics, must face in periods before roughly 1800 CE: the first is a question of our own attitudes toward the "deep" past of the preindustrial era; the second concerns available sources; and, finally, the third is the challenge of terminology.

Ideological Misconceptions about Preindustrial European Environmental History

One of the greatest obstacles to preindustrial environmental history is ideological. As Andrew Shryock and Daniel Smail have recently pointed out, over the last generation, there has been a tendency among historians and others who study the past from the perspectives of the humanities and social sciences to focus on the recent past. We believe the same is true of environmental history. For example, in a survey of the articles published in the last five years (2015–19) in the premier North American journal *Environmental History*, there are only five pieces dedicated to the pre-1800 period; and of those five, only one concerns pre-1600 Europe. Partially, this is a result of publication patterns; it is worth noting that there are few environmental history articles published in such top-tier general premodern journals as the *Journal of Medieval and Early Modern Studies*. The reality is this: many preindustrial environmental history articles are published in journals with narrower foci, such as *Landscape Studies*, *Agricultural History*, and *Water History*, as well as edited collections and monographs. The unfortunate result is that other environmental historians do not see preindustrial studies, and the cycle continues. In addition, key concepts that we use in this volume, such as "sustainability" and "conservation," are generally regarded as modern developments. As Shryock and Smail show, the focus on the more recent past narrows our historical perspective and, we believe, cripples our understanding of the whole human past, blocking from view patterns of change that have unfolded, as in the case of the human–nature interactions implicated in the management of resources, over long expanses of time.[1]

This emphasis on the more recent human past has, in turn, encouraged the persistence of a number of interrelated oversimplifications about past relations between humans and their environments. Such broad-brush generalizations are almost as numerous as the evidence to support them is weak. For example, some writers have suggested that long before the rise of industrialization, western cultures already evinced a domineering or "exploitative" attitude toward the natural world. In this view, the "seeds" of current environmental problems were sown as early as the Book of Genesis, which gave humans "dominion" over nature. Lynn White Jr., who proposed this dubious hypothesis in 1967, went on to argue that although western culture's environmentally destructive potential was implanted early on, it only gradually manifested, first becoming noticeable during Europe's medieval economic expansion, and then finally emerging more fully with the seventeenth-century Scientific Revolution and the eighteenth- and nineteenth-century Industrial Revolution.[2]

Subsequent scholars have largely abandoned White's "dominion" theory, recognizing instead that a multiplicity of modes of thinking about and interacting with nature can be found in premodern times. Still, a tendency persists to assume a progression among such modes of thinking and acting. David J. Herlihy suggested that an early medieval fear of nature as a hostile force gradually gave way by the Central Middle Ages (ca. 1000–1300 CE) to greater confidence in the human ability to shape nature positively, before the crises of the Late Middle Ages (1300–1500 CE) returned western cultures to a predominantly fearful mode.[3] Positing a prevailing mentality to entire historical periods oversimplifies preindustrial interactions with the natural world. Scholars such as Ellen Arnold and Paolo Squatriti have persuasively argued that early medieval peoples were not unequivocally afraid of nature; they could—and did—harness environment narratives for political gain.[4] Furthermore, as Richard Unger has noted, these "stage" theories leave in place the related assumption that White's thesis exemplified in that "some essential feature of Western thinking created the precondition for an assault on the natural world."[5]

In a similar vein, many scholars and members of the public continue to believe that in the distant past, before the emergence of modern science and industry, people coexisted peacefully with nature. Steeped in a strain of nostalgia endemic to environmental history, declensionist views of nature cast humans as agents of unrelenting, cumulative environmental destruction and past human interactions with nature as inherently less destructive and selfish. This "arcadian" view of nature, as Donald Worster called it, was popular in literature and art from the late eighteenth century on and was picked up by the early environmental movement of the mid-twentieth century.[6]

Declensionist narratives also lean heavily on modern definitions of environmental crisis, which, as Richard Hoffmann has pointed out, neither reflect the complexity of environmental change nor adequately account for the historical realities of premodern management and scarcity.[7] Although cataclysmic events grab and hold our attention, they may not reflect historical responses to change. As Paolo Squatriti has written, "some of the most celebrated monographs of environmental history adopt a chiding tone and use past cataclysms as the rods with which to chastise profligate and environmentally callous people today."[8] Further, such approaches set up an unproductive dichotomy that sees humans as either "Nature's rulers or victims."[9]

Closely related to declensionist narratives are romanticized views that a pristine form of nature or "wilderness" existed in the preindustrial past. The hope that we can stem the tide of ecological degradation by searching such a past for solutions is intimately tied to modern environmentalist movements.[10] These views are relevant to this volume because of an enduring popular belief that preindustrial peoples could not use "marginal" or "waste" landscapes effectively or even avoided them altogether, which has been thoroughly debunked.[11] For example, while beginning in the seventeenth century, reformers considered many heathlands as mismanaged, degraded environments, Maïka de Keyzer's chapter shows that these judgments often bore little relation to local realities.

Moreover, there is a link between views of preindustrial management practices as being either harmonious or insufficiently adaptive and colonialism. As Emmanuel Kreike explains, there is still a tendency to assume that "Western or 'modern' people" actively change nature, in contrast to "non-Western or 'premodern' peoples," who live within "the realm of Nature."[12] Such ideas enable views of premodern natural resource management as less complex, less exploitative, or less useful than modern, colonial, or capitalist modes of management. The contributors to this volume emphasize the practices, behaviors, and policies relating to environmental interaction in European populations themselves, thereby complementing scholarly attention to the environmental dimensions of European colonialism and overseas interactions.[13]

Sources: An Embarrassment of Riches

While scholars of preindustrial periods typically grapple with relatively limited sources, they are more abundant than is often assumed. Preindustrial Europe in the second millennium of the Common Era was not in any sense exceptional in the degree to which it valued or practiced the sustainable use of natural resources. Farmers the world over, for example, have long recognized and responded to the problem of maintaining soil fertility, whether by practicing shifting cultivation, crop-rotation, or manuring.[14] But in our period, roughly 1100–1800 CE, sustainable practices are particularly well documented—much more so than might appear from perusing the current literature of environmental studies.

Most contributors to *Conservation's Roots* find supporting evidence for these sustainability-oriented practices in the rapidly accumulating results of landscape archaeology, historical ecology, and other paleo-scientific approaches to past environments.[15] A few of the chapters incorporate paleo-scientific evidence more fully into their analyses, including Dolly Jørgensen's study of medieval pig-grazing and Péter Szabó's essay on traditional woodland management. Overall, however, our focus is on human artifacts, and especially written documents. The authors concentrate on documents that concern human uses of and impacts on nature, rather than on the literary, religious, or philosophical texts that illuminate cultural understandings of nature.[16]

In some regions, by the eighth century CE, surviving texts already include not only political and religious genres, like historical chronicles and saints' lives, but also many documents of practice, such as estate inventories, land deeds, and judicial decisions.[17] In most parts of Europe, however, it is not until after about 1000 CE that the sources for the study of human interactions with nature improve greatly, both in quantity and quality. Such documents include land transactions (gifts, sales, exchanges, mortgages, etc.), judicial records (dispute settlements, arbitrations, inquests, marriage contracts, wills, post-mortem inventories, etc.), business dealings (partnerships, insurance contracts, receipts of sale, bank deposits, etc.), account books, charters of liberties, and tax rolls.[18] Literacy expanded further from the fourteenth century, when paper began to replace more expensive parchment as a writing tool, and then even more with the advent of widespread printing from about 1500 CE.[19] All of these expansions in documentation and literacy over the last half-millennium or so of Europe's preindustrial era did not mean that most people could read and write well. Rather, these changes created conditions in which large numbers of people acquired at least the skills needed for "pragmatic literacy"—that is, the ability to understand and use written instruments well enough to help them in managing their affairs and in asserting their claims to property and resources.[20]

Contributors Abigail P. Dowling and Sébastien Poublanc both provide a numerical accounting of the size of their respective archives and demonstrate the volume and detail of information available in this period. Dowling's research is based on samples from thousands of account books and receipts that have survived from a period of just twenty-seven years in the fourteenth century from thirteen different administrative districts in northeastern France. Poublanc, who studies the royal program of reform that was carried out between 1665 and 1685 in the forests of the southern French region of Languedoc, draws on an even more impressive archive, which includes "twenty-seven linear meters of archived material, comprising 2,000 court cases, 1,000 forest maps, and information about 1,623 forests."

Given this kind of abundance, it is not surprising that this book's contributors each focus on just a few types of sources. Many of the authors of *Conservation's Roots* concentrate on the decrees, regulations, and legislation issued by villages, towns, and larger territorial authorities. Prescriptive in character, these sources express goals, ideals, and what people should do, rather than trying to record what people actually did. Such rules and commands nonetheless provide us with a snapshot of what those who claimed the authority to issue them were trying to do. Furthermore, in the centuries under study here, there was as yet very little in the way of learned traditions of resource management—at least for "uncultivated" resources that reproduced on their own, like most woods, pastures, wild game, and aquatic resources. Decrees, authoritative plans, and legislation on these matters often lacked, therefore, learned models to imitate. Instead, they tended to promote guidelines that leaned heavily on customary or community standards of best practice, and thus may be said to reflect local common practice. Perhaps the best examples of this dependence are the fishing regulations studied by Richard Hoffmann and Michael Zeheter: these rules were so technically detailed, varied, and subject to small adjustments over time that both authors argue persuasively that only the fishermen themselves could possibly have devised them.

Finally, we should note that in the agrarian world of preindustrial Europe, much of what was recorded in the above types of practical document inevitably concerns the usage of land and other natural resources. Scholars embarking on the exploration of these subjects do not have to start from scratch; they are able to build on long and continuing traditions of research on these sources by geographers, socioeconomic historians, and many others.[21] In addition to demonstrating the complexity and diversity of preindustrial European resource practice, we hope that this volume will help to make the potential of this embarrassment of riches for environmental history somewhat more familiar.

The Challenge of Terminology in Early Environmental History

Terminology is also a challenge for preindustrial environmental history, because there is no fully satisfactory set of general terms with which to describe how preindustrial people understood and interacted with the natural world. While we could use such labels as "proto-ecological" or "proto-environmental," one can call early ideas and efforts "environmental" only cautiously and as shorthand, given that this term had not yet acquired its modern meaning. Before the post-World War II rise of modern ecology and environmentalism, there was no unified and compelling way of thinking about the human relationship with the natural environment in its current, comprehensive sense.[22]

Similarly, use of the terms "conservation" and "sustainability" for the preindustrial period requires caution. One approach that many historians have taken, and which is adopted here, is to use both terms with their broad contemporary meanings. These strongly overlap, with both terms sharing the core meaning, "to maintain something (or someone) in good condition."[23] The two words also have, however, somewhat different implications. In current usage, "sustainability" is the more general term, referring to the capacity of humans to meet their needs over the long term, while maintaining the natural environment in as good condition as possible.[24] In contrast, "conservation" refers more specifically to "active management" to protect natural areas or to manage resources sustainably.[25]

Although currently these terms may be used in similar ways across the globe, before about 1980, usage varied a great deal nationally and even regionally. Thus, many words in the environmental historian's lexicon still bear the legacy of these diverse traditions. In the nineteenth and early twentieth centuries in many northern European countries, the earliest movements to protect aspects of the natural world, such as British "preservation" or German "nature protection" (*Naturschutz*), were primarily cultural, aesthetic, and patriotic in focus, and had little or nothing to do with sustaining resources for economic use over the long term.[26] In contrast, the US "conservation" movement was based squarely on what Germans and other Europeans called the "sustained yield" management of natural resources.[27] This potential for confusion is lessened, however, by this book's focus on the preindustrial period, before "nature protection" movements had emerged, and when instead an emphasis on something very much like "sustained yield" was widely shared.

Moreover, the terms "to sustain" and "to conserve" both have ancient origins in the Latin verbs *sustinere* and *conservare*,[28] and by the later Middle Ages, both these terms or their vernacular equivalents, at least in such Romance languages as French, could be used to refer to the management of natural resources.[29] Already by the late fifteenth and early sixteenth centuries, both the French and Spanish monarchies began issuing a series of edicts that mandated the "conservation" (French: *la conservation*; Spanish: *la conservación*) of forests, game, and aquatic resources.[30] By then, several other terms, many of them also derived from Latin,[31] were even more commonly used for the protection and careful management of natural resources. For example, as Sébastien Poublanc discusses in his chapter, the early modern French state was keen on the idea of setting aside, or "reserving" (from classical Latin *reservare*), areas for the growth of timber: *la réserve*—a term and idea that had direct parallels in Venetia, in northern Italy.[32] Further, as Sara Morrison shows in her chapter on Sherwood Forest, forms of the term "to preserve," deriving from the medieval Latin *preservare*, were commonly used in early modern English laws that targeted the protection of woodlands. Finally, as several of this book's chapters show, the protean term "forest" (medieval Latin *forestis*) was and still is found across Europe. Throughout the period covered by this book, many areas called "forests" continued to be managed under the special, natural-resources-oriented jurisdiction that rulers and lords claimed based on the medieval tradition of forest law.[33]

Of course, the early use of terms like "conservation" or "forest" should not be assumed to suggest the existence of a broader social and intellectual campaign of conservation or sustainable management analogous to those that began to develop from the mid-eighteenth and nineteenth centuries.[34] Rather, we suggest that using these terms in their broad senses is appropriate and helpful in highlighting conservative resource practices in preindustrial eras. Further, by using a shared set of terms, if not the ideological background to the concepts, we hope to better engage with other environmental scholars.

The Deep Roots of Conservation

> Having been informed about the thefts and abuses committed in the waters and forests of our realm . . . which things constitute a great detriment and diminution of the public good . . . we have ordered a search of our accounting office for the ancient edicts concerning the reform, upkeep, and conservation of our waters and forests and also in the matter of hunting . . . and our counselors have found them to be very useful for the conservation of our forests. (Edict of King Francis I of France, March, 1515)[35]

The quotation above is suggestive of the long tradition of environmental planning and regulation by lordships, villages, principalities, and states, which can be documented in detail as far back as the Central Middle Ages. One of this book's aims is to show that King Francis was correct to assert that such early efforts were indeed "useful for the conservation" of a wide range of natural resources, including woodlands, pastures, fresh water, fish, and wild game. We argue that the agreements, plans, orders, and regulations that record these efforts represent some of the earliest and most important sources for the practices that today we might refer to under such labels as "conservation" or "sustainable management."

The basic ideas underlying this argument are not new, even if claims similar to ours have traditionally focused only on some types of resources and have often been limited to one or a few specific regions or nations. But already in 1864, George P. Marsh noted that, at least for forests, efforts to preserve natural resources were widespread and had begun quite early: "The severe and even sanguinary legislation, by which some of the governments of mediaeval Europe . . . protected the woods, was dictated by a love of the chase, or the fear of a scarcity of fuel and timber. The laws of almost every European state more or less adequately secure the permanence of the forest."[36] Modern scholars have often been hesitant, however, to accept these claims at face value. There is in fact some justification for this hesitation, especially when it comes to the ideas and intentions underlying these measures. For example, Clarence Glacken memorably suggested that these early regulations should not be understood even as "unwitting conservation."[37] It is certainly true that early efforts to conserve resources usually had different goals from those of the more well-known projects of conservation or sustainable management that began to develop from about 1700 CE, when concerns grew that certain resources, especially woodlands, might be depleted on regional or national scales. These fears gradually led to a greater focus on the preservation of a particular resource in and of itself, which became a salient feature of the nineteenth-century movements of "conservation" and "sustained yield" management. Instead, most earlier conservation efforts were framed more narrowly. Typically lacking a broad approach to entire territories on regional or national scales, they aimed both to affirm the authority of lords, communities, and rulers and to safeguard the specific rights of those who used certain resources. Early plans and regulations often focused more on the rights of particular parties than on resources or territories.

Yet before dismissing these early measures too quickly on these grounds, it should be noted that they had implications and effects that frequently went beyond the protection of rights in the narrowest sense, in at least four ways. First, it is vital to remember that most preindustrial agrarian societies were, as Patricia Crone puts it, "dominated by scarcity." Scarcity was a "hallmark" of such societies less because of any absolute, ecological limits on available resources than because of the typically low levels of productivity and very high degrees of socioeconomic inequality. Beggars and the truly indigent poor could easily account for up to 10 percent of the population, while up to half or more of the people in these societies often struggled, even in good years, to provide modest levels of food, fuel, clothing, and housing.[38] Yet outside infrequent but unfortunately recurrent episodes of famine, war, and epidemic disease, scarcity was experienced by most people as a routine but manageable fact of life, and mostly at the scales of the household, village, or local district. Both Maïka de Keyzer's and Oliver Auge's chapters provide particularly compelling examples of how people coped over the long term, often with at least some success, with acute scarcities of such resources as fertile soils and fuel. Scarcity also forced a long-term approach to the maintenance of natural resources even for elites. Abigail P. Dowling's chapter follows the ways in which elite estates, monuments to excess, also required careful management to assure the continuous supply of quotidian and luxury goods that supported aristocratic culture.

A second reason not to dismiss early conservation efforts is the fact that access to most resources was subdivided among many parties and concerned multiple, overlapping types of usage. As a practical matter, the safeguarding of rights often amounted, therefore, to the preservation of the resource itself. Even outside of lands managed explicitly as "commons," a topic to which we shall return below, villagers often enjoyed some usage rights in the woodlands, pastures, and waters that lords claimed as parts of their domains. The chapters by Dolly Jørgensen and Richard Keyser document this type of layering of rights and usage early on in the medieval forests of Normandy and England. Péter Szabó's chapter reveals that usage rights related to such traditional practices as coppicing and wood-pasture became so deeply ingrained that Czech woodlands actually evolved to fit them. Although these practices and the multiple-use approach fell out of favor under the influence of scientific forestry in the twentieth century, recent conservationists have begun efforts to restore them within what is now understood to represent a long process of coevolution.

Third, at least by the Late Middle Ages (ca. 1300–1500), many regulations to conserve resources began to be framed more broadly so that they were gradually applied to more and more extensive territories, whether at the level of the region, principality, or kingdom. Thus, in her chapter, Eva Jakobsson traces legislation to maintain the free-flowing character of Swedish rivers to the oldest surviving book in Swedish, the thirteenth-century *Regional Law of Västergötland*, which, as its names suggests, records the legal customs that in theory applied across this large region in south-central Sweden. By the seventeenth century, Swedish monarchs decreed that a similar principle should be applied to all of Sweden's rivers. Similarly, Cristina Arrigoni Martelli details a centralizing trend in hunting regulations, which by the fifteenth century resulted in the replacement of dozens of local, municipal statutes by those of such regional powers as the duchies of Lombardy and Tuscany.

Finally, during this same period, justifications for regulating natural resources frequently went beyond narrow protections for the rights of specific parties to make claims about serving the "public good," as seen, for example, in this section's opening quotation. Beginning as early as the thirteenth century, this newly assertive approach to rule-making seems to have first targeted such problems as local flood control, urban sanitation, and, as Richard Hoffmann's chapter shows in abundant detail, the safeguarding of fish stocks in "public" rivers and lakes.[39] Of course, this kind of ideological packaging, which clearly has political motivations, is by no means an essential criterion for sustainable management. Nonetheless, its prevalence suggests that interest in preserving resources was widespread.

Environmental historians have overlooked or tended to downplay the importance of much of the early evidence for conservation presented in this book. The problem with minimizing the central- and late-medieval phases of the development of sustainable management and resource conservation is not merely a matter of dating these phenomena. A more accurate chronology also affords us a better understanding of what motivated preindustrial conservation and the kinds of contexts in which it emerged. This chronology reveals that scholars have told only part of the story by focusing heavily, sometimes almost exclusively, on the familiar and eye-catching demands of early modern and nineteenth-century industries, metropolitan cities, powerful nation–states, early scientific investigations, and colonial enterprises. The evidence presented in this book shows that resource-conserving and sustainability-oriented plans and practices were much more deeply rooted and had more diffuse origins than these standard accounts would suggest.

A fuller history of preindustrial conservation and sustainable management must recognize the central importance of the needs of villages and relatively small urban centers for pasture, game, small wood, lumber, fish, and fresh water. Doing so would shift attention away from the warnings of scarcity and plans for action that elite administrators working near the centers of power began to issue from the late seventeenth century onward. This book argues instead that early conservation's most typical contexts were local lordships, various kinds of commons, and small principalities and emerging states. This is a crucial fact to grasp: preindustrial conservation and sustainable management emerged first and were most securely anchored at the local level. Although great lords and rulers and their educated administrators played important roles in expanding and enforcing the necessary rules, the customs and techniques on which the rules were based derived from local knowledge and were implemented primarily by villagers, woodmongers, fishermen, merchants, and other nonelites. They worked to conserve and sustain resources within a broadly shared set of customary assumptions, which dictated that careful management was necessary in order to provide for human needs.

Preindustrial Conservation and Sustainable Management: More Than Ideas

Although some recent scholars have shown interest in the prehistory of the eighteenth- and nineteenth-century beginnings of modern conservation and sustainable management, so far, most research of this kind focusing on developments within Europe has concentrated on the cultural and intellectual history of the early modern period (ca. 1500–1800). This work offers important insights for understanding how natural resources were managed in the preindustrial period, even if it focuses less than we do on the kinds of evidence presented in this book.

The most prominent hypothesis for the origins of modern ideas about conservation or sustainability within preindustrial Europe traces these ideas to early modern forest administrations, and thus may for convenience be referred to as the "state forestry" model.[40] By the time modern environmentalism began to emerge in the 1960s, some scholars traced the precursors of modern conservation or sustainable yield management to the later 1600s and early 1700s, when a few leading administrators and writers addressed fears of shortages of wood or timber on large, often national scales, and promoted equally large-scale solutions. The key players in this story are a trio of men who were close to the centers of power in France, England, and the German principality of Saxony: Jean-Baptiste Colbert (1619–1683), John Evelyn (1620–1706), and Hans Carl von Carlowitz (1645–1714). All three became concerned that wood and timber shortages would undermine, respectively, French and English naval power or the business of mining in Saxony. Through their policies or books—including Colbert's reform of royal forest administration (1661–85) and his authorship of Louis XIV's Forest Ordinance of 1669, Evelyn's *Sylva, or a Discourse of Forest Trees and the Propagation of Timber in his Majesties Dominions* (1664), and Carlowitz's *Sylvicultura Oeconomica* (1713)—all three promoted as solutions to the problem of timber shortages the planting, preserving, and sustainable harvesting of trees.[41]

Yet far from being isolated geniuses that developed these ideas on their own, these men should rather be seen as important links in a long process of development that not only continued after them, but that also reached back centuries earlier, as three of this book's chapters help us to understand. To begin with Colbert, it is apparent that his

"Great Reformation," as it became known, was largely traditional in its concrete silvicultural guidelines; it innovated primarily in its centralizing vision and will to enforce.[42] Yet as Sébastien Poublanc shows in masterful detail, even the centralization for which these reforms are chiefly remembered was at first, in the late seventeenth century, often an illusion. Rather than being able to impose their own vision in southern France's Languedoc region, where local traditions of woodland management were thriving, royal foresters had to engage in the kinds of long, drawn-out negotiations and compromises that were typical of early modern governance.

Similarly, Sara Morrison's study of Sherwood Forest in the seventeenth century reveals that John Evelyn's famous call for the widespread planting of timber for naval needs had very little immediate impact.[43] She shows that Sherwood's diverse woodlands were already managed in highly sustainable, mostly traditional ways, and that where new experiments in "plantation forestry" were carried out, such as in the Forest of Dean, they ran into difficulties and for the next century did not make any significant contribution to naval provisioning.

Although in his book of 1713 Carlowitz goes beyond his French and English predecessors by promoting more systematic and explicit plans for ensuring the sustainability of timber and other woodland resources,[44] Oliver Auge's chapter shows that already by the sixteenth century, the need to sustain woodlands for the sake of posterity was becoming a prominent motif for some German writers. Moreover, his research demonstrates that in the principality of Schleswig-Holstein, people confronted a scarcity of woodlands through a wide range of adaptive responses, including the production of firewood through coppicing, the use of alternative energy sources like peat and cow dung, and regulations requiring replanting whenever trees were cut down. In sum, Colbert, Evelyn, and Carlowitz are notable primarily because they made already-traditional fears about shortages of woodland resources more well-known, and they framed both the problems they discussed and the solutions they proposed on large, state or national scales. These authors built squarely on traditions in woodland management that reached back to the Middle Ages.[45]

Furthermore, even though these three men and even earlier figures had become concerned about using up woodland resources, these worries lacked crucial elements of modern forms of "sustainability thinking" that we now take for granted. As Paul Warde has recently argued in the *Invention of Sustainability*, the modern idea that "whole societies" risk "undermining [their] environmental foundations," which would not be described with the term "sustainability" until the 1970s, coalesced only in the nineteenth century. The emergence of this idea depended not only on awareness of possible scarcities on a wide scale, but also on several other gradual shifts in thinking about agronomy, chemistry, demography, and state-level economic growth, as well as a new interest in and capacity to calculate total stocks of resources. Perhaps most crucially, agronomists and others became more aware of the chemical mechanisms behind soil fertility and its loss.[46]

Thus, one must avoid too easily assimilating to our modern concept of sustainability the virtually omnipresent efforts that preindustrial people made to keep a farm, for example, going. As far as we can tell, the basic motives for sustaining resources were not ecological, but rather social and economic.[47] Although many aspects of agrarian life during the long centuries of the preindustrial era exhibited a "tendency towards sustainability," it seems clear that, as Warde puts it, the "operative rules" of agrarian life "regulated the impact of one person's actions on others, and were not concerned with ecological sustainability itself." Instead, these norms "were preoccupied with neighborly relations rather than ecological management." They were part of an array of "'homeostatic' forces [that] exerted a balancing force on preindustrial societies." Their ultimate goal was predictability, which it was hoped would help protect the interests of the farming household against "the caprice of one's neighbors," and this across the agrarian landscape.[48]

Although we would agree with these insights concerning most preindustrial actors' conscious intentions and abstract ideas, in this book, we instead emphasize the aggregate effects that many traditional limitations on practice and regulations could have. These effects were especially salient in contexts such as "commons," where resources were held as common or shared property. In the 1980s, Elinor Ostrom and other social scientists developed an influential theoretical model of "common property" regimes.[49] Also known simply as "commons," what Ostrom refers to as "common property" or "common pool" resource regimes should not be confused with conditions of "open access," which allow anyone to use a resource. Garrett Hardin made precisely this mistake in his famous essay on the "Tragedy of the Commons," when he explained why a hypothetical sheep pasture that was held in common would inevitably be overgrazed.[50] In contrast to resources left open to use by all comers, in a regime of "common property," only those with vested rights, such as the members of a village community, have access.

These stakeholders were, collectively, both the owners and users of such shared resources, and thus had a strong interest in their sustainable management over the long term.

Furthermore, even if such resources as local waters and nonarable lands like pastures and woods were parts of the larger agro-ecosystem,[51] the evidence suggests that the prevailing agrarian outlook was applied somewhat differently in these less intensively worked areas. Preindustrial people often distinguished between "cultivated" and "uncultivated" lands, a contrast founded on the different levels of labor applied to each, by opposing cultivated fields (*ager*, *terra arabilis*) and "uncultivated" or "waste" spaces (*saltus*, *terra inculta*, *vastum*). These latter, "uncultivated" waters and lands were also those that were most often shared among multiple users and subject to many different modes of exploitation. It was precisely in these contexts that early efforts to preserve and protect resources were expressed most clearly, especially when such spaces were governed as commons.

Several of this book's chapters contribute to the ongoing reappraisal of Europe's historical commons, which across the last several centuries of the preindustrial period often remained both widespread and resilient.[52] Three chapters, those by Maïka de Keyzer, Michael Zeheter, and Eva Jakobsson, focus the most on commons' governance—in very different contexts. De Keyzer studies how villagers in sandy heathlands maintained their extensive common lands, whose fragility they well understood, from the Late Middle Ages right down to the nineteenth century. Zeheter examines commons that developed in an at least equally challenging context: the Lake Constance fisheries. Governing aquatic resources over an expanse of more than 500 square km of water might seem nearly impossible in preindustrial circumstances, especially given that from the fourteenth to the nineteenth century the lake was subject to a shifting variety of rival political authorities, including city-states, the Habsburg Holy Roman Emperors, and the Swiss Confederacy. Yet all these authorities deferred to the lake's professional fishermen and their regulations, thereby avoiding any apparent tragedy for almost five hundred years. Jakobsson's chapter tackles a unique commons: the free-flowing character of rivers. By defending the customary principle that one-third of a river's flow must be held open, undiminished and unobstructed, preindustrial Swedes balanced the competing demands of fishers, mill-owners, and farmers.

These examples may suffice to show that the analysis of the *idea* of sustainability in the preindustrial world by no means exhausts the interest of this topic for historians. Out of an abundance of caution, one might avoid the use of modern terms like "sustainability" and "conservation" altogether, for example by defining our topic as that of "homeostatic forces." But such a change in terminology would accomplish little. Instead, the contributors to this volume use "sustainability" and "conservation" advisedly, without assuming all of their modern meanings.

The Plan of *Conservation's Roots*

The goal of this volume is to bring together case studies from all over preindustrial Europe to deepen our understanding of resource management itself and the ways in which it reflected and shaped larger sociocultural developments. A few shared premises set this volume apart from similar collections. The first of these is the commitment, as explained above, to applying the perspectives and methods of environmental history to the centuries prior to the massive transitions associated with the industrial revolution that began in Britain and Europe in the late eighteenth and nineteenth centuries. Although there is now a growing body of scholarship on the environmental history of preindustrial societies, much of this research has focused on the impact of Europe's overseas colonialism,[53] the implications of landscape manipulation for elite political power,[54] or, as just explained, the origins of our modern attitudes toward nature.

Second, this volume is distinctive because of its broad chronology and Europe-wide scope. In the last two decades, scholars have published geographically targeted or resource-specific studies; however, this volume represents one of the first collective scholarly efforts to explore the concrete, practical methods and approaches that preindustrial populations in Europe adopted to manage such naturally regenerating resources as pastures, wild game, woodlands, fish, and flowing fresh water, and situate them within their historical contexts. Finally, all of the contributors to this volume tackle matters of resource management while asking one of the central questions posed by modern environmental historians: Just how "wisely," "conservatively," or "sustainably" did premodern Europeans manage their resources?

Richard Keyser is Senior Lecturer in Legal Studies in the Center for Law, Society, and Justice at the University of Wisconsin-Madison, where he teaches classes on American legal history, law and environment, and related topics. Trained in European history (PhD Johns Hopkins University, 2001), he has published articles on the legal, religious, and environmental history of medieval France and England. His 2009 essay on "The Transformation of Traditional Woodland Management" won the American Society for Environmental History's Alice Hamilton prize.

Abigail P. Dowling is Assistant Professor of History at Mercer University in Macon, Georgia, US. She received her PhD in Medieval Studies and History at the University of California, Santa Barbara (2014). She teaches classes on environmental history and material culture in Greece, Rome, and the Middle Ages. She has published articles on ecology in classical and medieval farming manuals and the socio-political symbolism of resource management in medieval parks. She is currently working on a book on resource management and transport in early fourteenth-century Artois and an article on humorism in the political rhetoric of the post-Roman western kingdoms.

Notes

1. Andrew Shryock and Daniel Lord Smail, eds., *Deep History: The Architecture of Past and Present* (Berkeley: University of California Press, 2011), 3–20.
2. Lynn T. White Jr., "The Historical Roots of Our Ecologic Crisis," *Science* 155 (1967): 1203–7.
3. David J. Herlihy, "Attitudes toward the Environment in Medieval Society," in *Historical Ecology: Essays on Environment and Social Change*, ed. Lester J. Bilsky (London: Kennikat Press, 1980), 101–15; John Aberth, *An Environmental History of the Middle Ages: The Crucible of Nature* (New York: Routledge, 2013), 5–9; Richard Hoffmann, *An Environmental History of Medieval Europe* (New York: Cambridge University Press, 2014), 94–112.
4. Ellen F. Arnold, *Negotiating the Landscape: Environment and Monastic Identity in the Medieval Ardennes* (Philadelphia: University of Pennsylvania Press, 2013); Paolo Squatriti, "The Floods of 589 and Climate Change at the Beginning of the Middle Ages: An Italian Microhistory," *Speculum* 85 (2010): 799–826.
5. Richard Unger, "Introduction: Hoffmann in the Historiography of Environmental History," in *Ecologies and Economies in Medieval and Early Modern Europe: Studies in Environmental History for Richard C. Hoffmann*, ed. Scott G. Bruce (Boston: Brill, 2010), 15.

6. Donald Worster, *Nature's Economy: History of Ecological Ideas*, 2nd ed. (New York: Cambridge University Press, 1994), 3–25; Paul Warde, "The Environmental History of Pre-industrial Agriculture in Europe," in *Nature's End: History and the Environment*, ed. Sverker Sörlin and Paul Warde (New York: Palgrave Macmillan, 2009), 71.
7. Richard C. Hoffmann, Nancy Langston, James C. McCann, Peter C. Perdue, and Lise Sedrez, "AHR Conversation: Environmental Historians and Environmental Crisis," *The American Historical Review* 113, no. 5 (2008): 1431–65.
8. Paolo Squatriti, "Introduction: Natures Past and Present Environmental Histories," in *Natures Past: The Environment and Human History*, ed. Paolo Squatriti (Ann Arbor: The University of Michigan Press, 2006), 4.
9. Emmanuel Kreike, *Environmental Infrastructure in African History* (New York: Cambridge University Press, 2013), 1.
10. William Cronon, "The Trouble with Wilderness; or, Getting Back to the Wrong Nature," in *Uncommon Ground: Rethinking the Human Place in Nature*, ed. William Cronon (New York: W. W. Norton & Co, 1995), 69–90.
11. Petra J. E. M. Van Dam, "Sinking Peat Bogs: Environmental Change in Holland, 1350–1550," *Environmental History* 6, no. 1 (2001): 32–46; Karl Jacoby, *Crimes Against Nature: Squatters, Poachers, Thieves and the Hidden History of American Conservation*, rev. ed. (Berkeley: University of California Press, 2014).
12. Kreike, *Environmental Infrastructure*, 2; Philip J. Deloria, *Playing Indian* (New Haven: Yale University Press, 1998), 154–80.
13. For example, see Alfred Crosby, *Ecological Imperialism: The Biological Expansion of Europe, 900–1900*, 2nd ed. (New York: Cambridge University Press, 2004); Richard Grove, *Green Imperialism: Colonial Expansion, Tropical Island Edens, and the Origins of Environmentalism, 1600–1860* (New York: Cambridge University Press, 1995); and John Richards, *The Unending Frontier: An Environmental History of the Early Modern World* (Berkeley: University of California Press, 2003).
14. Joachim Radkau, *Nature and Power: A Global History of the Environment*, trans. Thomas Dunlap (New York: Cambridge University Press, 2008), 11–19, 36–45, 55–85.
15. For example, see Naomi J. Sykes, "Hunting for the Anglo-Normans: Zooarchaeological Evidence for Medieval Identity," in *Just Skin and Bones? New Perspectives on Human-Animal Relations in the Historical Past*, ed. Aleksander Pluskowski (Oxford: BAR International Series, 2005), 73–80; Roberta Gilchrist, *Gender and Archaeology: Contesting the Past* (New York: Routledge, 1999); and Péter Szabó, "Historical Ecology: Past, Present, and Future," *Biological Reviews. Cambridge Philosophical Society* 90 (2015): 997–1014.
16. For the classic distinction between these three approaches to environmental history (ideas about nature, human uses of nature, and human impacts on

nature), see Donald Worster, "Appendix: Doing Environmental History," in *The Ends of the Earth: Perspectives on Modern Environmental History*, ed. Worster (New York: Cambridge University Press, 1988), 289–307.

17. Warren C. Brown, et al., eds., *Documentary Culture and the Laity in the Early Middle Ages* (New York: Cambridge University Press, 2013).
18. Joel T. Rosenthal, ed., *Understanding Medieval Primary Sources* (New York: Routledge, 2012).
19. Robert A. Houston, *Literacy in Early Modern Europe: Its Growth, Uses and Impact, 1500–1800*, 2nd ed. (New York: Routledge, 2002), 141–71.
20. Michael T. Clanchy, *From Memory to Written Record: England, 1066–1307*, 3rd ed. (Malden, MA: Wiley-Blackwell, 2013); Richard Britnell, ed., *Pragmatic Literacy East and West, 1200–1330* (Woodbridge: Boydell Press, 1997); Mathias Schmoeckel and Werner Schubert, eds., *Handbuch zur Geschichte des Notariats der europäischen Traditionen* (Baden-Baden: Nomos, 2009).
21. Georges Duby, *Rural Economy and Country Life in the Medieval West*, trans. Cynthia Postan (Columbus: University of South Carolina Press, 1968; repr. Philadelphia: University of Pennsylvania Press, 1998); Fernand Braudel, *Civilization and Capitalism, 15th–18th Century*, vol. 1: *The Structures of Everyday Life*, trans. Siân Reynold (New York: Perennial Library, 1985); Carlo M. Cipolla, *Before the Industrial Revolution: European Society and Economy, 1000–1700*, 3rd ed. (New York: W. W. Norton & Co., 1994); and Bruce Campbell, *English Seigniorial Agriculture, 1250–1450* (New York: Cambridge University Press, 2000).
22. Worster, *Nature's Economy*, 342–87; Radkau, *Nature and Power*, 250–60; and Paul Warde, Libby Robin, and Sverker Sörlin, *The Environment: A History of the Idea* (Baltimore: Johns Hopkins University Press, 2018).
23. *Oxford English Dictionary Online* (New York: Oxford University Press, 2018), http://www.oed.com: "Conserve, v.;" "conservation, n.;" "sustain, v.;" and "sustainable, adj."
24. Ulrich Grober, *Sustainability: A Cultural History* (Totnes, UK: Green Books, 2012), 155, 174, 181–86; and Jeremy Caradonna, *Sustainability: A History* (New York: Oxford University Press, 2014).
25. "Conservation," UN World Conservation Monitoring Centre, http://www.biodiversitya-z.org/content/conservation; and, in Europe, see the Bern Convention on the Conservation of European Wildlife and Natural Habitats, in force since 1982: "Bern Convention," Council of Europe, https://www.coe.int/en/web/bern-convention/.
26. James Winter, *Secure from Rash Assault: Sustaining the Victorian Environment* (Berkeley: University of California Press, 1999); Thomas Lekan, *Imagining the Nation in Nature: Landscape Preservation and German Identity, 1885–1945* (Cambridge, MA: Harvard University Press, 2004); Frank Uekoetter, *The Green and the Brown: A History of Conservation in Nazi Germany* (New York: Cambridge University Press, 2006), xiii–vx, 17–21.

27. Henry E. Lowood, "The Calculating Forester: Quantification, Cameral Science, and the Emergence of Scientific Forestry Management," in *The Quantifying Spirit in the Eighteenth Century*, ed. Tore Frängsmyr, H. L. Heilbron, and Robin E. Rider (Berkeley: University of California Press, 1990), 337–42; Grober, *Sustainability*; and compare: Samuel P. Hays, *Conservation and the Gospel of Efficiency: The Progressive Conservation Movement 1890–1920* (Cambridge, MA: Harvard University Press, 1959), 28–33.
28. *Oxford Latin Dictionary*, 2nd ed., 2 vols., ed. P. G. W. Glare (New York: Oxford University Press, 2012). See Grober, *Sustainability*, 19, 37–43, 46, 51–54, who does not examine early environmental practices.
29. For the French verbs *soutenir* and *conserver* and the noun *la conservation*, see *Dictionnaire Historique de la Langue Française*, 2 vols., ed. Alain Rey (Paris: Le Robert, 1992).
30. For a French example, see the next section here below. For the Spanish case, see John T. Wing, *Roots of Empire: Forests and State Power in Early Modern Spain, c. 1500–1750* (Boston: Brill, 2015), 2–3, 19–29, 38, 44, 54–56, 68, etc.
31. For all of the Latin terms cited in this passage and below, see Jan F. Niermeyer and C. Van de Kieft, eds., *Mediae Latinitatis Lexicon Minus*, rev. ed., 2 vols., ed. Johannes W. J. Burgers (Boston: Brill, 2002).
32. Karl Appuhn, *Forest on the Sea: Environmental Expertise in Renaissance Venice* (Baltimore: Johns Hopkins University Press, 2009), 94–111.
33. See the chapter by Dolly Jørgensen and all five chapters of the book's third section on woodlands.
34. Such campaigns had clearly begun by the mid-eighteenth century; see Jeremy Caradonna, "Conservationism Avant la Lettre? Public Essay Competitions on Forestry and Deforestation in Eighteenth-Century France," in *Invaluable Trees: Cultures of Nature, 1660–1830*, ed. Laura Auricchio, Elizabeth H. Cook, and Giulia Pacini (Oxford: Voltaire Foundation, 2012), 39–54.
35. François-André Isambert et al., eds., *Recueil général des anciennes lois françaises depuis l'an 420 jusqu'à la Révolution de 1789*, 29 vols. (Paris: Belin-Leprieur, Plon, 1826–1833), vol. 12: 49–74: "*Nous deuement advertis et informez des pilleries, larrecins, et abus qui se font aux eaues et forests de nostre royaume, les quelles choses . . . reviennent au grand détriment et diminution du bien de la chose publique. . . . À cette cause . . . avons envoyé quérir en nostre chambre des comptes les anciennes ordonnances faictes sur la réformation, entretenement et conservation de nos eaues et forests et aussi sur le faict des chasses, lesquelles . . . les gens de nostre conseil . . . nous ont rapporté icelles ester très utiles et proffitables pour la conservation desdites forests.*"
36. George P. Marsh, *Man and Nature, or Physical Geography as Modified by Human Action* (New York: Charles Scribner and Co., 1864), 279. More recently, see Charles R. Young, "Conservation Policies in the Royal Forests of Medieval England," *Albion* 10, no. 2 (1978): 95–103; and James Latham, "The Development of Forest Conservation in Europe," in *Europe's Changing Woods*

and Forests: From Wildwood to Managed Landscapes, ed. Keith J. Kirby and Charles Watkins (Boston: CAB International, 2015), 309–25.

37. Clarence Glacken, *Traces on the Rhodian Shore: Nature and Culture in Western Thought from Ancient Times to the End of the Eighteenth Century* (Berkeley: University of California Press, 1967), 318–41, quoted at 326; cf. 485–97.
38. Patricia Crone, *Pre-industrial Societies: Anatomy of the Pre-Modern World*, 2nd ed. (London: Oneworld Publications, 2013), 16, 204, and passim; Cipolla, *Before the Industrial Revolution*, 7–22; Astrid Kander, Paolo Malanima, and Paul Warde, *Power to the People: Energy in Europe Over the Last Five Centuries* (Princeton: Princeton University Press, 2013), 14, 37–43, 90–103.
39. See also Dolly Jørgensen, "'All Good Rule of the Citee': Sanitation and Civic Government in England, 1400–1600," *Journal of Urban History* 36, no. 3 (2010): 300–15; and Hoffmann, *Environmental History*, 263–78.
40. We omit here another important hypothesis about the source of these ideas, which may be called the "colonial" model, and which was pioneered by Grove, *Green Imperialism*. As Grove and others have shown, ideas deriving from colonial contexts had negligible impact on sustainable practices within Europe before ca. 1750.
41. Glacken, *Traces on the Rhodian Shore*, 485–97; Keith Thomas, *Man and the Natural World: Changing Attitudes in England, 1500–1800* (New York: Penguin, 1983), 197–202; Simon Schama, *Landscape and Memory* (New York: Knopf, 1995), 142–84; Grober, *Sustainability*, 63–85; Caradonna, *Sustainability*, 32–54; and Kieko Matteson, *Forests in Revolutionary France: Conservation, Community, and Conflict, 1669–1848* (New York: Cambridge University Press, 2015), 34–38, 50–68.
42. Andrée Corvol, *L'homme et l'arbre sous l'Ancien Régime* (Paris: Economica, 1984), 187–264; *Les Eaux et Forêts du 12e au 20e siècle* (Paris: Centre National de la Recherche Scientifique, 1987), 131–65, 434, and 593 (showing traditional views); Jérôme Buridant, *Espaces forestiers et industrie verrière, XVIIe–XIXe siècle* (Paris: L'Harmattan, 2005), 291–364; and Martine Chalvet, *Une histoire de la forêt* (Paris: Éditions du Seuil, 2011), 162–63.
43. John Evelyn, *Sylva: Or, a Discourse on Forest Trees and the Propagation of Timber in his Majesty's Dominions*, 4th ed., 2 vols., ed. John Nisbet (London: A. Doubleday & Co., 1908). On Evelyn's intellectual influence, see Schama, *Landscape and Memory*, 159–74; Paul Warde, *The Invention of Sustainability: Nature and Destiny, c. 1500–1870* (New York: Cambridge University Press, 2018), 61–64, 74–75.
44. Hans Carl von Carlowitz, *Sylvicultura oeconomica* (Leipzig: Johann F. Braun, 1713), repr. and ed. Harald Thomasius and Bernd Bendix (Remagen-Oberwinter: Verlag Kessel, 2013). See Grober, *Sustainability*, 80–88; Warde, *Invention of Sustainability*, 166–71.
45. Warde, *Invention of Sustainability*, 58–101. Recent studies of state forestry dating from the fifteenth and sixteenth centuries include Corrine Beck, *Les eaux et forêts en Bourgogne ducale (vers 1350–vers 1480): Société et biodiversité*

(Paris: L'Harmattan, 2008); Appuhn, *Forest on the Sea*; and Wing, *Roots of Empire*.
46. Warde, *Invention of Sustainability*, 5–12, 333–35.
47. Joachim Radkau, "Wood and Forestry in German History: In Quest of an Environmental Approach," *Environment and History* 2 (1996): 64; and Joachim Radkau, "'Nachhaltigkeit' als Wort der Macht: Reflexionem zum methodischen Wert eines umweltpolitischen Schlüsselbegriffs," in *Umwelt und Herrschaft in der Geschichte. Environnement et pouvoir: Une approche historique*, ed. François Duceppe-Lamarre and Jens Ivo Engels (Munich: R. Oldenbourg, 2008), 132.
48. Warde, "Environmental History," 72, 76, and 88; Warde, *Invention of Sustainability*, 4–5.
49. Elinor Ostrom, *Governing the Commons: The Evolution of Institutions for Collective Action* (New York: Cambridge University Press, 1990).
50. Garrett Hardin, "The Tragedy of the Commons," *Science* 162 (1968): 1243–48.
51. Donald Worster, "Transformations of the Earth: Toward an Agroecological Perspective in History," *The Journal of American History* 76, no. 4 (1990): 1093–94; and Hoffmann, *Environmental History*, 148–49, 155–58, and passim.
52. Martina de Moor, Paul Warde, and Leigh Shaw-Taylor, eds., *The Management of Common Land in North West Europe, c. 1500–1850* (Turnhout: Brepols, 2002); Christopher Rodgers, Eleanor A. Straughton, Angus J. L. Winchester, and Margherita Pieraccini, *Contested Common Land: Environmental Governance Past and Present* (Washington, DC: Earthscan, 2011); and Tine De Moor, *The Dilemma of the Commoners: Understanding the Use of Common-Pool Resources in Long-Term Perspective* (New York: Cambridge University Press, 2015).
53. See above, in the section on ideological misconceptions.
54. L. M. Cantor and J. Hatherly, "The Medieval Parks of England," *Geography* 64, no. 283 (1979): 71–85; Constance Hoffman Berman, "Medieval Agriculture, the Southern French Countryside, and the Early Cistercians: A Study of Forty-Three Monasteries," *Transactions of the American Philosophical Society* 76, part 5 (1986): i–179; Robert Liddiard, ed., *The Medieval Park: New Perspectives* (Macclesfield: Windgather Press, 2007).

Bibliography

Aberth, John. *An Environmental History of the Middle Ages: The Crucible of Nature.* New York: Routledge, 2013.

Appuhn, Karl. *Forest on the Sea: Environmental Expertise in Renaissance Venice.* Baltimore: Johns Hopkins University Press, 2009.

Arnold, Ellen F. *Negotiating the Landscape: Environment and Monastic Identity in the Medieval Ardennes.* Philadelphia: University of Pennsylvania Press, 2013.

Beck, Corrine. *Les eaux et forêts en Bourgogne ducale (vers 1350–vers 1480): Société et biodiversité*. Paris: L'Harmattan, 2008.

Berman, Constance Hoffman. "Medieval Agriculture, the Southern French Countryside, and the Early Cistercians: A Study of Forty-Three Monasteries." *Transactions of the American Philosophical Society* 76, part 5 (1986): i–179.

"Bern Convention." Council of Europe. Retrieved 9 October 2019. https://www.coe.int/en/web/bern-convention/.

Braudel, Fernand. *Civilization and Capitalism, 15th–18th Century*. Vol. 1: *The Structures of Everyday Life*. Translated by Siân Reynold. New York: Perennial Library, 1985.

Britnell, Richard, ed. *Pragmatic Literacy East and West, 1200–1330*. Woodbridge: Boydell Press, 1997.

Brown, Warren C., Marios Costambeys, Matthew Innes, and Adam J. Kosto, eds. *Documentary Culture and the Laity in the Early Middle Ages*. New York: Cambridge University Press, 2013.

Buridant, Jérôme. *Espaces forestiers et industrie verrière, XVIIe–XIXe siècle*. Paris: L'Harmattan, 2005.

Campbell, Bruce. *English Seigniorial Agriculture, 1250–1450*. New York: Cambridge University Press, 2000.

Cantor, Leonard M., and J. Hatherly. "The Medieval Parks of England." *Geography* 64, no. 283 (1979): 71–85.

Caradonna, Jeremy. "Conservationism Avant la Lettre? Public Essay Competitions on Forestry and Deforestation in Eighteenth-Century France." In *Invaluable Trees: Cultures of Nature, 1660–1830*, edited by Laura Auricchio, Elizabeth H. Cook, and Giulia Pacini, 39–54. Oxford: Voltaire Foundation, 2012.

_____. *Sustainability: A History*. New York: Oxford University Press, 2014.

Chalvet, Martine. *Une histoire de la forêt*. Paris: Editions du Seuil, 2011.

Cipolla, Carlo M. *Before the Industrial Revolution: European Society and Economy, 1000–1700*. 3rd ed. New York: W. W. Norton & Co., 1994.

Clanchy, Michael T. *From Memory to Written Record: England, 1066–1307*. 3rd ed. Malden, MA: Wiley-Blackwell, 2013.

Coates, Peter. *Nature: Western Attitudes Since Ancient Times*. Berkeley: University of California Press, 1998.

"Conservation." UN World Conservation Monitoring Centre. Retrieved 9 October 2019. http://www.biodiversitya-z.org/content/conservation.

Corvol, Andrée. *L'homme et l'arbre sous l'Ancien Régime*. Paris: Economica, 1984.

Crone, Patricia. *Pre-industrial Societies: Anatomy of the Pre-Modern World*. 2nd ed. London: Oneworld Publications, 2013.

Cronon, William. "The Trouble with Wilderness; or, Getting Back to the Wrong Nature." In *Uncommon Ground: Rethinking the Human Place in Nature*, edited by William Cronon, 69–90. New York: W. W. Norton & Co, 1995.

Crosby, Alfred. *Ecological Imperialism: The Biological Expansion of Europe, 900–1900*. 2nd ed. New York: Cambridge University Press, 2004.

Deloria, Philip J. *Playing Indian*. New Haven: Yale University Press, 1998.

de Moor, Martina, Paul Warde, and Leigh Shaw-Taylor, eds. *The Management of Common Land in North West Europe, c. 1500–1850.* Turnhout: Brepols, 2002.
de Moor, Tine. *The Dilemma of the Commoners: Understanding the Use of Common-Pool Resources in Long-Term Perspective.* New York: Cambridge University Press, 2015.
Dictionnaire Historique de la Langue Française, 2 vols., edited by Alain Rey. Paris: Le Robert, 1992.
Duby, Georges. *Rural Economy and Country Life in the Medieval West.* Translated by Cynthia Postan. Columbus: University of South Carolina Press, 1968; reprinted at Philadelphia: University of Pennsylvania Press, 1998.
Evelyn, John. *Sylva: Or, a Discourse on Forest Trees and the Propagation of Timber in his Majesty's Dominions.* 4th ed. 2 vols., edited by John Nisbet. London: A. Doubleday & Co., 1908.
Gilchrist, Roberta. *Gender and Archaeology: Contesting the Past.* New York: Routledge, 1999.
Glacken, Clarence. *Traces on the Rhodian Shore: Nature and Culture in Western Thought from Ancient Times to the End of the Eighteenth Century.* Berkeley: University of California Press, 1967.
Grober, Ulrich. *Sustainability: A Cultural History.* Totnes, UK: Green Books, 2012.
Grove, Richard. *Green Imperialism: Colonial Expansion, Tropical Island Edens, and the Origins of Environmentalism, 1600–1860.* New York: Cambridge University Press, 1995.
Hardin, Garrett. "The Tragedy of the Commons." *Science* 162 (1968): 1243–48.
Hays, Samuel P. *Conservation and the Gospel of Efficiency: The Progressive Conservation Movement 1890–1920.* Cambridge, MA: Harvard University Press, 1959.
Herlihy, David J. "Attitudes toward the Environment in Medieval Society." In *Historical Ecology: Essays on Environment and Social Change*, edited by Lester J. Bilsky, 101–15. London: Kennikat Press, 1980.
Hoffmann, Richard. *An Environmental History of Medieval Europe.* New York: Cambridge University Press, 2014.
Hoffmann, Richard, Nancy Langston, James C. McCann, Peter C. Perdue, and Lise Sedrez. "AHR Conversation: Environmental Historians and Environmental Crisis." *The American Historical Review* 113, no. 5 (2008): 1431–65.
Houston, Robert A. *Literacy in Early Modern Europe: Its Growth, Uses and Impact, 1500–1800.* 2nd ed. New York: Routledge, 2002.
Isambert, François-André, et al., eds. *Recueil général des anciennes lois françaises depuis l'an 420 jusqu'à la Révolution de 1789*, 29 vols. Paris: Belin-Leprieur, Plon, 1826–1833.
Jacoby, Karl. *Crimes Against Nature: Squatters, Poachers, Thieves and the Hidden History of American Conservation.* Revised ed. Berkeley: University of California Press, 2014.
Jørgensen, Dolly. "'All Good Rule of the Citee': Sanitation and Civic Government in England, 1400–1600." *Journal of Urban History* 36, no. 3 (2010): 300–15.

Kander, Astrid, Paolo Malanima, and Paul Warde. *Power to the People: Energy in Europe Over the Last Five Centuries*. Princeton: Princeton University Press, 2013.

Kreike, Emmanuel. *Environmental Infrastructure in African History*. New York: Cambridge University Press, 2013.

Latham, James. "The Development of Forest Conservation in Europe." In *Europe's Changing Woods and Forests: From Wildwood to Managed Landscapes*, edited by Keith J. Kirby and Charles Watkins, 309–25. Boston: CAB International, 2015.

Lekan, Thomas. *Imagining the Nation in Nature: Landscape Preservation and German Identity, 1885–1945*. Cambridge, MA: Harvard University Press, 2004.

Les Eaux et Forêts du 12e au 20e siècle. Paris: Centre National de la Recherche Scientifique, 1987.

Liddiard, Robert, ed. *The Medieval Park: New Perspectives*. Macclesfield: Windgather Press, 2007.

Lowood, Henry E. "The Calculating Forester: Quantification, Cameral Science, and the Emergence of Scientific Forestry Management." In *The Quantifying Spirit in the Eighteenth Century*, edited by Tore Frängsmyr, H. L. Heilbron, and Robin E. Rider, 315–42. Berkeley: University of California Press, 1990.

Marsh, George P. *Man and Nature, or Physical Geography as Modified by Human Action*. New York: Charles Scribner and Co., 1864.

Matteson, Kieko. *Forests in Revolutionary France: Conservation, Community, and Conflict, 1669–1848*. New York: Cambridge University Press, 2015.

Niermeyer, Jan F., and C. Van de Kieft, eds. *Mediae Latinitatis Lexicon Minus*. Revised ed. 2 vols., edited by Johannes W. J. Burgers. Boston: Brill, 2002.

Ostrom, Elinor. *Governing the Commons: The Evolution of Institutions for Collective Action*. New York: Cambridge University Press, 1990.

Oxford English Dictionary Online. New York: Oxford University Press, 2018. Retrieved 9 October 2019. http://www.oed.com.

Oxford Latin Dictionary. 2nd ed. 2 vols., edited by P. G. W. Glare. New York: Oxford University Press, 2012.

Radkau, Joachim. "'Nachhaltigkeit' als Wort der Macht: Reflexionem zum methodischen Wert eines umweltpolitischen Schlüsselbegriffs." In *Umwelt und Herrschaft in der Geschichte. Environnement et pouvoir: Une approche historique*, edited by François Duceppe-Lamarre and Jens Ivo Engels, 131–36. Munich: R. Oldenbourg, 2008.

_____. *Nature and Power: A Global History of the Environment*. Translated by Thomas Dunlap. New York: Cambridge University Press, 2008.

_____. "Wood and Forestry in German History: In Quest of an Environmental Approach." *Environment and History* 2 (1996): 63–76.

Richards, John. *The Unending Frontier: An Environmental History of the Early Modern World*. Berkeley: University of California Press, 2003.

Rodgers, Christopher, Eleanor A. Straughton, Angus J. L. Winchester, and Margherita Pieraccini. *Contested Common Land: Environmental Governance Past and Present*. Washington, DC: Earthscan, 2011.

Rosenthal, Joel T., ed. *Understanding Medieval Primary Sources*. New York: Routledge, 2012.

Schama, Simon. *Landscape and Memory*. New York: Knopf, 1995.

Schmoeckel, Mathias, and Werner Schubert, eds. *Handbuch zur Geschichte des Notariats der europäischen Traditionen*. Baden-Baden: Nomos, 2009.

Shryock, Andrew, and Daniel Lord Smail, eds. *Deep History: The Architecture of Past and Present*. Berkeley: University of California Press, 2011.

Squatriti, Paolo. "The Floods of 589 and Climate Change at the Beginning of the Middle Ages: An Italian Microhistory." *Speculum* 85 (2010): 799–826.

_____. "Introduction: Natures Past and Present Environmental Histories." In *Natures Past: The Environment and Human History*, edited by Paolo Squatriti, 1–15. Ann Arbor: The University of Michigan Press, 2006.

Sykes, Naomi J. "Hunting for the Anglo-Normans: Zooarchaeological Evidence for Medieval Identity." In *Just Skin and Bones? New Perspectives on Human-Animal Relations in the Historical Past*, edited by Aleksander Pluskowski, 73–80. Oxford: BAR International Series, 2005.

Szabó, Péter. "Historical Ecology: Past, Present, and Future." *Biological Reviews. Cambridge Philosophical Society* 90 (2015): 997–1014.

Thomas, Keith. *Man and the Natural World: Changing Attitudes in England, 1500–1800*. New York: Penguin Books, 1983.

Uekoetter, Frank. *The Green and the Brown: A History of Conservation in Nazi Germany*. New York: Cambridge University Press, 2006.

Unger, Richard. "Introduction: Hoffmann in the Historiography of Environmental History." In *Ecologies and Economies in Medieval and Early Modern Europe: Studies in Environmental History for Richard C. Hoffmann*, edited by Scott G. Bruce, 1–21. Boston: Brill, 2010.

Van Dam, Petra J. E. M. "Sinking Peat Bogs: Environmental Change in Holland, 1350–1550." *Environmental History* 6, no. 1 (2001): 32–46.

von Carlowitz, Hans Carl. *Sylvicultura oeconomica*. Leipzig: Johann F. Braun, 1713. Reprinted and edited by Harald Thomasius and Bernd Bendix. Remagen-Oberwinter: Verlag Kessel, 2013.

Warde, Paul. "The Environmental History of Pre-industrial Agriculture in Europe." In *Nature's End: History and the Environment*, edited by Sverker Sörlin and Paul Warde, 70–92. New York: Palgrave Macmillan, 2009.

_____. *The Invention of Sustainability: Nature and Destiny, c. 1500–1870*. New York: Cambridge University Press, 2018.

Warde Paul, Libby Robin, and Sverker Sörlin. *The Environment: A History of the Idea*. Baltimore: Johns Hopkins University Press, 2018.

White, Lynn T., Jr. "The Historical Roots of Our Ecologic Crisis." *Science* 155 (1967): 1203–7.

Wing, John T. *Roots of Empire: Forests and State Power in Early Modern Spain, c. 1500–1750*. Boston: Brill, 2015.

Winter, James. *Secure from Rash Assault: Sustaining the Victorian Environment*. Berkeley: University of California Press, 1999.

Worster, Donald. "Appendix: Doing Environmental History." In *The Ends of the Earth: Perspectives on Modern Environmental History*, edited by Worster, 289–307. New York: Cambridge University Press, 1988.

_____. *Nature's Economy: A History of Ecological Ideas*. 2nd ed. New York: Cambridge University Press, 1994.

_____. "Transformations of the Earth: Toward an Agroecological Perspective in History." *The Journal of American History* 76, no. 4 (1990): 1087–1106.

Young, Charles R. "Conservation Policies in the Royal Forests of Medieval England." *Albion* 10, no. 2 (1978): 95–103.

PART I

Multiple-Use Resource Management in Preindustrial Societies

Pigs, Parks, Game, and Heathlands

CHAPTER 1

Controlling Pigs in Countryside and City for Sustainable Medieval Agriculture

Dolly Jørgensen

> They claimed the right to send all their herds of pigs into the two forests, but the count did not wish to allow them more than a customary two herds. The two parties eventually compromised on four herds which could always be sent into the forests, but not into the count's fenced-off enclosure. However, if the count sent his own herds of pigs or other herds into the fortified enclosure, then the monks should also be allowed to do so. The monks were to be allowed to graze all other animals in the forests.
>
> —David Bates, ed., *Regesta Regum Anglo-Normannorum*, no. 262

This agreement between the abbey of Saint-Wandrille and count William of Evreux in 1074 in Normandy provides a tantalizing glimpse into the medieval practice of controlling pigs for sustainable medieval agriculture. Both parties wanted to take advantage of fattening their pigs on acorns falling from the oak trees during the fall season before slaughter. Customary rights for using the woodland were already established by the time of this conflict. But as the account above reveals, use rights could be a complicated matter with real environmental implications. If too many pigs rooted in the forest, they could damage the forest ecosystem, harming both alternative woodland uses such as wood gathering and deer grazing and the future use of the woodland as fodder ground. Count William had created a fenced-off enclosure, most likely a designated area for deer hunting. In this protected area, the vegetation could grow higher than in the surrounding wood-pasture areas to provide adequate shelter and fodder for the deer. Swine could feed within the enclosure, but the count wished to limit this practice to avoid damage to the deer habitat. Feeding pigs in the forest required deliberate control to ensure both short- and long-term viability of the resource.

Domesticated pigs (*Sus scrufa*) were an integral component of medieval agriculture. Although archaeological evidence indicates that pork was less consumed than beef by weight and that average consumption decreased over the Middle Ages (ca. 500–1500 CE), pork consumption in medieval Europe was common.[1] Pigs are particularly attractive to raise because of their high fecundity (with at least two litters per year of up to fifteen piglets each), their omnivorous diet (pigs are not selective eaters and can, like humans, consume both meat and vegetable matter), and the amenability of pork to preservation (pork is easily salted, smoked, or dried). Recipes for preserved pork such as blood pudding, chitterlings, sausage, and salted pork cuts are commonly recorded in medieval cookbooks, such as one written for a fourteenth-century knight's Parisian household.[2] Boiled pork was commonly served to monks in colder months, from autumn through early spring.[3] Pigs were the only livestock kept solely for meat production in medieval Europe.

Because of traits making upkeep and consumption relatively easy, pigs were kept by people both in rural and urban settings.[4] For example, the *Domesday* survey compiled in England in 1086 often recorded rural woodland size based its swine carrying capacity or the amount of rent paid in swine, which could range from a few up to hundreds, and even thousands.[5] Beyond the rural space, owning one or two pigs was quite common for city dwellers: a taxation roll from 1297 for Reading, England, shows 30 of the 102 assessed persons owned pigs, yet only six people had more than two of them.[6] An archaeological study of Turku, Finland, indicated that large numbers of pigs could be kept by urban individuals who apparently operated a "self-sufficient pig-raising system."[7] Pigs were integral to medieval food production in both rural and urban settings.

This chapter examines medieval strategies for controlling pigs in both countryside and city, with a focus on England and neighboring Normandy. There were differing challenges to managing pigs in a sustainable manner in rural settings versus urban ones. But in both cases, people attempted to control damage to productive environments, whether those were grain fields, wooded pastures, woodlands, or the cityscape. The chapter is organized around the limitations placed on where, when, and how many pigs were allowed to maintain the sustainability of the environment and pig keeping as a practice.

Where?

Pigs in the countryside are particularly visible in medieval grants of usufruct rights in woodlands and wood pastures. This primarily comes in the form of *pannage* (in medieval Latin *pasnagio*), which conferred the right to feed pigs in an area in exchange for a number of fattened pigs. Pannage rights were regularly included in monastic charters and grants, which is our most consistent source for the practice. These documents explicitly state where pigs could go and whether or not pannage would be assessed for that pasturage. For example, Robert Count of Mortain gave a grant to abbey of Marmoutier allowing that "wherever the count's pigs go into the forest, the monks' pigs shall do likewise without payment of pannage."[8] Nigel the Viscount likewise gave the church of Saint-Sauveur-le-Vicomte the right for pigs to graze in the forest and a named enclosure, "and for them to wander wherever Nigel's pigs go."[9] What is interesting about both of these cases is that the gift specifies parity—that the pigs of the religious house are allowed to go in the same places as the lord's pigs. This clause could have sustainability thinking behind it because it concentrates the pigs in certain areas.

These concentration areas would likely have been managed woodland, rather than dense forest. Although grants of rights often include the word "forest," the areas controlled by kings, counts, and viscounts had many larger open areas than the word "forest" indicates in modern usage. In medieval usage, forest was a legal term rather than a term for woodland—a medieval forest could be almost completely devoid of trees because labeling something as a "forest" was a means of establishing a legal jurisdiction. Owning a forest as legal entity came with the ability to set policies and rules about which resources (like pannage) could be used and how they could be used.[10] Medieval records are not always clear with giving names to pasturage areas. For example, when King Henry I gave the right for the monks of St. Mary of Bec to pasture their swine within his forest, the specific area was called a *nemus*, which technically means woodland, but might in this case be a more open woodland grove.

Pigs in particular would have done well in medieval wood pasture. In a managed woodland, particularly one with large pollarded oak trees, there is no need to avoid consuming all available acorns, seedlings, and other understory growth. Pollarding, which is the cyclical practice of cutting off the branches of a grown tree above the grazing height of animals and then allowing the branches to grow back, especially in a wood pasture setting, is an effective way of increasing available swine food.[11] The acorn-producing potential of individual oaks is associated with the exposure of the crown to light, so oaks growing in open wood pasture produce more acorns per tree than those in closed-canopy forests. In addition, three-quarters of oak buds occur on current-year shoots, meaning that even recently pollarded trees should not have significantly decreased acorn production.[12] At the same time, the open spaces created by pollarding trees on fields also encourages grass growth, which can also be consumed by foraging pigs.[13] Only oaks over twenty years old will produce a significant number of acorns, so while it is not necessary to cultivate young trees, care must be taken to keep old trees alive.[14] Large productive trees were especially valued and conserved, as we see in the Laws of Ine issued in seventh century Wessex, which set the fine for illegally cutting down a tree that could "shelter 30 swine" at twice the amount of a regular tree.[15]

It is possible that pig grazing was also combined with other woodland management techniques such as coppicing. In coppicing, trees that resprout from the base, such as willows, are cut low down in order to maximize the production of smaller branches, which can be harvested on a short rotation of three to seven years as firewood and fencing material. Although those trees would not have been particularly useful for the pigs, coppicing also tends to be practiced in an open wood pasture setting, so grasses would abound in the area. The pigs of Nottingham are known to have foraged in an area called "Coppice" outside of the Swine Bar town gate, which gives evidence of this practice in the fourteenth century.[16]

The practice of grazing pigs in wood pasture is well attested in medieval images, which show pigs gorging on acorns in the autumn to fatten up before the winter slaughter (Figure 1.1). These illustrations, which were primarily created to represent the agricultural task for October or November in manuscript calendars, often show wood pasture (rather than dense forest) as the feeding location. In these images, the swineherd is often involved in knocking down acorns, rather than just letting the pigs feed on what was naturally there.[17] This appears to have been common practice: the thirteenth-century scholar Bartholomew the Englishman commented on this imagery choice, saying that a rustic beating the oak was appropriate for the month for fattening swine.[18] Acorns knocked from the trees could even be collected in baskets and used as fodder at a later time.[19] Because these areas were managed

Figure 1.1. Feeding pigs on acorns in November. Book of Hours made in France by the workshop of the Bedford Master, first half of the fifteenth century. Getty Ms. Ludwig IX 6, fol. 11. Image released into the public domain under Creative Commons Attribution 4.0 International License by the Getty Museum.

wood pasture, natural regeneration of trees by acorns would have been unnecessary. Regular pollarding would have kept each tree healthy for hundreds of years.

As in the example of Saint-Wandrille at the start of this chapter, restrictions also separated pig grazing from other uses, such as deer parks. Medieval woodlands had many potential uses as sources of income for farming communities, including timber cutting, firewood collection, and wildlife hunting.[20] An 1115 confirmation of a grant to the church of Abingdon in England, for example, allowed the monks to graze their pigs throughout an area of woods, except within a named deer park.[21] Another grant to the Norman abbey of Lessay from the late eleventh century also distinguished the deer park from other grazing areas held by the donor: "Outside the park, the monks are to have pasture for all their animals and for all their pigs, as well as pasture for the animals of the servants who take care of the monks' herd; if the pigs are sent to graze inside the park, the monks shall have the right freely to graze one hundred pigs from Martinmas until Lent."[22] The distinction of inside and outside of the park was specifically intended to avoid negative effects on deer habitat. Deer need more dense forest for cover and browsing, and because pig grazing tends to create extended grassy glades, pig habits can be counterproductive for deer. In this gift, an unlimited number of pigs could be grazed outside the park; but there was a tight restriction on the herd size inside the park. The timing of the grazing inside the park was also restricted to minimize interference with deer, a subject to which we will return in the section "When?" below. Segregating the woodland uses increased the overall efficiency of acorns, grass, deer, and fattened pigs. Grazing practices that limited the location of pigs to wood pasture or forested areas without deer created a long-term ecologically viable system.

The common assumption by historians has been that pigs must have roamed the medieval streets, because towns made statements against the practice. City governments did issue statements against pigs wandering the towns, often assessing high fines and the seizure of the pig as penalties, including York in 1301, London in 1388, and Coventry in 1421.[23] London even named "swine killers" who were authorized by the government to look for stray pigs and kill them (keeping the meat), or give them back to the owner in return for a 4 pence fine.[24] The fear of wandering swine could be well-founded: when the Norwich city leaders restricted roaming pigs in their city in 1354, they did it because pigs had wrecked gardens, damaged houses, dug up bodies in the graveyard, attacked residents, and even killed a child.[25] But rather than seeing wandering pigs as the norm, we should understand them as the exception.

In the cityscape, pig movements were regulated by city governments intent on limiting damage to goods and property within the town boundaries.[26] These kinds of restrictions appear early in city records; for example, in 1297 the London council ordered that "pig-sties that are in the streets shall be speedily removed; and that no swine shall be found in the streets, on pain of forfeiture thereof, in aid of making the walls and gates."[27] In this by-law, both the location of the pigs themselves and their habitations (pigsties) were restricted. These two elements of pig control would be repeated in other city laws throughout Europe in the Middle Ages.

The urban pigsty is a critical element of understanding how pigs were controlled for sustainability. Sties served as both the home and feeding place of urban pigs. In 1277, the inhabitants of London were told by the city council that pigs could only be fed either in the open away from the King's Highway or indoors.[28] Archaeological investigations of tooth microwear have also indicated that urban pigs were in general stall fed rather than fed via foraging or rooting. Pigs were fed at troughs placed outside their sties, a practice that appears often in late medieval images of the prodigal son feeding the swine, such as in the finely detailed painting in the Södra Råda church in Sweden in which three swine stick out their heads from a wooden building to feed in two troughs being filled by the son.[29] The Grandes Heures d'Anne de Bretagne dated 1505–15 also includes an image of pigs feeding in a trough overseen by two swineherds as the illustration for the month of November calendar (see Figure 1.2).[30] In troughs, pigs were fed on milling and brewing waste, as well as other organic food wastes.[31] These practices would have been a good way to recycle otherwise wasted caloric materials into edible meat. This would have also maximized the sustainability of

Figure 1.2. Feeding pigs at a trough in an illustration for November. *Book of Hours*, late fifteenth century. Angers Bibl. mun., ms. 0134, fol. 11. Image in public domain.

keeping the pigs within the town boundaries, since they would have needed less grazing.

Many cities mandated that pigs be kept in their stalls except on two occasions. The first was when they went out to nearby fields and woodlands for extra feed. In this case, the pigs were not alone, but rather in the presence of a swineherd who was responsible for collecting the animals and herding them out to specific feeding locations.[32] An illustration in one copy of *Le Chevalier errant* dated to 1400 shows an urban swineherd leaning on his staff while keeping an eye on a group of pigs in the marketplace.[33] Some images show the pigs being led by a rope so that they would not stray from the swineherd (Figure 1.3). Taking pigs out to feed on grain stubble left in the fields would have been a reasonable way to make use of the edible material. But there could be problems caused by pig damage to unharvested crops; thus, the Coventry government ordered the swineherd to drive the pigs only into empty land or marshy areas, rather than productive meadows or

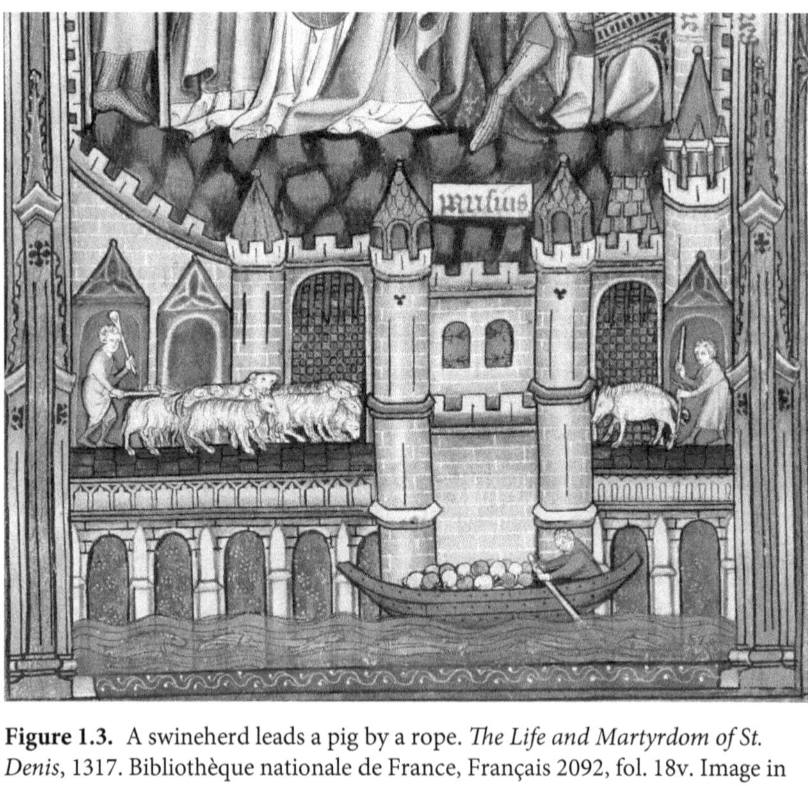

Figure 1.3. A swineherd leads a pig by a rope. *The Life and Martyrdom of St. Denis*, 1317. Bibliothèque nationale de France, Français 2092, fol. 18v. Image in the public domain.

fields near the city.[34] Pigs were often ringed to keep damage to productive land at a minimum. Ringing involved putting a wooden or iron ring or bar through the pig's snout, which kept them from digging with the snout while yet allowing them to feed.[35] This would have prevented root damage to plants and trees. Local governments mandated the use of rings, as evidenced in fines levied on three men who had "left their pigs unrung with iron" in Eynsham in 1296, an order for two men to ring their pigs in Ealing in 1492, and a command to the tenants of South Hamingfield manor to ring their pigs in 1501.[36]

The second time pigs were regularly let out was when the sties were being cleaned. The cities of Lynn and Norwich, for example, both required pig sties to be cleaned on Saturday afternoon, so that was the only time pigs were permitted in the streets.[37] The Norwich documents are very specific, ordering each person who has pigs to take them out of their stalls every Saturday between noon and the evening to clean the pens, during which time the animals must be kept under control, and then return the pigs to their pigsty immediately after the cleaning.[38] Swineherds would have been required during the cleaning to provide that watchful eye.

When?

Pigs were most certainly put out in wooded areas with oak or beech trees to feed on acorns or beech mast during the fall months. This is attested by the many visual images of swineherds knocking acorns out of the trees to feed their herd, as well as written evidence.[39] Domesday Book, for example, employs a standard formulation of "wood for *x* swine" (*silva ad x porcos* or *silva de x porcis*) in its taxation assessment, with the Shropshire entries using the even more specific "wood for fattening (*incrassandis*) *x* swine" phrase. In his foundational work on social and agrarian conditions in rural England, Reginald Lennard argues: "Extensive pig-farming was a normal feature of districts in which there was much glandiferous woodland, and in some of these must have played a really important part in the economy. Especially notable is the region between the escarpment of the Chilterns and the London area . . . [where] twenty places were credited in 1086 with wood for a thousand swine or more."[40] Henry C. Darby, who compiled an extensive study of Domesday geography, drew the same conclusion as Lennard that wood's economic value in the eleventh century was often tied to swine feeding.[41] Oliver Rackham has countered this claim by arguing that fall acorn or beech crops can fail, thus these numbers cannot represent actual swine in the woodland;[42] we do indeed have one entry in the Anglo-Saxon Chronicle for 1116 that records a complete failure of the harvest. Although that can happen, oak silviculture studies show that despite year-to-year fluctuations in acorn crops, moderate crop sizes are just as common as extreme years.[43] This means that although high productivity "mast" years are limited, acorns are typically available every year.

Moreover, there is nothing in the Domesday Book entries, nor in most accounts of pannage gifts, that specifies pannage as a fall activity. Pannage is actually defined as the payment (of either money or in-kind) levied for feeding pigs in a wooded area. Although the fee was collected in conjunction with the fall feeding on acorns, the woodland feeding may have taken place throughout the year. The seasonality has been assumed by historians based on the idea that the only time pigs would be in the woodlands is for fattening on acorns.[44] Yet, pigs cannot feed on acorns alone. Even when acorns are available, grazing swine will consume large quantities of grass, roots, and plant rhizomes.[45] The Roman writer Palladius, who was later translated into Middle English verse, noticed this and commented that pigs thrived in woodlands because of both acorns and the grass beneath the trees.[46] If we look at evidence of feeding practices, we see that pigs were put out to graze in open wood pasture for much of the year.

The grant to Lessay discussed previously specified that pigs were permitted in the enclosure from Martinmas (11 November) until Lent (if they mean the beginning of Lent, this would be anytime from 4 February to 10 March, depending on the year). This is interestingly *not* when acorns or beech mast would have been on the ground. According to a fourteenth-century husbandry handbook titled *Seneschaucy*, owners were encouraged to provide food directly to their herd only during severe frost; the rest of the time, the pigs should find their own food in forests, wood pastures, marshes, or untilled fields.[47] Piggeries, according to this text, could be set up for permanent use in marshes or woodlands, a practice that is also attested in the Grande Heures d'Anne de Bretagne image mentioned earlier. We also have documents stating that pannage fees were collected whether there was mast or not,[48] another indicator that pannage may have had a wider pig pasturage implication than just acorn feeding. Orderic Vitalis even mentions that Ralph of Conches gave a monastic foundation "land for pannage in such a way that their tenants should pay nothing for the first pannage but should pay for the second and third pannages."[49] This would appear to indicate that pannage could be a series of events—again hinting at pannage as meaning times of pig grazing rather than acorn-feeding time. This body of evidence emphasizes the year-round nature of pig pasturage.

The times that pigs were allowed in grain fields rather than woodland had to be carefully controlled, especially early in the season before the vegetation was established and late in the season when the harvest was neigh. Fines were assessed if an owner was negligent and let his pigs into grain fields at harvest time, such as a man in 1276 in Wiltshire who had to repay the crop damage caused by his herd.[50] The animals might, however, be taken to feed on fields already harvested where they could scrounge up leftover vegetation. One example of this is at Basingstoke, where the pigs were allowed in the "Lent" field after 29 September and in the winter field from 8 November.[51] Later evidence from the sixteenth century shows continued concern about keeping pigs out of grain crops in the summer months, specified, for example, as 25 May to harvest in Latton, Wiltshire, in 1596.[52]

How Many?

The sharing of woodland resources would appear to have been the common modus operandi, so inevitably conflicts arose over how many pigs could feed in a given area. The Visigothic laws included an entire title devoted to the pasturage of swine and how to resolve potential conflicts. The law code anticipated a dispute between persons who held an oak woodland in common but had different numbers of pigs feeding there. In that case, the common woodland could be divided so that each owner would have an equal-sized tract of land. The fee for pasturage (set at 10 percent) would then be set on each herd separately. This law shows an awareness of managing the commons for the environment. Because pigs could have a negative impact on the environment if too many were grazed in the same area, this law permits the owner with fewer pigs to conserve his woodland and avoid it being damaged by the other owner's large herd.[53]

Limitations on the number of pigs in particular areas appear to have been commonly set. In the opening vignette, the monks of Saint-Wandrille wanted to send all of their pigs into the woodland, but the count wanted to allow only the "customary" two herds. The parties reached a settlement permitting the monks to send four herds into a designated part of the woodland. This case demonstrates that "custom" had been active in the area to dictate how many pigs could be in the woodlands, which functioned as common in spite of being "held" by a lord. While the number of pigs was raised to resolve the conflict, these additional pigs were not allowed to go wherever they wanted. This evidence indicates that the parties understood the environmental consequences of pannage and acted to balance resource demands.

Within cities, on the contrary, restrictions on how many pigs a particular individual could own do not appear in the records. The numbers kept would have been limited by practical space and food concerns. Taxation lists give us an indication of pig keeping. In Lynn at the end of the thirteenth century, one resident had twenty-three pigs and piglets, while another had twelve piglets. Of all persons assessed in the tax rolls of both Lynn and Reading, 40 percent owned livestock, with pigs being the most commonly owned.[54] These numbers indicate that a few people had fairly large piggeries within town walls, and many households owned one or two pigs. This made the regulation of pig movement through requirements to be in sties rather than on the street and to have a swineherd particularly important. Some cities paid a city swineherd who collected individual pigs from their owners and took them out together to feed outside the walls. Pigs would have been branded to ensure they went back to the rightful owner.[55]

Conclusion

The conflict between the Abbey of Saint-Wandrille and count William of Evreux over their pigs nearly resulted in bloodshed. A duel with champions had been proposed before the legal settlement was finally reached in 1074. The agreement included compromise—more of the monks' pigs were allowed on the Count's land, but the herd was excluded from the count's fenced-in hunting sections. Custom played a part in establishing the norms for pig grazing, but so did this legal ruling. Pig management was serious business and thus could be escalated to the highest levels. The King of England (and Duke of Normandy) William I issued the document resolving the conflict, indicating the importance of good land and resource management. Swine could be as precious as pearls.

Without control, pigs were potentially destructive to the environment and future livelihoods of medieval farmers as well as townspeople. As seen in the examples discussed in this chapter, legal measures were taken to control where pigs could be, when pigs could be put out to graze, and how many pigs could be in a given area. When individuals violated legal or customary standards, they were penalized with fines or even the loss of their valuable animals.

The practices implemented aimed to make pig rearing compatible with other land uses in the long term. Open wood pasture with pollarded or even coppiced trees provided grazing areas that were valuable even beyond the short acorn season. The caloric potential of stubble in harvested fields, brewery dregs, and other food wastes could be turned into pork by allowing swine to feed on them. Keeping urban pigs in pens, feeding them at troughs, and sending them out only under the watchful eye of a swineherd constituted a constellation of practical control methods to limit pig damage while maximizing that meat production. These were not measures intended to preserve "nature," but they were intended to ensure the long-term viability of agricultural practice, and thus the communities that practiced them.

Dolly Jørgensen is Professor of History at University of Stavanger, Norway, specializing in environmental history. Her scholarship is unconstrained by typical periodization boundaries: she is just as comfortable writing about eleventh-century forest management or fifteenth-century urban sanitation as she is writing about twentieth-century offshore oil operations or contemporary efforts to resurrect extinct animal species. She has published several articles on medieval swine management practices.

Notes

1. Umberto Albarella, "Pig Husbandry and Pork Consumption in Medieval England," in *Food in Medieval England: Diet and Nutrition*, ed. Christopher M. Woolgar, D. Serjeantson, and T. Waldron (Oxford University Press: Oxford, 2006), 72–87; Julie Hamilton and Richard Thomas, "Pannage, Pulses and Pigs: Isotopic and Zooarchaeological Evidence for Changing Pig Management Practices in Later Medieval England," *Medieval Archeology* 56 (2012): 234–59.
2. Nicole Crossley-Holland, *Living and Dining in Medieval Paris: The Household of a Fourteenth-Century Knight* (Cardiff: University of Wales Press, 1996), 131.

3. Barbara Harvey, *Living and Dying in England 1100–1540: The Monastic Experience* (Oxford: Clarendon Press, 1993), 51–53.
4. For a deeper discussion of controlling medieval urban pigs, see Dolly Jørgensen, "Running Amuck? Urban Swine Management in Late Medieval England," *Agricultural History* 87 (2013): 429–51.
5. Henry C. Darby, "*Domesday* Woodland," *The Economic History Review* 3, no. 1 (1950): 23–24.
6. Barbara Dodwell, "Reading Records (3): Taxation Roll, 1297," *Berkshire Archaeological Journal* 60 (1962): 101–12.
7. Auli Tourunen, *Animals in an Urban Context: A Zooarchaeological Study of the Medieval and Post Medieval town of Turku*, PhD dissertation, University of Turku, 2008.
8. David Bates, ed., *Regesta Regum Anglo-Normannorum: The Acta of William I (1066–1087)* (Oxford: Oxford University Press, 1998), no. 205. Hereafter abbreviated *RRAN I*.
9. Bates, *RRAN I*, no. 260.
10. See Dolly Jørgensen, "The Roots of the English Royal Forest," in *Anglo-Norman Studies 32*, ed. Chris Lewis (Woodbridge, UK: Boydell Press, 2010), 114–28.
11. Dolly Jørgensen, "Pigs and Pollards: Medieval Insights for UK Wood Pasture Restoration," *Sustainability* 5 (2013): 387–99; Dolly Jørgensen and Peter Quelch, "The Origins and History of Medieval Wood-pastures," in *European Wood-pastures in Transition: A Social-ecological Approach*, ed. Tibor Hartel and Tobias Pleininger (New York: Routledge, 2014), 55–69.
12. Paul S. Johnson, Stephen R. Shifley, and Robert Rogers, *The Ecology and Silviculture of Oaks* (Wallingford: CABI Publishing, 2002), 55.
13. See Jørgensen, "Pigs and Pollards" for a discussion of the ecosystems created by pig feeding.
14. Jules N. Pretty, "Sustainable Agriculture in the Middle Ages: The English Manor," *Agricultural History Review* 38 (1990): 16.
15. The Laws of Ine, capitulary 44, in Frederick L. Attenborough, ed. and trans., *The Laws of the Earliest English Kings* (New York: Russell & Russell, 1963).
16. Stephen Alsford, "Tending of livestock," Florilegium website, http://users.trytel.com/~tristan/towns/florilegium/community/cmfabr27.html.
17. Jørgensen, "Pigs and Pollards," 389–91. Three images of the practice of knocking down acorns are also reproduced in that article.
18. Herbert Meritt, "Beating the Oaks: An Interpretation of Christ 678–9," *American Journal of Philology* 66 (1945): 6.
19. Vincent of Beauvais, *Speculi Maioris*, Book 6, Volume 2, Chapter 148 (Venice, Italy: Dominicum Nicolinum, 1541). See also the discussion in Jørgensen, "Pigs and Pollards," 391.
20. Pretty, "Sustainable Agriculture."
21. "Abingdon Abbey," *Regesta* 1089, The Charters of William II and Henry I, version dated 23 September 2013, https://actswilliam2henry1.files.wordpress.com/2013/09/h1-abingdon-2013-1.pdf, 162.

22. Bates, *RRAN I*, no. 175.
23. Michael Prestwich, ed., *York Civic Ordinances, 1301* (York: St. Anthony's, 1976), 16; 12 Richard II. A.D. 1388. Letter-Book H. fol. ccxxxvii, translated in Henry T. Riley, ed., *Memorials of London and London Life: In the XIIIth, XIVth and XVth Centuries* (London: Longmans, Green and Co., 1868), 509; Mary Dormer Harris, ed., *The Coventry Leet Book: or Mayor's Register, Containing the Records of the City Court Leet or View of Frankpledge, AD 1420–1555, with Divers Other Matters*, 4 vols. (London: Kegan Paul, Trench, Trübner, 1907–1913), 1: 27.
24. 20 Edward I. A.D. 1292. Letter-Book C. fol. ii, in Riley, *Memorials of London*, 28. This was a task that was later allocated to the Surveyor of streets and lanes in the fourteenth century: Reginald R. Sharpe, ed., *Calendar of Letter-books preserved among the archives of the City of London at the Guild Hall, Letter-book H, 1375–1399* (London: J. E. Francis, 1907), 355.
25. William Hudson and John C. Tingey, trans. and eds., *The Records of the City of Norwich*, 2 vols. (Norwich: Jarrold, 1906–1910), 2: 205–6.
26. For a deeper discussion of controlling medieval urban pigs, see Jørgensen, "Running Amuck?" 429–51.
27. 25 Edward I. A.D. 1297. Letter-Book B. fol. xxxiii, in Riley, ed., *Memorials of London and London Life*, 35.
28. Reginald R. Sharpe, ed., *Calendar of Letter-books preserved among the archives of the City of London at the Guild Hall, Letter-book A, 1275–1298* (London: J. E. Francis, 1899), 217.
29. A color reproduction of the wall painting is found in Janken Myrdal, *Boskapsskötseln under medeltiden: En källpluralistisk studie* (Stockholm: Nordiska museets förlag, 2012), 249. See Jørgensen, "Running Amuck?" 435–39 for more examples of medieval illustrations of sties and troughs. In Dundee in the 1560s, residents were scolded for having swine troughs at their doors, under stairs, and in open windows on the street: Catherine Smith, "A Grumphie in the Sty: An Archaeological View of Pigs in Scotland, from their Earliest Domestication to the Agricultural Revolution," *Proceedings of the Society of Antiquaries of Scotland* 130 (2000): 714.
30. BNF Latin 9474, folio 14.
31. Smith, "A Grumphie in the Sty," 716; Hamilton and Thomas, "Pannage, Pulses and Pigs," 235; Albarella, "Pig Husbandry and Pork Consumption," 79.
32. Jørgensen, "Running Amuck?" 437. Swineherds were also necessary for pigs kept in completely rural settings to control movement and damage to neighboring crops and property: Jamie Kreiner, "Pigs in the Flesh and Fisc: An Early Medieval Ecology," *Past and Present* 236 (2017): 8–10.
33. BNF Français 12559, folio 167.
34. Dormer Harris, *The Coventry Leet Book*, 1: 27.
35. Jørgensen, "Running Amuck?" 17–18.
36. These examples come from Warren O. Ault, "Open-Field Husbandry and the Village Community: A Study of Agrarian By-Laws in Medieval

England," *Transactions of the American Philosophical Society* 55, no. 7 (1965): 27.

37. Hudson and Tingey, *The Records of the City of Norwich*, 2: 205–6; Stephen Alsford, trans., "Lynn By-laws," laws dated 29 September 1331 and early fifteenth century, Florilegium website, http://users.trytel.com/~tristan/towns/lynnlaws.html.
38. Hudson and Tingey, *The Records of the City of Norwich*, 2: 205–6.
39. See Jørgensen, "Pigs and Pollards" for visual examples, as well as C. L. ten Cate, *Wan god mast gift: Bilder aus der Geschichte der Schweinezucht im Walde* (Wageningen, Netherlands: Centre for Agricultural Publishing and Documentation, 1972).
40. Reginald Lennard, *Rural England 1086–1135: A Study of Social and Agrarian Conditions* (Oxford: Oxford University Press, 1959), 256.
41. Darby, "*Domesday* Woodland."
42. Oliver Rackham, *Trees and Woodland in the British Landscape: The Complete History of Britain's Trees, Woods & Hedgerows*, rev. ed. (London: Phoenix Press, 1990), 48–49.
43. See Walter D. Koenig and Johannes M. H. Knops, "The Behavioral Ecology of Masting in Oaks," in *The Ecology and Silviculture of Oaks*, ed. Paul S. Johnson, Stephen R. Shifley, and Robert Rogers (Oxon UK: CABI Publishing, 2002); and Cathryn H. Greenberg and B. R. Parresol, "Dynamics of Acorn Productivity by Five Species of Southern Appalachian Oak," in *Oak Forest Ecosystems*, ed. William J. McShea and William M. Healy (Baltimore: Johns Hopkins University Press, 2002).
44. Charles J. Cox, *The Royal Forests of England* (London: Methuen & Co., 1905), 41, defines pannage as applying from the Triumph of the Holy Cross (September 14) to Martinmas (although he incorrectly identifies this as November 18 instead of November 11).
45. Vicente Rodríguez-Estévez, Anton Garcia, Francisco Peña, and Antonio G. Gómez, "Foraging of Iberian Fattening Pigs Grazing Natural Pasture in the Dehesa," *Livestock Science* 120 (2009): 135–43.
46. Barton Lodge, ed., *Palladius, On husbondrie* (London: Early English Text Society, 1879).
47. Dorothea Oschinsky, ed., *Walter of Henley and Other Treatises on Estate Management and Accounting* (Clarendon Press: Oxford, 1971), chapter viii, 285.
48. Jean Birrell, "Common Rights in the Medieval Forest: Disputes and Conflicts in the Thirteenth Century," *Past & Present* 117 (1987): 40.
49. Marjorie Chibnall, ed. and trans., *The Ecclesiastical History of Orderic Vitalis*, vol. 3 (Oxford: Oxford University Press, 1972), 127.
50. Ault, "Open-Field Husbandry," 28
51. Ault, "Open-Field Husbandry," 28
52. Ault, "Open-Field Husbandry," 28
53. Samuel P. Scott, trans., *The Visigothic Code (Forum Judicum)* (Boston: Boston Boko Company, 1910), 298–99. Book VIII, Title V, Law II.

54. Alsford, "Tending of Livestock."
55. Ten Cate, *Wan god mast gift*, 120–23.

Bibliography

"Abingdon Abbey." *Regesta* 1089. The Charters of William II and Henry I. Version dated 23 September 2013. https://actswilliam2henry1.files.wordpress.com/2013/09/h1-abingdon-2013-1.pdf. Accessed 11 February 2020.

Albarella, Umberto. "Pig Husbandry and Pork Consumption in Medieval England." In *Food in Medieval England: Diet and Nutrition*, edited by Christopher M. Woolgar, D. Serjeantson, and T. Waldron, 72–87. Oxford: Oxford University Press, 2006.

Alsford, Stephen. "Tending of Livestock." Florilegium website. Retrieved 21 January 2020. http://users.trytel.com/~tristan/towns/florilegium/community/cmfabr27.html.

_____, trans. "Lynn By-laws," Florilegium website. Retrieved 21 January 2020. http://users.trytel.com/~tristan/towns/lynnlaws.html.

Attenborough, Frederick L., ed. and trans. *The Laws of the Earliest English Kings*. New York: Russell & Russell, 1963.

Ault, Warren O. "Open-Field Husbandry and the Village Community: A Study of Agrarian By-Laws in Medieval England." *Transactions of the American Philosophical Society* 55, no. 7 (1965): 1–102.

Bates, David, ed. *Regesta Regum Anglo-Normannorum: The Acta of William I (1066–1087)*. Oxford: Oxford University Press, 1998.

Birrell, Jean. "Common Rights in the Medieval Forest: Disputes and Conflicts in the Thirteenth Century." *Past & Present* 117 (1987): 22–49.

Chibnall, Marjorie, ed. and trans. *The Ecclesiastical History of Orderic Vitalis*. Vol. 3. Oxford: Oxford University Press, 1972.

Cox, Charles J. *The Royal Forests of England*. London: Methuen & Co., 1905.

Crossley-Holland, Nicole. *Living and Dining in Medieval Paris: The Household of a Fourteenth-Century Knight*. Cardiff: University of Wales Press, 1996.

Darby, Henry C. "*Domesday* Woodland." *The Economic History Review* 3, no. 1 (1950): 21–43.

Dodwell, Barbara. "Reading Records (3): Taxation Roll, 1297." *Berkshire Archaeological Journal* 60 (1962): 101–12.

Dormer Harris, Mary, ed. *The Coventry Leet Book: or Mayor's Register, Containing the Records of the City Court Leet or View of Frankpledge, AD 1420–1555, with Divers Other Matters*. 4 vols. London: Kegan Paul, Trench, Trübner, 1907–1913.

Greenberg, Cathryn H., and B. R. Parresol. "Dynamics of Acorn Productivity by Five Species of Southern Appalachian Oak." In *Oak Forest Ecosystems*, edited by William J. McShea and William M. Healy. Baltimore: Johns Hopkins University Press, 2002.

Hamilton, Julie, and Richard Thomas. "Pannage, Pulses and Pigs: Isotopic and Zooarchaeological Evidence for Changing Pig Management Practices in Later Medieval England." *Medieval Archeology* 56 (2012): 234–59.

Harvey, Barbara. *Living and Dying in England 1100–1540: The Monastic Experience.* Oxford: Clarendon Press, 1993.

Hudson, William, and John C. Tingey, trans. and eds. *The Records of the City of Norwich.* 2 vols. Norwich: Jarrold, 1906–1910.

Johnson, Paul S., Stephen R. Shifley, and Robert Rogers. *The Ecology and Silviculture of Oaks.* Wallingford: CABI Publishing, 2002.

Jørgensen, Dolly. "Pigs and Pollards: Medieval Insights for UK Wood Pasture Restoration." *Sustainability* 5 (2013): 387–99.

______. "The Roots of the English Royal Forest." In *Anglo-Norman Studies 32,* edited by Chris Lewis, 114–28. Woodbridge, UK: Boydell Press, 2010.

______. "Running Amuck? Urban Swine Management in Late Medieval England." *Agricultural History* 87 (2013): 429–51.

Jørgensen, Dolly, and Peter Quelch. "The Origins and History of Medieval Wood-pastures." In *European Wood-pastures in Transition: A Social-ecological Approach,* edited by Tibor Hartel and Tobias Pleininger, 55–69. New York: Routledge, 2014.

Koenig, Walter D., and Johannes M. H. Knops. "The Behavioral Ecology of Masting in Oaks." In *The Ecology and Silviculture of Oaks,* edited by Paul S. Johnson, Stephen R. Shifley, and Robert Rogers. Oxon, UK: CABI Publishing, 2002.

Kreiner, Jamie. "Pigs in the Flesh and Fisc: An Early Medieval Ecology." *Past and Present* 236 (2017): 1–42.

Lennard, Reginald. *Rural England 1086–1135: A Study of Social and Agrarian Conditions.* Oxford: Oxford University Press, 1959.

Lodge, Barton, ed. *Palladius, On husbondrie.* London: Early English Text Society, 1879.

Meritt, Herbert. "Beating the Oaks: An Interpretation of Christ 678–9." *American Journal of Philology* 66 (1945): 1–12.

Myrdal, Janken. *Boskapsskötseln under medeltiden: En källpluralistisk studie.* Stockholm: Nordiska museets förlag, 2012.

Oschinsky, Dorothea, ed. *Walter of Henley and Other Treatises on Estate Management and Accounting.* Oxford: Clarendon Press, 1971.

Prestwich, Michael, ed. *York Civic Ordinances, 1301.* York: St. Anthony's, 1976.

Pretty, Jules N. "Sustainable Agriculture in the Middle Ages: The English Manor." *Agricultural History Review* 38 (1990): 1–19.

Rackham, Oliver. *Trees and Woodland in the British Landscape: The Complete History of Britain's Trees, Woods & Hedgerows.* Rev. ed. London: Phoenix Press, 1990.

Riley, Henry T., ed. *Memorials of London and London Life: In the XIIIth, XIVth and XVth Centuries.* London: Longmans, Green and Co., 1868.

Rodríguez-Estévez, Vicente, Anton Garcia, Francisco Peña, and Antonio G. Gómez. "Foraging of Iberian Fattening Pigs Grazing Natural Pasture in the Dehesa." *Livestock Science* 120 (2009): 135–43.

Scott, Samuel P., trans. *The Visigothic Code (Forum Judicum)*. Boston: Boston Boko Company, 1910.

Sharpe, Reginald R., ed. *Calendar of Letter-books preserved among the archives of the City of London at the Guild Hall, Letter-book A, 1275–1298*. London: J. E. Francis, 1899.

_____, ed. *Calendar of Letter-books preserved among the archives of the City of London at the Guild Hall, Letter-book H, 1375–1399*; London: J. E. Francis, 1907.

Smith, Catherine. "A Grumphie in the Sty: An Archaeological View of Pigs in Scotland, from their Earliest Domestication to the Agricultural Revolution." *Proceedings of the Society of Antiquaries of Scotland* 130 (2000): 705–24.

Ten Cate, C. L. *Wan god mast gift: Bilder aus der Geschichte der Schweinezucht im Walde*. Wageningen, Netherlands: Centre for Agricultural Publishing and Documentation, 1972.

Tourunen, Auli. *Animals in an Urban Context: A Zooarchaeological Study of the Medieval and Post Medieval Town of Turku*. PhD dissertation, University of Turku, 2008.

Vincent of Beauvais. *Speculi Maioris*. Venice, Italy: Dominicum Nicolinum, 1541.

CHAPTER 2

Sustainability and Natural Resource Management at Hesdin, Artois, France, 1302–1329

Abigail P. Dowling

> To Symon Pikart to cut wood below the hunting stand [for bow-and-stable shooting] so that madame may better view the beasts [deer] in chase and make bundles of firewood from the branch [and] to put the large firewood for monsieur [the cart driver] in a place to be carried away by car, 8 l. 16 s.
>
> —Entry in 1325 Works Account for the Countess Mahaut d'Artois (1302–29) for the cost to cut down and cart away firewood from the Park at Hesdin

Some of the earliest studies of medieval landscape and, by extension resource use, were studies of elite parks. Parks could vary widely in size from only a few hundred to well over a thousand acres. Often referred to as "deer parks," medieval parks did indeed house deer; however, that name is more appropriate for elite parks in the Early Middle Ages (ca. 800–1000 CE). By the Central Middle Ages (ca. 1000–1300 CE), elite parks likely included an elaborate system of fishponds, an assortment of water fowl (e.g., swans and heron), other game birds (such as pheasant), and a castle with windows offering carefully crafted vistas onto the park. In literature, the elite park has a long history, with mentions to hunting parks going back to at least Xenophon in the fourth century BCE. By the Central Middle Ages, a genre that we might call "estate manual" had developed to direct elite landowners how best to shape and use their land and its natural bounty. The most elaborate parks, such as the one I will be discussing at Hesdin, had follies, hermitages, and extensive horse and hound stables. Crucially, the park was only one element of a large elite estate, which would almost always include farmland, additional unwalled woodlands, and perhaps even an animal reserve, such as for rabbits.

While it is true that medieval park studies have predominantly addressed questions of power and display, since at least the 1980s, landscape historians and archaeologists have considered how elites exploited their natural resources to provide crucial social capital, most notably luxury food stuffs, such as venison, other game, and fish for the lordly table. Recent research has shown that parks and their associated landscapes, woodland, meadow, and stream were managed as multi-purpose and productive spaces.

The quotation from the extensive and highly detailed accounts of the influential lord, Mahaut d'Artois, Countess of Artois and Peer of France (ruled 1302–29), demonstrates the intersection between elite cultural practices (hunting) and resource use (firewood) that embody the multipurpose character of the medieval park and broader estate.[1] A close-reading of those same account books demonstrates that Mahaut, like her elite peers elsewhere in Europe, as we see in Cristina Arrigoni Martelli's chapter in this volume, engaged in a pattern of sustainable practices geared toward effective use and long-term supply of mundane and luxury natural resources at her estates, including within her most celebrated "pleasure park," Hesdin. Conservative practices defied park boundaries; retainers harvested practical resources, such as wood, from within the park walls as well as outside them. Further, infrastructure, such as mills and limekilns, were placed strategically all over the estate, including within the park. Although bounded by a wall, the park was not managed separately from the rest of the estate. In addition to the park, unbounded woodlands, for hunting and for wood products, and a hunting reserve were common features of aristocratic estates in the Central Middle Ages. François Duceppe-Lamarre, French archaeologist and scholar of Hesdin, has called this the "hunting triad" and urges us to look beyond the boundaries of the park to consider management of the estate as a whole.[2] When we take a more holistic approach, we see that Mahaut's resource management presses past the boundaries of the park and parallels practices that form the backbone of our current philosophy of conservation.

By centering my analysis of Mahaut's accounts on the questions of long-term supply and the stability of an estate—that is park, woodlands, and reserve—as a system, we can see that Mahaut and her retainers employed short- and long-term scales as well as balanced the needs of local and distant resources to maintain the estate's lands and resources as a continuously producing system. Woodland practices at Hesdin make clear that Mahaut's management was developed to create local, long-term supply with minimal additional cost and low waste. Medieval scholars have frequently posited that this kind of conservative behavior occurred and was part of the pattern of orally transmitted knowledge now called "Traditional Ecological Knowledge;"[3] however, we rarely have the opportunity to see these practices so clearly executed in our written sources outside literature. Finally, I conclude the chapter with a short reflection on how the natural resource management of elite estates may productively relate to later precepts of conservation as defined by nineteenth- and twentieth-century conservationists.

Hesdin: A Park and Estate

Mahaut's Artois, which is situated in modern-day northeastern France, provides a rich case study to reconstruct the practices of late medieval resource exploitation in a controlled estate setting under the auspices of an elite rather than in a town or commons, as we see in many of the other valuable contributions in this volume. Mahaut ruled Artois independently for twenty-seven years after the successive and untimely deaths of her father, Robert II, Count of Artois, and her husband, Otto, Count of Burgundy, in the French war against the Flemish. Her park at Hesdin is one of the most famous and influential of all medieval parks.

Hesdin's park, which is simply called the "Park" in the accounts, was established in 1295 by Mahaut's father, Robert II, Count of Artois (1250–1302).[4] It enclosed roughly 2,000 acres of fields, woodlands, marshes, and ponds. Accounts tell us that Mahaut's retainers exploited the Park's natural resources to its fullest; workers harvested stone, timber and firewood, fruit, deer, rabbits, and even produced lime mortar from within the bounds of the Park;[5] however, Hesdin's Park is best known for its technical marvels, such as the fur-covered monkey automata.[6] Scholarship on Hesdin's Park overwhelmingly focuses on the genesis of these marvels, with little attention given to the many ways in which Mahaut managed the production of resources, luxury and mundane, within the boundaries of the Park and its adjacent lands.[7]

Mahaut hosted the English king and queen Edward II and Isabelle of France (daughter of King Philip IV of France, and Mahaut's distant cousin) at Hesdin in 1313.[8] As Stephen Mileson argues, Hesdin's style of park, which was itself heavily influenced by earlier park models, from classical Roman to Islamic and Norman Sicilian,[9] directly shaped Edward II's park at Windsor.[10] Amanda Richardson has posited that Edward II's visit to Hesdin spurred renewed English royal interest in large, elaborate parks in general, as Edward III continued his father's emparkment program.[11] Hesdin's influence on northern European parks likely continued until its destruction by Charles V, Holy Roman Emperor (1519–1556) as part of his wars with France. Given Hesdin's contemporary influence, it makes an excellent case study; it is likely that similar patterns of resource management existed in other, contemporaneous parks.

Mahaut's archive contains an astounding number of minutely detailed financial accounts, which Sharon Farmer has posited is directly due to Mahaut's father's regency in Sicily in the late thirteenth century and the influence of Muslim accounting practices and a parallel rise in account details at the royal court.[12] Mahaut's "treasury"—as it is still frequently called because it is such a precious and rich resource—is so full that nearly the entire first floor of the departmental archive of Pas-de-Calais is dedicated to her documents, which number in the thousands. The pre-Revolutionary branch of the archive is named for the countess because her documents form the backbone of the collection. It is impossible to quantify the number of documents in her collection because it contains boxes upon boxes of receipts; the number of pieces certainly approaches ten thousand. Her accounts address everything from the number of bolts of cloth-of-gold purchased for Mahaut's household and retinue to when the grass in her celebrated park at Hesdin was trimmed or how many pests of each type (otter, bird of prey, polecat, etc.) were killed by the exterminator on retainer in the park and adjacent woodlands that comprised the estate at Hesdin.

Within Mahaut's archive, there are regular, trice yearly accounts for each of the thirteen administrative districts, called bailiwicks (*baillis*), that were prepared by a retainer called the "General Receiver." Mahaut's archive is nearly intact. It has only minor documentary losses, especially during the years of the Baron's Revolts, in which low-ranking French noblemen rebelled against the kings and local lords (1314–19).[13] Mahaut's nephew, also named Robert of Artois, used the revolts to attempt to take control of the county. Mahaut temporarily lost several of her estates during the period and was embroiled in court battles and scandal until her death in 1329. The General Receiver accounts concern the entire bailiwick and record any form of income, which was most frequently in the form of rent. Logically, expenditures in the general bailiwick accounts focus on areas directly controlled by the countess. Thus, most of the accounts concern the park, any nearby woodlands, and game reserves. In addition to the regular bailiwick accounts, Mahaut's archive includes frequent, more granular accounts that cover resource extraction and management: Works (i.e., construction), Woods and Waters, and Hunting. For this chapter, I rely on both the General Receiver and more specific accounts.

By far, the most plentiful and best-documented resource in Hesdin the bailiwick—confusingly, the park, estate, and administrative district all carry the same name—is wood. Partially, this is artifactual because woodland nurtured Hesdin the estate's many animal denizens and was an important part of the natural appeal of the park, hunting woodland, and reserve favored by medieval elites.[14] In addition to wood, elite woodland management also encompassed social hunting (from stands, as we saw in the opening quotation), hunting for status meat, and the joy and repose derived from sculpting greenery so that one could observe animals living among the foliage of the park, as described in the estate manuals of the time.[15] Although the many virtues of a pleasing landscape and social capital of luxury meats were crucial to Hesdin the park's purpose,[16] I focus here on wood rents and use for infrastructure construction and maintenance at the estate of Hesdin to understand the ways in which estate management was conducted holistically to balance short- and long-term social and economic needs at the estate level.

Wood Rents

It is well-known that lords commonly leased the rights to woodland for income in addition to household and estate use.[17] Mahaut was no different: her woodland leases generated substantial sums in many of the wooded bailiwicks of Artois. But more importantly for this study, the account entries include additional information about the lessee, what stage of the lease the current payment was for, and the length of the lease. When we look at lease length and frequency and aggregate income, we can discern patterns that show adaption to regular and extraordinary resource needs.

In the early years of Mahaut's reign, wood leases comprised a significant amount of Hesdin's income. Wood rents generated nearly 47 percent of the total average yearly income for the bailiwick. Wood rent income started to decline in early 1303 and stayed low for the rest of the decade. While wood rent income recovered somewhat by 1306, gross wood rents and sales were still only about 60 percent of what they were in 1301 (and in previous years under Count Robert II, who died in 1302). In 1308, woodland rent income declined suddenly again and stayed at about that same percentage until the 1320s—a sign that Mahaut stopped leasing. Other scholars have remarked on the precipitous decline, but no satisfactory explanation has been tendered.[18] I argue that the change reflects a wartime supply problem: there was not enough wood immediately available to harvest because it had already been taken, and perhaps was even overharvested, to provide for the troops in Flanders and bolster defenses at home.

In 1302 to 1305, King Philip IV of France (r. 1285–1314) embroiled Mahaut and Artois in a long-running and ultimately ineffectual war with Flanders. Flemish combatants regularly raided her districts on the border with Flanders: Lens, Aire, and Fampoux. But woodland all over Artois was requisitioned. Even though the war was entirely of the king's making, Mahaut's accounts document the substantial outlay of resource necessary to protect Artois and its armies. Nonthreatened woodlands, such as those in Hesdin—roughly 75 kilometers from the active raids—provided invaluable wood to make weapons, arrows, and the scaffolding and ladders to fortify castle walls during the war.[19] Even though Hesdin was never in any danger of Flemish raids, in 1302, the bailiff of Hesdin recorded costs to fortify Hesdin's castle that necessitated actively cutting down woodlands. For example, an entry "for the needs of the war" specified that workers "cut down lumber [*mairien*] from 'the Little Forest' [*Forestel*] for the castle's defensive scaffolds [*hourdes*] and to make ladders [*eschieles*] for fortification." Anne Hagopian van Buren surmised that "the Little Forest" was not contiguous to the park, but very close, and to the south of the town of Hesdin, which was itself at the southernmost point of the Park.[20] In an additional entry in the carpentry section for the same task, the bailiff recorded the additional labor cost to "cut down the lumber [*mairien*], smooth it out, and build the named scaffolds." An additional entry specifies that the scaffolds required that they use a full *arpent*[21] of wood from the Little Forest. "To complete the wickerwork [*cloie*] of the scaffolds, they gathered 4,400 switches [*vergue*]," which took two carts and three horses two days to move to the construction site. Unfortunately, unlike the scaffolds themselves, the account does not detail where the switches came from; however, the lack of recorded cost and short transport time demonstrate the switches were harvested from the estate itself. Later in the same account, there is a cost recorded to carry willow (*saulon*), the usual source of wicker, from the Park. Far from being exclusively for recreation as scholars have often assumed, the Park supplied the raw materials for war preparation.

An additional wartime preparation entry from the same roll records the building of two crossbows and the towers on which they were mounted. In addition to a number of costs to affix the large crossbows, they "sawed out the *courbes* [the curved portion of the bow]." The wood for the crossbows was taken from oaks (*caisnes*) of an unspecified origin.[22] There is no reference to what type of wood is used for the tower structure; however, it is reasonable to assume it and the bow wood were from within the estate because no transportation costs were listed. In general, this pattern of felling trees as they were needed generally assured a consistent supply of timber; although in the Late Middle Ages and Early Modern period, it will be targeted by reformers, as Sébastian Poublanc demonstrates in this volume.[23] Yet, in this case, it appears that they used too much wood in the construction of these various war preparations to continue regular harvest for some years.

In 1324 and 1325, woodland income still only made up 16 percent of Hesdin's total revenue (versus nearly 50 percent before the war), which would be in line with my hypothesis that the woodlands were harvested so much during the war years that they could not be let out for a number of years to allow for regrowth. In 1325 and 1326, Mahaut signed new agreements to sell and rent out additional woodland that boosted the total income derived from woodland those years to 33 percent, which doubled the percent of income deriving from wood rents after the war years (although of course this income was substantially lower than earlier prewar woodland rental income).[24] Unfortunately, there is no direct evidence to connect a new harvesting agreement in 1326 to post-war regrowth, and the accounts are imprecise on the question of species except in rare cases, and only then when the trees were felled; however, post-war regrowth is a plausible explanation for the jump in woodland income roughly twenty years after the exceptional wartime use. I posit that as the trees regrew, likely with care not recorded in the accounts, Mahaut could rerent out the woodlands for harvest that had been used to provide for the war effort.

The new 1325–26 contract used the same terms implemented in the prewar years: a short three-year contract cycle with payments three times a year. Although not explicitly stated, other scholars have demonstrated that short rent cycles facilitated the necessary supervision to prevent over-grazing and felling, which were on longer cycles.[25] It was common for these leases to be for small wood, which would preserve long-term timber growth.[26] Thus, the fact that even likely small wood leases were impaired strengthens the argument that the wartime impacted the long-term supply of older trees for lumber *and* the ability to harvest small and dead wood (used for kindling and firewood, which I discuss later in this chapter).

Medieval elites and their woodland managers planned rents to account for small-wood production (primarily used for fuel, but also for some small construction projects) and long-term timber growth for estate use (for large-scale construction) in addition to the commercial rental incomes. Despite the clear supply problem evinced by the catastrophic drop in woodland rents during and for two decades after the war, infrastructure repair and even new projects (for example, the water mills I will discuss shortly) continued in the Park and the estate as a whole. Hesdin's local wood harvest for estate use was not impacted by the war, unlike the raided regions on the border with Flanders, which saw catastrophic declines in rents and construction.[27] The most likely conclusion is that Hesdin's woodlands were harvested to the point that it took nearly twenty years for them to recover the prewar level of leasing, but that management in the interim years prioritized consistent estate supply.

Water Mills

Mills were crucial infrastructure in the late medieval estate and provide an excellent example of mixed local wood use. While many lords and towns used mills to power machinery, on lordly estates, mills were used primarily for grinding grains for food and beverage preparation.[28] Further, mills needed constant repair, so they occur frequently in the accounts and make it easy to track Hesdin's local wood use. The accounts demonstrate that despite the precipitous decline in woodland rent income, wood harvesting for new infrastructure and repairs appears to have been unaffected. It is crucial to note that for Mahaut, mills were themselves also a source of additional rents, another common elite practice to generate income, but that did not compensate for the lost woodland revenues during and after the war.[29]

For "sustainability" purposes, mills represent a nexus of multiple forms of resource management. They required extensive labor and diverse wood and wood-derived resources necessary to build and maintain their complex structures. Further, lords generally controlled multiple mills in any given estate, and Mahaut was no exception. Because the mills only occur incidentally in the accounts for repairs or to record rent income, it is difficult to estimate how many were operating in a given time frame. But it is safe to estimate that the bailiwick of Hesdin, which retained the best records, had at least three mills operating at any given time, at least one of which was in the Park.

Because references are so plentiful, even a quick skim of Mahaut's archive reveals dozens of entries on mill construction at Hesdin alone. Indeed, there are so many entries, not just at Hesdin but at all of Mahaut's estates, that it is impossible to cover their full extent here. A brief survey focusing on the entries that document construction and repair of mills can help us understand natural resource management practices. The best documented are water mills at Mahaut's at Hesdin and in the cities of Saint-Omer and Tournehem in the marshy north of the county near the English Channel. The information in the accounts is not sufficient to judge the type of watermill, although it would most likely be the more powerful vertical rather than the horizontal waterwheel.[30] Regardless, both types of watermills required a steady, strong flow of water, which was often regulated artificially with great difficulty, a water wheel, axel, cogwheel, arm, trundle, and millstone.

Water mills were largely made of timber. The accounts record the frequent replacement of these elements, from the wheels themselves (which required diverting the watercourse) to the arms, rims of interior wheels, and sprockets. While it was normal to name the type of work done (e.g., "To Master Simon, the carpenter, to make a pair of small wheels for the mill of the Grand Pond"), in some cases the accounts specify that the wood was cut locally.[31] Like with the entries for wartime fortification, in rare cases, the accounts even indicate from which woodland and what type of wood was used.

For example, preserved in a badly burned manuscript from Tournehem (a wooded and riverine area in the northeast of Artois), the bailiwick account records the cost to a Lambert Estienenin to cut down and cart oaks (*kaisnes*) to place in the mill.[32] The account reports no cost for the trees themselves, hinting that they were taken from "stores" rather than purchased. A more elaborate entry from Hesdin in 1325—the year before Mahaut started re-leasing the woodlands used for wartime supply—describes the costs for laborers to build a new mill and canals in the Park. Unfortunately, the accounts do not specify the location within the Park, as they sometimes do. The "new mill" was built from the water level up, from stone foundations to a new wheel and grinding mechanism. In this case, the account specifically records that wood, willow (*sauls*), from the Little Forest, the open woodland to the southeast of the walled park, is used to make the canals and the sluice gates for the new mill.[33] Willow, a tree that grows well in riparian conditions, appears to have been used for the submerged portions of the mill and waterways, likely due to its resistance to water damage. Willow was also harvested and employed for other aqueous uses, such as fish traps.[34] It was also likely the source of the wickerwork for the scaffolds constructed at the castle in 1302, which was discussed at the beginning of this chapter. Considering that the other named wood, oak, appears to have been employed exclusively for the above-ground arms of the mill, it is clear that the master carpenters understood and exploited the benefits of each wood type—behavior that economic historians have pointed out for decades but is worth noting is a pattern in line with the current conceptions of conservation and sustainability.

The majority of descriptions of mill work occur in the Works accounts. These accounts record total wages, by the week and by type, for the enumerated tasks. Works accounts rarely list the cost for the wood building materials, and it is likely that any wood listed that has no material cost recorded, even if the origin is unstated, derived from woodlands under Mahaut's control. The widespread practice of recording the costs only of externally purchased timber would explain why the majority of costs listed in the accounts are for the labor to move, fashion, and install timber. The kind of specificity of wood use we saw in the opening examples during the fortification of the castle is extraordinary. When exceptional costs for wood purchases are recorded, the accounts note the charge explicitly. For example, a singular charge in the nearly one hundred entries on mills I surveyed for this chapter describes a "large [*grant*] board [*ais*] of oak cut by Mikiel le Moy" to install in the mill.[35] It is worth noting that is also the only combination of "grant" and "ais" in the data set, which intimates it was a remarkably large piece of timber. Entries, such as this one "to cut and piece curved wood pieces [*courbes*] and other wood for the mill, 19 sous" occur with greater frequency. Even in the bailiwick accounts, which cover income and expenses more broadly than the works accounts, entries recording labor but no material, or material with no purchase cost, far outnumber those with additional resource cost.[36] Recording only extraordinary purchases of materials and using internal resources "silently" was common practice in late medieval households.[37] If Mahaut followed that pattern, which she most likely did, most of her substantial mill infrastructure was maintained through estate-sourced and managed wood and timber, which again parallels modern principles of sustainable management.

Mills also necessitated repair and replacement of additional materials, such as millstones, various small iron pieces, and stone. While the millstones and stones for the foundations can be quarried without any wood beyond incidentals, the creation of iron pieces and mortar required fuel (either wood itself or wood in the processed form of charcoal). Although the creation process is not described, "ironworks" are a common section in the works accounts and various iron pieces are "remade" at the mills regularly.[38] Mortar shows up even more frequently in the accounts. For example, in 1322, a Jehan de Waus was directed to cart mortar (*cauffour*) to the "Woods between Two Waters," west of the Park at Hesdin "to cover" the mill there (it is unclear *what* they are covering).[39] In general, every construction site would have a limekiln nearby, as mortar cannot travel far without drying out.

Archaeologist François Duceppe-Lamarre has located at least three major kiln sites both in and near the Park at Hesdin, although the accounts suggest that temporary limekilns would also pop up near major construction sites at a distance. Just as with modern brick buildings, mortar had to be replaced periodically. In Aire (on the eastern border with Flanders), an interior cogwheel at a mill had to be "re-mortared" (*rencauchie)* in place.[40] Of course, other brick and mortar buildings required the same construction and maintenance. For example, in the winter of 1310, when laborers were charged with "making bundles of branches of wood to burn in the castle of Madame and to make a limekiln to do the work on the walls of the new manor."[41] In addition to the creation of bricks and mortar, the kiln worker also made charcoal from wood unsuited to carpentry.[42] The household used some charcoal,[43] but it would also have been used in various industries. Although not planned in the same manner as the rent cycle for wood resource production, and certainly not at the level of production of nineteenth- or twentieth-century forestry, in the harvesting of small wood for fuel and small tasks, medieval people systematically—one might even say, sustainably—used all productive elements possible with as little waste as possible.

Other Carpentry and Fuel

Oliver Rackham maintains there was (and is) a sharp distinction between lumber, wood used for beams and construction, and wood that was used for firewood, mending, and other projects. In the accounts, the authors rarely use the term "wood" or "woodland" (*bos* or *bois*) to indicate firewood. Instead, they call it "dead wood" (*morbos*) or just "wood" (*laigne*). *Morbos* likely references fallen, dead wood, while *laigne* suggests green wood that would be cured for use as firewood.[44] Timber was the most valuable wood product because it consumed the trunk of the tree. But it takes years to grow a timber tree. Other useful wood was gathered from fallen branches, coppices, suckers, and the branches of trees felled for timber. Unfortunately, the language in the accounts is not precise enough to make a clear distinction. Sometimes the accounts specify "lumber" (*mairien*), and sometimes they say "wood" (*bos*), the most general term, but go on to describe its use in construction. Sometimes they call the wood very specifically "planks" (*plankes*). Although both timber and wood were renewably sourced, the small branches used to cut pieces of wood regrew faster, making it significantly less expensive and still an excellent source of smaller building materials.

Craftsmen used small wood for everyday repairs of other buildings.[45] Local wood products were employed to create floorboards, posts, and other materials for areas all over the estate from rooms within the castle[46] to the chapel[47] to gateposts on the fence[48] and to outdoor pavilions.[49] On one occasion, workers made posts from a tree in the park felled specifically for this purpose, which were then transported to a workshop in Arras for finishing.[50] As we saw with the "large planks" cut on demand from the Park's oak, timber was a luxury wood resource that elicited extraordinary measures. In this case, it was intended for a large building owned by Mahaut in the administrative center of Arras. Further, far from respecting the boundary of the Park, Mahaut's laborers used wood sources from anywhere in the area, including the Park, which we often see as exclusively for pleasure, to provide the wood necessary for their carpentry. On occasion, men from the neighboring villages even utilized general "wood for carpentry" derived from woodland around Hesdin to make tools for their work, such as when they took smaller cuts of wood to make "pitchforks for the vines."[51] Another entry in the same account specifies that the wood is used for the handle of a pitchfork.

Wood also provided fuel for the other purposes in the kitchen, house, and industry. The trees both in the Park and in the woodlands just outside the walls generated a lot of dead wood (*morbos*)—small naturally occurring dried sticks—that were bundled together and used as kindling and firewood. Hesdin's woodlands also produced firewood (*laigne*). Laborers harvested firewood from outside the Park for both the kitchen and the houses.[52] In fact, Hesdin's harvest of firewood (both dead and green) was so plentiful that it was sold by the cartload nearly every week from the Forest of Hesdin for varying amounts; although, unfortunately, we do not know to where it was carted.[53] Even during the lean years in rent income, Hesdin's firewood production was healthy; cartloads were sometimes delivered only on the promise of repayment.[54] Finally, woodlands were multiuse and produced more than wood products, such as hay, which was used to feed animals.[55] Now that we have clearly seen a systematic pattern of management to privilege local, estate-based use during and after a period of extraordinary need, it is worth considering if we should call that systematic management sustainable and categorize it as conservation. And further, if we do, what it may help us accomplish.

Some Reflections on Using Sustainability and Conservation to Study the Preindustrial Era

As the introduction to this volume sketches out, "sustainability" and eventually "conservation" were clearly defined by the eighteenth century and further refined in the nineteenth century to mean "continuous, resilient, and sustainable use."[56] Both terms were rapidly folded into the political cultures of the major Western countries, in particular, what will become Germany, France, and the United States. For the most part, the medieval environmental history subfield has skirted the question of whether the widespread strategic, planned reuse of materials seen at Mahaut's Hesdin could constitute a form of sustainable practice as defined by these modern intellectuals. Instead, premodern scholars have focused on the aesthetic and sociopolitical value of elite resource management.

It is well known that "pleasure parks," like Hesdin, were aristocratic playgrounds that hosted desirable natural resources, such as venison, and demonstrated wealth and prestige. As many scholars have shown, elites manipulated landscapes to reinforce their wealth, placement in the social hierarchy, and political authority.[57] But parks were not exclusively for pleasure or entertainment. Resources in and around parks were managed to support the quotidian workings of an elite's holdings, such as building infrastructure and economic solvency, which did boost sociopolitical value but were pragmatically necessary; however, evidence fine-grained and systematic enough to reconstruct premodern management practices, such as that in Mahaut's archive, is uncommon even for medieval environmental historians. Mahaut's documents provide an extraordinarily rich view of day-to-day estate and park activities and their impacts.

A recent essay in the *American Journal of Botany* by Erin Beller et al. on historical ecology helpfully establishes the defining principles and research scales of which environmental history is a part and emphasizes how history actively complements environmental studies. The authors suggest that principles of historical ecology "might address human-environment interactions and feedbacks, define the scope of 'history' in the context of ecological change, delineate the spatial and temporal scale and resolution of historical ecology studies and temporal continuity, address the importance of place, address theories of ecological and landscape transformation and disturbance, and reflect on its application to current and future ecosystems."[58] Beller et al. break environmental studies down into four "scales"—population, community, ecosystem, and landscape—and emphasize no one study needs to address all scales. Thus, historians should helpfully occupy the community and ecosystem levels, where they can demonstrate how resource-use patterns have shaped long-term resource availability and how patterns created long-term legacies.[59] Further, Beller et al. emphasize that the authors and their resulting studies should "draw inspiration from existing ecological principles," including sustainable land-use management.[60] In essence, it is helpful for historians to work with terms that may not be native to the historical period studied in order to draw historical parallels and build interdisciplinary bridges.

In this vein, I did not seek to argue in this chapter that Hesdin or medieval elites aimed for "sustainability" or were "conservationists" in the strict, modern, economic view of development. Rather, I argue that by looking at Mahaut's management through the lenses of sustainability and conservation, we see that she and her retainers viewed and treated the environment as a complex, exploitable system that prioritized local resources, both from taxes in kind and estate production, and sought to limit waste through repurposing, such as in the quotation that opened this essay. Mahaut ordered a tree to be cut so she might "see the chase" better, as in the opening quotation, and it was recorded that the wood from that clearing should also find use as firewood. By focusing on the growth, harvesting, transportation, and exploitation patterns of Mahaut's natural resources rather than by focusing exclusively on her economic or political motivations—which were legion—we see that medieval people implicitly understood their environments as interdependent systems (ecologies) to be manipulated for personal (or community, in the case of other studies in this volume) long-term gain.

As the Introduction noted, local pressures (e.g., the realities of transport and the slow-growing nature of trees) drove a mixture of short- and long-term management rather than any abstract sense of preservation. Further, medieval management was dependent on the vicissitudes of local physical and political environments. Emergencies, such as war, could disrupt the careful balance between short-term and long-term supply; however unfortunate for Mahaut and her leaseholders, the unusual interruption in leases is fortunate for scholars—the extraordinary entries demonstrate that Mahaut and her retainers had an awareness of and an intentional resource adaptation in response to changing external forces, in this case, an extraordinary need for woodland during the Franco-Flemish Conflict of the first decade of the fourteenth century. To meet the singular needs required by the war and balance consistent production of timber and small wood, Mahaut and her retainers limited leases, freeing up woodland for the supply of internal timber, small wood, and fuel.

Mahaut's accounts clearly show that medieval resource practices shared some commonalities with the modern principles of sustainability and conservation. I would even suggest it checks off two of the categories of "conservation" as defined by the great US American environmentalist, Gifford Pinchot: development of infrastructure and the prevention of waste.[61] She engaged in a pattern of conservative practices geared toward most effective use and a balance of short- and long-term supply, the self-same practices that would later form the backbone of modern conservationism. To me, that suggests it is worth our time to consider further how these modern concepts might illuminate the complex systems of premodern management and challenge us to think differently about the development of management in Europe and globally.

Abigail P. Dowling is Assistant Professor of History at Mercer University in Macon, Georgia, US. She received her PhD in Medieval

Studies and History at the University of California, Santa Barbara (2014). She teaches classes on environmental history and material culture in Greece, Rome, and the Middle Ages. She has published articles on ecology in classical and medieval farming manuals and the sociopolitical symbolism of resource management in medieval parks. She is currently working on a book on resource management and transport in early fourteenth-century Artois and an article on humorism in the political rhetoric of the post-Roman western kingdoms.

Notes

1. Here is the original text for the translated epigraph above: "Praimes a Symon Pikart pour abatre bos dessous le bersoire pour che que madame peust mieu voir les bestes courre et faire fagos de le menne raine mettre le grosse legne pour monsieur en lieu la bon peust kerkier a karete viii lb. xvi s." Archives départementales Pas-de-Calais, Dainville, Arras (AdPdC), Series A441.5.
2. François Duceppe-Lamarre, "La fonction cynégétique des espaces boisés médiéveaux à rravers l'exemple des cervidés et lagomorphes (Nord-Pas-de-Calais, XIe-XVe Siècles)," *Anthropozoologica* 28 (1998): 37–38.
3. Fikret Berkes, Johan Colding, and Carl Folke, "Rediscovery of Traditional Ecological Knowledge as Adaptive Management," *Ecological Applications* 10, no. 5 (2000): 1251–62; C. R. Menzies, *Traditional Ecological Knowledge and Natural Resource Management* (Lincoln: University of Nebraska Press, 2006).
4. Marguerite Charageat, "Le parc d'Hesdin, création monumentale du XIIIe siècle," *Bulletin de la société de l'histoire de l'art français* (1950): 94–106.
5. Abigail P. Dowling, "Landscape of Luxuries: Mahaut d'Artois's (1302–1329) Management and Use of the Park at Hesdin," in *Rural Space in the Middle Ages and Early Modern Age: the Spatial Turn in Premodern Studies*, ed. Albrecht Classen (Berlin: De Gruyter, 2012), 367–88.
6. Anne Hagopian Van Buren, "Reality and Literary Romance in the Park of Hesdin," in *Medieval Gardens: Dumbarton Oaks Colloquium on the History of Landscape Architecture*, ed. Elizabeth B. MacDougall (Washington, DC: Dumbarton Oaks Library and Collection, 1986), 129.
7. For an example of a detailed study, see Amanda Richardson, "'Hedging, Ditching and Other Improper Occupations': Royal Landscapes and their Meaning Under Edward II and Edward III," *in Fourteenth Century England*, ed. J. Hamilton (Woodbridge: Boydell and Brewer, 2006), 26–42.
8. Sharon Farmer, "Aristocratic Power and the 'Natural' Landscape: The Garden Park at Hesdin, c. 1291–1302," *Speculum* 88, no. 3 (July 2013): 645.
9. Farmer, "Aristocratic Power," 644–46.
10. Stephen A. Mileson, *Parks in Medieval England* (Oxford: Oxford University Press, 2009), 92.

11. Amanda Richardson, "'The King's Chief Delights': A Landscape Approach to the Royal Parks of Post-Conquest England," in *The Medieval Park: New Perspectives*, ed. Robert Liddiard (Macclesfield, England: Windgather Press, 2007), 37.
12. Farmer, "Aristocratic Power," 644–46; Jean Dunbabin, *The French in the Kingdom of Sicily, 1266–1305* (New York: Cambridge University Press, 2011), 109–15.
13. Elizabeth A. R. Brown, "Reform and Resistance to Royal Authority in Fourteenth-Century France: The Leagues of 1314–1315," in *Politics and Institutions in Capetian France*, ed. Elizabeth A. R. Brown (Hampshire, England: Variorum, 1991), 109–37.
14. Farmer, "Aristocratic Power," 646–48.
15. Pietro de Crescenzi, *Ruralia Commoda: Das Wissen Des Vollkommenen Landwirts Um 1300*, ed. Will Richter and Reinhilt Richter-Bergmeier, Editiones Heidelbergenses, no. 25 (Heidelberg: C. Winter, 1995), 3: 3.
16. L. M. Cantor and J. Hatherly, "The Medieval Parks of England," *Geography* 64, part 2, no. 283 (April 1979): 71–85; John Fletcher, *Gardens of Earthly Delight: The History of Deer Parks* (Oxford: Windgather Press, 2011); Naomi J. Sykes, "Taking Sides: The Social Life of Venison in Medieval England," in *Breaking and Shaping Beastly Bodies: Animals as Material Culture in the Middle Ages*, ed. Aleksander Pluskowski (Oxford: Oxbow Books, 2007), 149–60.
17. Oliver Rackham, *Trees & Woodland in the British Landscape: The Complete History of Britain's Trees, Woods, & Hedgerows*, rev. ed. (London: Phoenix Press, 1990), 76–83.
18. Bernard Delmaire, ed., *Le compte général du receveur d'Artois pour 1303–1304: édition précédée d'une introduction à l'étude des institutions financières de l'Artois aux XIIè-XIVè siècles*, ed. Des Lettres et des Beaux-Arts de Belgique Commission Royale d'Histoire Academie Royale des Sciences (Brussels: Palais des Academies, 1977), lxxvii.
19. AdPdC 180.3. Partial transcription available in Jules-Marie Richard, *Inventaire-sommaire des archives départementales antérieures à 1790, Pas-de-Calais, Archives Civiles—Série A*, 2 vols. (Arras: Imprimerie de la Société du Pas-de-Calais, 1887), 1: 192.
20. Van Buren, "Reality and Literary Romance," plate 2.
21. The *arpent* was a unit of land measurement approximate to an acre.
22. AdPdC 180.3. Partial transcription available in Richard, *Inventaire-sommaire*, 1: 192. Delmaire, *Compte général*, lxxvii.
23. Rackham, *Trees & Woodland*, 69.
24. AdPdC A436.1 and A451.3.
25. Rackham, *Trees & Woodland*, 67; Richard Keyser, "The Transformation of Traditional Woodland Management: Commercial Sylviculture in Medieval Champagne," *French Historical Studies* 32, no. 3 (2009): 353–84.

26. Rackham, *Trees & Woodland*, 62–66.
27. This is part of an unpublished book manuscript I am writing. It was presented as "Mapping Nature through Use: Natural Resource Exploitation during the Attack and Rebuilding of Saint-Omer (1302–09)" (Sewanee Medieval Colloquium, Sewanee, TN, 2016).
28. John Langdon, *Mills in the Medieval Economy: England 1300–1450* (Oxford: Oxford University Press, 2004), 2.
29. AdPdC A198.6.
30. Langdon, *Mills*, 73; Richard Holt, *The Mills of Medieval England* (London: Blackwell Publishers, 1988), 121.
31. AdPdC A190.1
32. AdPdC A198.6.
33. AdPdC A441.5.
34. de Crescenzi, V.39.4.
35. AdPdC A 438.1.
36. AdPdC A441.
37. Malcolm Vale, *The Princely Court: Medieval Courts and Culture in North-West Europe* (Oxford: Oxford University Press, 2001), 78.
38. AdPdC A198.6.
39. AdPdC A410.2.
40. AdPdC A177.1.
41. AdPdC A277.
42. AdPdC A441.5.
43. For example, in the daily account for the winter of 1315 for "sacs of carbon," AdPdC A329.
44. Frédéric Eugène Godefroy, and P. Godefroy, *Dictionnaire de l'ancienne langue française et de tous ses dialectes du 9e au 15e siècle* (Paris: F. Vieweg, 1881), 4: 701.
45. Rackham, *Trees & Woodlands*, 67–70.
46. AdPdC A441.5.
47. AdPdC A446.3.
48. AdPdC A441.5.
49. AdPdC A441.5.
50. AdPdC A441.5.
51. AdPdC A477.1.
52. AdPdC A441.5.
53. AdPdC A477.1.
54. AdPdC A477.1.
55. AdPdC A441.5.
56. Hans Carl von Carlowitz, *Sylvicultura Oeconomica, oder hauswirthliche Nachricht und naturmässige Anweisung zur wilden Baumzucht* (Leipzig: Braun, 1713), as cited in Jeremy L. Caradonna, *Sustainability: A History* (Oxford: Oxford University Press, 2014), 35–38. George Perkins Marsh, *Man and Nature*, repr. ed., ed. David Lowenthal (Seattle: University of Washington

Press, 2003 [1864]), 28–29; Gifford Pinchot, *The Fight for Conservation* (New York: Doubleday, Page & Co., 1910), 42, 44, 49.

57. Oliver Creighton, *Designs Upon the Land: Elite Landscapes of the Middle Ages* (Woodbridge: Boydell & Brewer, 2009).
58. Erin Beller et al. "Toward Principles of Historical Ecology," *American Journal of Botany* 104, no. 5 (May 2017): 648.
59. Beller, et al., "Principles of Historical Ecology," 647.
60. Beller, et al., "Principles of Historical Ecology," 648.
61. Pinchot, *The Fight*, 42, 44, 49.

Bibliography

Beller, Erin, et al. "Toward Principles of Historical Ecology." *American Journal of Botany* 104, no. 5 (May 2017): 645–48.

Berkes, Fikret, Johan Colding, and Carl Folke. "Rediscovery of Traditional Ecological Knowledge as Adaptive Management." *Ecological Applications* 10, no. 5 (2000): 1251–62.

Brown, Elizabeth A. R. "Reform and Resistance to Royal Authority in Fourteenth-Century France: The Leagues of 1314–1315." In *Politics and Institutions in Capetian France*, edited by Elizabeth A. R. Brown, 109–37. Hampshire, England: Variorum, 1991.

Cantor, L. M., and J. Hatherly. "The Medieval Parks of England." *Geography* 64, part 2, no. 283 (April 1979): 71–85.

Caradonna, Jeremy L. *Sustainability: A History*. Oxford: Oxford University Press, 2014.

Charageat, Marguerite. "Le parc d'Hesdin, création monumentale du XIIIe siècle." *Bulletin de la société de l'histoire de l'art français* (1950): 94–106.

Creighton, Oliver. *Designs Upon the Land: Elite Landscapes of the Middle Ages*. Woodbridge: Boydell & Brewer, 2009.

Dark, Petra. *The Environment of Britain in the First Millennium AD*. London: Bristol Classical Press, 2000.

de Crescenzi, Pietro. *Ruralia Commoda: Das Wissen Des Vollkommenen Landwirts Um 1300*. 4 vols., edited by Will Richter and Reinhilt Richter-Bergmeier. Editiones Heidelbergenses, no. 25. Heidelberg: C. Winter, 1995.

Delmaire, Bernard, ed. *Le compte général du receveur d'Artois pour 1303–1304: édition précédée d'une introduction à l'étude des institutions financières de l'Artois aux XIIè-XIVè siècles*, edited by Des Lettres et des Beaux-Arts de Belgique Commission Royale d'Histoire Academie Royale des Sciences. Brussels: Palais des Academies, 1977.

Dowling, Abigail P. "Landscape of Luxuries: Mahaut d'Artois's (1302–1329) Management and Use of the Park at Hesdin." In *Rural Space in the Middle Ages and Early Modern Age: The Spatial Turn in Premodern Studies*, edited by Albrecht Classen, 367–88. Berlin: De Gruyter, 2012.

Duceppe-Lamarre, François. "La Fonction cynégétique des espaces boisés médiéveaux à rravers l'exemple des cervidés et lagomorphes (Nord-Pas-de-Calais, XIe-XVe Siècles)." *Anthropozoologica* 28 (1998): 37–38.
Dunbabin, Jean. *The French in the Kingdom of Sicily, 1266–1305.* New York: Cambridge University Press, 2011.
Farmer, Sharon. "Aristocratic Power and the 'Natural' Landscape: The Garden Park at Hesdin, c. 1291–1302." *Speculum* 88, no. 3 (July 2013): 644–80.
_____. "La Zisa/Gloriette: Cultural Interaction and the Architecture of Repose in Medieval Sicily, France, and Britain." *Journal of the British Archaeological Association* 166, no. 1 (September 2013): 99–123.
Fletcher, John. *Gardens of Earthly Delight: The History of Deer Parks.* Oxford: Windgather Press, 2011.
Godefroy, Frédéric Eugène, and P. Godefroy. *Dictionnaire de l'ancienne langue française et de tous ses dialectes du 9e au 15e siècle.* Paris: F. Vieweg, 1881.
Hoffmann, Richard C. *An Environmental History of Medieval Europe.* New York: Cambridge University Press, 2014.
Holt, Richard. *The Mills of Medieval England.* London: Blackwell Publishers, 1988.
Keyser, Richard. "The Transformation of Traditional Woodland Management: Commercial Sylviculture in Medieval Champagne." *French Historical Studies* 32, no. 3 (2009): 353–84.
Langdon, John. *Mills in the Medieval Economy: England 1300–1450.* Oxford: Oxford University Press, 2004.
Marsh, George Perkins. *Man and Nature.* Reprint ed. Edited by David Lowenthal. Seattle: University of Washington Press, 2003 [1864].
Menzies, C. R. *Traditional Ecological Knowledge and Natural Resource Management.* Lincoln: University of Nebraska Press, 2006.
Mileson, Stephen A. *Parks in Medieval England.* Oxford: Oxford University Press, 2009.
Pinchot, Gifford. *The Fight for Conservation.* New York: Doubleday, Page & Co., 1910.
Pretty, Jules N. "Sustainable Agriculture in the Middle Ages: The English Manor." *The Agricultural History Review* 38, no. 1 (1990): 1–19.
Rackham, Oliver. *Trees & Woodland in the British Landscape: The Complete History of Britain's Trees, Woods, & Hedgerows.* Rev. ed. London: Phoenix Press, 1990.
Richard, Jules-Marie. *Inventaire-sommaire des archives départementales antérieures à 1790, Pas-de-Calais, Archives Civiles—Série A.* 2 vols. Arras: Imprimerie de la Société du Pas-de-Calais, 1887.
Richardson, Amanda. "'Hedging, Ditching and Other Improper Occupations': Royal Landscapes and their Meaning under Edward II and Edward III." In *Fourteenth-Century England,* edited by J. Hamilton, 26–42. Woodbridge: Boydell and Brewer, 2006.
_____. "'The King's Chief Delights': A Landscape Approach to the Royal Parks of Post-Conquest England." In *The Medieval Park: New Perspectives,* edited by Robert Liddiard, 27–48. Macclesfield, England: Windgather Press, 2007.

Sykes, Naomi J. "Taking Sides: The Social Life of Venison in Medieval England." In *Breaking and Shaping Beastly Bodies: Animals as Material Culture in the Middle Ages*, edited by Aleksander Pluskowski, 149–60. Oxford: Oxbow Books, 2007.

Vale, Malcolm. *The Princely Court: Medieval Courts and Culture in North-West Europe*. Oxford: Oxford University Press, 2001.

Van Buren, Anne Hagopian. "Reality and Literary Romance in the Park of Hesdin." In *Medieval Gardens: Dumbarton Oaks Colloquium on the History of Landscape Architecture*, edited by Elizabeth B. MacDougall, 115–34. Washington, DC: Dumbarton Oaks Library and Collection, 1986.

von Carlowitz, Hans Carl. *Sylvicultura Oeconomica, oder hauswirthliche Nachricht und naturmässige Anweisung zur wilden Baumzucht*. Leipzig: Braun, 1713.

Worster, Donald. "Transformations of the Earth: Toward an Agroecological Perspective in History." *The Journal of American History* 76, no. 4 (March 1990): 1087–1106.

CHAPTER 3

Eating Your Game and Having It Too

North-Central Italian Conservation of Game Animals and Birds, 1300–1550

Cristina Arrigoni Martelli

This chapter deals with the legislative strategies employed by late medieval North and Central Italian municipalities and ducal administrations for managing game animals and birds. These urban and lordly powers issued legislation aimed at regulating access and use of game resources in their territories. Aristocrats and local hunters attributed different sociocultural value to game animals and had different opportunities to access them, yet both were cognizant of the political value of the game as a resource. As hunters, dukes and local individuals possessed a concrete knowledge of their territories and of the activity of capturing animals and birds, which they had garnered through practice and observation. Hunting gave them a clear and functional vision of what is called ecological interdependence. Although they did not explicitly express their measures in an ecological light, their ecological clarity comes through when we read them through the lenses of sustainability and conservation.

The different value attributed to game affected the way it was hunted. Hunting by aristocrats was a performance involving raptors, packs of dogs, long chases on horseback, and entourages of huntsmen. The commercial hunters, on the other hand, exploited their local hunting landscapes efficiently with nets, snares, camouflage, and bait.[1] For the elites, the meat was not the goal. When they wanted their larders filled, elites sent their huntsmen to procure it. These men hunted in the same way as most commercial hunters did while enjoying legal exceptions, at least in Lombardy. Meat for the markets was the goal of the municipal hunters, citizens of individual townships, who hunted to sell to up-and-coming urban aristocracies keen on copying aristocratic behaviors and cultural attitudes but who were unable themselves to hunt.[2] To maintain their hunts, both townships and aristocratic lords responded to maintain game animal populations but with very different methods. This chapter will focus on two geographical areas of Italy: Fucecchio, a small, well-documented locality in Tuscany; and ducal Visconti-Sforza Lombardy, an area rich in hunting reserves and parks. It will look in detail at the strategies of those in a position to legislate on hunting. Then I will present those strategies in the wider and dynamic context of Italian landscapes as producers of natural resources and as politically contested areas.

Legal Context

In medieval Italy, as elsewhere in Europe, who used land and for what purposes was part and parcel of a dynamic process determined by demographic shifts and power politics. Over a five- to six-hundred-year period, parts of Italy witnessed dramatic landscape changes. In the ninth and tenth centuries, sparser populations and less centralized lordly powers had led to less intensive land use. Demographic increase, the rise of the municipalities (*comuni*), and the corollary increase in the value of agricultural land sped up the process of land reclamation. In Italy, this ultimately led to the creation of new villages (*borghi nuovi*). Urban development and the privileging of a commercial economy augmented the pressures on the countryside and those living in it. While game animals and their hunts had traditionally provided luxury rituals and products for kings, emperors, and nobility, there was increasingly now money involved in the commercial capture of game. All the while, reduced areas of pasture and woodlands, refuges to game populations, combined with increased human populations put pressure on the desirable beasts.

Hunting landscapes were often also spaces with additional vital uses, be it cutting of wood and timber, pasturage, or, in the case of aquatic birds, fishing. Legal protections for coexisting ownership and use rights were superimposed over these multiuse landscapes. This resulted in a complicated patchwork of rights of precedence, extraction, and access. One overarching distinction derived from Roman law was found between direct dominion (*dominium directum*) and useful dominion (*dominium utile*). The former was the prerogative of kings and private landowners; the latter a right of use obtainable through continuous use or specific agreement with the owner. Coveted or vital resources were thus redistributed and use right could even be upheld in court as a right of individuals who had invested years of work.[3]

Local users considered resources they extracted through prolonged use as essentially their own, despite claims to the contrary. In a dispute from 1276, Alberto and Bandinello Lese claimed that "they and their elders" had captured "birds" (ducks) on the river (for free) for almost twenty years and stated that their right to do this amounted to "almost ownership" of the duck hunting location(s), and this seemingly implied future use.[4] As long as demand for land and its resources was kept in check, such arrangements lasted; however, by the late fourteenth century, medieval Italy and its cities developed into voracious economic powerhouses casting increasingly wider nets over the countryside around them. Control of land and its resources required stronger tools than prolonged use and work. An indirect tax or fee (*gabella*) distributed annually through municipal auctions became the price to be paid by the highest bidder for the exclusive license to use a resource in a certain area. The town government, claiming superior ownership and lordship, now acted as the landlord of its territory's resources toward its citizens.

The decades after the Great Mortality of 1348 brought environmental and administrative change to hunting and other forms of resource extraction. On the environmental side, scholarship argues for a rebound in natural resources with a drastically reduced, albeit growing, human population. On the administrative side, what an individual or group previously obtained and retained for free through a state of "almost ownership" simply through use now increasingly had to be paid for each year. The indirect tax (*gabella*) was introduced slowly in Tuscany around 1200 as a substitute for direct and personal taxes. Fucecchio embraced this fiscal device in 1315. After 1348, with a reduced population base, the levying of taxes changed conceptually. No longer was the individual the focus of town administrators, but rather the remaining individuals' uses of resources and services. These became associated with a cost changing how work was viewed and organized.

In Fucecchio, not only was hunting now subject to a fee, town authorities disaggregated resource extraction activities that had previously been bundled in order to maximize revenues. In the fifteenth century, for example, fowling was separated from fishing.[5] Fucecchio's hunting taxes were now two, one of which was levied on the hunting of roe deer in the local hills of the Cerbaie—animals captured with nets and with an explicit commercial intent—but with no apparent limitations of other kinds. The second was levied on the capture of ducks, similarly with nets, in and near the banks of the nearby Gusciana River.[6] In both cases, the licenses were auctioned to the highest bidder in an annual event.

For all this, there is no doubt that hunting was a marginal element in Fucecchio's fiscal balance sheet. For example, in 1395, the income from the tax on deer hunting accounted for a mere 1.4 percent of the city's total revenue.[7] In the post 1348 economy, this still represented an attractive income for the town. Further, the resource-use tax may also be seen as part of the contemporary professionalization of Italian business practices, and it helped protect municipal citizenries, their livelihood, and town administrations. Threatened from the outside by the political schemes of larger cities and municipal neighbors, and from the inside by a falling population base, towns became more defensive of their resources and alert to their economic value and their potential as bulwarks against encroachment.[8] Using the entry fee as part of a selection process, towns became managers of their resources, which in turn made their use visible in the written record. It also resulted in the birth of a group of professional investors for whom long-term viability and monopoly of access were desirable.

This apparently small change in the way resources were allocated, from use-right to license, affected the internal organization of those who extracted them. Being a practitioner in a certain field was a lifelong vocation, and the practice of trades ran in families.[9] Hunting rights were now no longer a matter of past and present work; they were simply a matter of available cash. The investor needed a guarantee of pecuniary returns to balance added risks and expenses while attempting to organize labor and production in the face of competition. Those who could not pay the fee, but who still needed to work, had to put themselves under an investor—a vastly different situation from that of Alberto and Bandinello.

The resulting operational dynamics relied less on kin solidarity than on impersonal or meritocratic arrangements, especially as these were embodied in the type of contractual business partnership known as "society" (*societas*). Although these contracts were flexible, this new system resulted in what Malvolti calls a "loosening" of the "collective and corporative tradition" that had characterized groups or families that fished up to the second half of the thirteenth-century.[10] In watching the Tuscan hunters at work, it is possible to suggest a similar scenario.

While whatever changes took place upriver, or over the next set of hills, could affect local game populations substantially, these were beyond the control of local administrators and hunters. Game had to compete with such land uses as the cultivation of crops and the grazing of cattle.[11] Yet statutory prohibitions of certain techniques demonstrate that contemporaries understood which strategies would catch many animals. Hunting manuals and written sources that otherwise rely for much of their knowledge on the observation of common folk practice, however, also point to a perspicacious understanding of things that is not necessarily destructive.[12] Fucecchio's extant fiscal registers show that for at least fifty years running, hunting through the payment of the tax continued without any evidence of scarcities. This may have been because the deer population remained viable enough to hunt commercially; however, there is a recurrence of names among those who obtained the license, suggesting continuous individual and familial practice.[13] Further, the accounts indicate that these investors/hunters were involved with other extractive activities, such as water fowling, fishing, or pasturage, which took place in the very same locations as the hunts. The inability to control that which lay beyond one's political grasp did not negate ecological knowledge based mostly on continuous practice and observation.

A relatively small group of men consistently tied their fortunes to diversified investment in nonagricultural resource extraction in contiguous environments, even to the point of serving as warrants for one another.[14] Potentially competing activities, such as deer hunting and pasturage, could thus be coordinated, and knowledge and use of productive potential habitats maximized. Complementary seasonality between, for example, fishing and fowling benefitted efficient management of resources, time, and equipment. Perhaps these arrangements also offset losses between extractive activities. Evidence from the visible and lucrative fishing tax record shows increasing ties and even overlap between who obtained the fishing right and public officials.[15] Being specialists of their environment carried political power, as those able to access hunting as a resource were fewer, better off, and better connected. These men were increasingly monopolizing access to land in ways different from that of simple use. It may well be that the hunting tax continued to be desirable because hunting continued and was lucrative. It may also have been desirable simply because it allowed a few individuals free, predictable access all through the year to a desirable area for other purposes—for example, pasturage. As recent scholarship is also discovering, the secret behind both environmental and political resilience in Tuscany may have been a reasonably wise use of a multiplicity of small-scale, locally managed resources.[16] The hunting legislation put in place by towns on their territories, on both sides of the Black Death watershed, was geared at making such resilience possible.

Municipal Practices: Fucecchio and Its Hunts

The medieval municipality of Fucecchio was situated 35 km west of Florence on a knoll between what were then the vast wetlands of the plain of Lucca (*Padule di Fucecchio*), the southern edge of the central Apennines, the Cerbaie hills rich in oaks and chestnuts, and the Gusciana River (now Canale Usciana), which drained the said wetlands and flowed into the Arno River. Towns in this area display a representative sample of municipally sponsored sustainable measures geared toward local game. Hills and wet areas, the former unsuited for intensive agriculture, made for a small hunters' paradise. The documented targets were the roe deer (*Capriolus capriolus*) that lived in the Cerbaie hills and the "ducks" (mallards and widgeon, respectively *Anas plathirynchos* and *Anas penelope*), which populated the Gusciana River and its banks. Various sources sporadically mention "birds" that were not aquatic, likely quail, partridges, and pheasants (*Coturnix coturnix* and *Perdix perdix*, *Phasianus cholchicus*). Small passerines such as thrushes, larks, and starlings *(Turdidae*, *Alaudidae*, and *Sturnidae*) as well as boar (*Sus scrofa*) and bear (*Ursus actos*), were also hunted, but receive sporadic mention in the sources.

Fucecchio's internal matters were regulated by a collection of laws produced by the municipal administration: the customary "Statute" (*Statuto*). This had been the practice for North-Central Italy's many independent municipalities (*liberi communi*) starting around 1100. The statute was a foundational piece of legislation that covered subjects considered indispensable to the functioning of self-governed municipalities, as well as matters crucial to local landscape uses and resources, including hunting, and could be integrated by "resolutions" (*deliberazioni*)—ad hoc and binding decisions by the municipal council. When hunting regulations appear in statutory documents, they consist of limitations to access, amounts of game taken, locations, seasons, techniques, and sometimes even the organization of the hunters. Recent research for Italy shows that communal lands, or "commons," owned by the municipality and used in various guises by its members were vital for resource supply as well as community building.[17] It was therefore important for a municipality to carefully balance different uses and users.

A closer look at the regulations helps shed light on what contemporaries considered good management practices and what malpractice legislators saw as threats to supply. In 1340, Fucheccio's statute stated that nobody could hunt "birds" with "nets." Such devices captured many birds at once, and so were preferred by commercial hunters; however, if injudiciously used, nets were potentially destructive of local bird populations. A similar concern may have informed Volterra's prohibition of the so-called "small shield" (*scudicciolo*). This camouflaging device was a crude mask with eyeholes, sometimes red in color, that the hunter held up to his face to break up the human profile, hide the gleam of his eyes, and thus get closer to the unsuspecting birds, for example to throw a net over them. Aimed at maintaining local bird populations, a chapter in the Statute of 1393 prohibited the capture of birds in and around the Gusciana River in July, August, and September when the birds had their young. About thirty years later, in 1426, an addition to the statutes informs us that if anybody wanted to hunt "ducks" in the Gusciana River with a technique that involved collapsible nets and bait strung out around a water source—a technique known as the *panthera* and in use beyond Tuscany—then the interested individuals had to register with the town notary in the month of January, many months ahead of their actual hunts. No further information is available, but the town seems to have been trying to limit (or at least anticipate) the number of hunters exploiting the aquatic resource and so the number of captured animals.

Further limitation on the catch may be gleaned by Fucecchio's (and other town's) prohibition against bringing and selling game animals and fish captured in the city's territory to markets outside of Fucecchio—what was caught within the territory was to be sold within it also.[18] This shielded Fucecchio's fish and game from the demands of the big city markets, such as Florence's *Mercato Vecchio*. The profits to be made in a place of much higher demands (and deeper purses) may have led Fucecchiese hunters to capture much larger amounts of animals than those they could expect to sell locally. In 1391, in nearby Santa Maria a Monte, townsmen wanting to hunt pheasants, partridges, and quail had to register with the notary ahead of hunting season as well as pay a municipal fee consisting of four fat quail. These hunts could take place only in September, likely in an attempt to lessen the hunting pressure. The Statute for Santa Maria a Monte also dictated that a citizen who wanted to fowl on the Gusciana River was allowed only one "position" (*stato*) in order to do so. As stated in the 1391 statutory regulations for Santa Maria a Monte, the "positions" of the fowlers using the municipal river banks had to be at least 300 fathoms, so 400–600 meters, apart. Capping the total number of hunters limited the number of captured ducks.[19]

In the case of the 1291–1340 Statute of Sambuca Pistoiese near Pistoia, controls on hunting took a slightly different form. When interested in hunting roe deer, this township's citizens, in order to make themselves known to the municipality, grouped themselves in hunting "societies" (*societates*); such contracts were increasingly common in all spheres of the commercialized Italian economy.[20] Two other such hunting societies notarized in 1370 and 1375 and from the locations Pieve di Santo Stefano and Pieve di Arliano in the environs of Lucca demonstrate how this contract applied to hunting.[21] Individual members contributed differently, in terms of labor and monetary investment, and were rewarded accordingly. Commercial hunting was certainly participating in the mechanisms of the larger Italian economy.

The statutes also reveal towns' attempts to manage complex ecological and cultural realities. Hunting had to fit in with many other uses of natural resources. For example, in the Statute of Fucecchio of 1410, while it was possible for anyone to hunt ducks in the Gusciana River, fowlers were to behave in ways not detrimental to the fishermen.[22] The cultural importance of fish gave their procurement precedence. On the downstream end of the Gusciana, Santa Maria a Monte's fowlers could hunt the banks but were not allowed to set up devices in the river, which could have created an impediment to navigation or caused siltation. As the Gusciana was the only outlet of the wetlands of the Paudle into the Arno, movement of goods and people may have been deemed more important than ducks, and thus, a balance had to be struck between the two activities. Italian statutory legislation abounds in injunctions meant to protect waterways, vines, and ripe crops from trampling local lords looking for lost falcons.[23] Yet legislation in both Santa Maria a Monte and Montevettolini also protects hunters from having game stolen from snares and nets. Multiple activities took place within the town's geographic jurisdiction (*contado*), with some, like fishing and fowling, overlapping. Towns were keen to maintain the use, and the income, from *all* their municipal resources. Balances had to be struck and priorities established—not to disallow some activities, but to maintain the viability of all.

Unpredictable Outside Threats

Municipalities could control their public territories and citizens but were otherwise at the mercy of those around them. How much that was the case depended on political factors. In Tuscany, the lengthy wars between Florence and Lucca prompted municipalities to put themselves under the protection of a larger city, which in Fucecchio's case was Florence. After 1330, the Statute of Fucecchio had to be approved by the dominant city and deposited in its archives. Given the cultural status of game animals and meats, there were further, perhaps unanticipated, consequences. Between 1374 and 1443, Fucecchio experienced the compulsory extraction of game.[24] In this period, the "Priors" (*Priori*), representatives of the guilds of the Florentine Republic, were Florence's highest political elite. Such men were under cultural obligation to consume only foods that matched their high status (and game was amongst them) and to offer gifts and celebrate their peers though festive meals.[25] Everywhere in medieval Europe, the fresh meat of game animals as well as items for the hunt, be it dogs or falcons, were integral to aristocratic rituals of duty, courtesy, and gift giving—often to an excess, as is reflected in the menus of elite banquets for special occasions.[26] Although the aristocrats were those in the best position to participate, a keen sense of social rank enabled similar social exchanges all through society.

Unable themselves to hunt due to their obligations, the Priors and other elites demanded a supply of fowl, roe deer, and boar from places such as Fucecchio. A parallel example appears in the well-documented life of the famous Prato cloth merchant, Francesco Datini, for whom food items were active objects in negotiating his lifelong interaction with the world around him.[27] Unable to hunt, but not in the position to expropriate, Datini obtained his game meats and other desirable foodstuffs from markets and a lively exchange economy. Monasteries similarly required gamey foods for special occasions or visitors.[28] While occasionally, and for reasons invisible to us, the Florentine elites paid for their game, other times they did not. Regardless, monetary compensation did little to lessen the pressure on game animal populations, which the municipality was trying to limit.[29]

While each tried to consume according to his means and abilities, even if that meant just a brace of thrushes, both quality and quantity was redirected toward the extremely rich.[30] These individuals were becoming hegemonic lords with the political ability to control vast areas of territory, procure game meats, and host hunts, thus amplifying their political authority. Cultural imperatives, connections, wealth, and political fiat gave these lords power as consumers, and hunters, but also as game managers. Yet the still very personal nature of power of the late Middle Ages also left them entangled in webs of relations requiring personal and ad hoc decisions.[31] While on the one hand, large areas were reserved to preserve game for the lords' hunts, reasons of political convenience sometimes required granting other noblemen permission to hunt otherwise protected animals or in otherwise protected areas. In the countryside, this shift was keenly felt in the negative. While the Priors' expropriations took place, Fucecchiesi hunters continued to pay the tax and acquire hunting rights to ply their trade. What effectively shut down all municipal management efforts, and commercial hunts in the area, was the emergence of an even stronger political player: the Medici Duchy.

Aristocratic Hunting

Unlike the Priors, aristocrats controlled the land, the hunts, and, thus, the game. Traditionally, prowess and success at the hunt were paramount to establishing the aristocratic warrior lord as entitled to rule, and it is practically impossible to overestimate the importance of hunting for European elites in the medieval and early modern periods. The Italians were no different, as documentation from the fifteenth century Visconti-Sforza courts of Lombardy and that of the sixteenth century Medici of Tuscany show. Young aristocratic men, and some women, were brought up in a hunting tradition centered around falconry and long chases on horseback with large retinues of horses, hounds, and huntsmen.[32] Raptors were used to hunt birds ranging from ducks and quail to herons, while the chase concentrated on (mostly) male exemplars of all deer species and boar. Compared to commercial hunting, which operated in an increasingly regulated municipal system, aristocratic activities appear unrestrained. Dukes produced their personal legislation through centralized administrations.[33] In Milan, career bureaucrats and a "police" force of locals (*campari*) enforced legislation.

In Tuscany, it fell to a judiciary body called *Gli Otto di Guardia e Balia*.[34] The protection of the lord's hunts and game excluded everyone else.

Italian aristocratic hunting reserved vast areas for exclusive elite use. In some cases, hunting reserves also had limitations on other activities, such as woodcutting or grazing of pigs in the stubble during the summer. Without special dispensation, any other hunts constituted poaching and were subject to heavy fines.[35] Italian social historian Massimo Montanari has claimed that the gradual expropriation of hunting by the elites forms the grand narrative of medieval and early modern Italian hunting.[36] Further, Lombard documentation testifies to aristocratic disregard of their own rules when advantageous, even when, for example, their bureaucrats acted like the thieves they were supposed to fine.[37] The complicated picture of Italian politics and hunting suggests that sometimes at least myopic decisions and crass short-term appropriation belied the rhetoric of good stewardship. The blatant overstretching of lordly permission to hunt, as happened when noblemen who had been allowed to hunt the (often nuisance) boar with crossbows trained their weapons at scarce deer instead, was socially and politically motivated rather than about the resources per se.[38]

Notwithstanding such political malfeasance, aristocratic strategies and legislation were generally sound. If properly maintained and respected, the regulations may have kept the elites hunting in the long term. Due to the magnitude of their territorial control, aristocratic measures would have been better suited both to the nature of wild game and local fluctuations in ecological variables such as cover and food than what the towns could muster. Despite their reputation for profligacy and ostentatious display, aristocrats were neither ignorant of the mechanisms of nature, nor of the connections between competing land uses and the well-being of game populations. To a large extent, this was due to the fact that they were actively and continuously involved in hunting.

Hunting was a privileged way to learn about the natural world in premodern communities. Based on continuous and prolonged hands-on practice, at times critical of established authority, such a tradition existed at both the courtly and popular level. At the popular level, information was transmitted orally, and by hands-on practice. The literate elites took to the pen to illustrate and share their findings. Among the masters of hunting erudition were Holy Roman Emperor Frederick II (ruled 1220–45), King Alfonso XI of Castile (r. 1312–50), and Gaston Phoebus, Count of Foix, in southern France (r. 1343–91). Based on personal experience and close observation, these men wrote deeply insightful hunting manuals.[39] These manuals demonstrate that despite a reputation as heedless, status-enamored consumers, elites from all over Europe could also be careful, knowledgeable, innovative, and passionate masters and advocates of things connected to hunting. Spending long hours in the outdoors, aristocrats and their huntsmen noticed ecological causalities and anthropogenic pressures.

Aristocratic Strategies

Hunting featured in legislation as the Italian dukes superimposed their urban-centered power over the existing geopolitical landscapes they sought to dominate.[40] By the fifteenth century, if not before, there was an established tradition for Italian cities to extend their economic and political control to their surrounding countryside (*contado*). As cities became parts of duchies, the dukes simply stepped into an established practice to centralize dominion in one large city, whether Milan or Florence. At the core of both Lombard and Tuscan ducal hunting legislation was the creation of areas of varying size, whether unwalled reserves or walled parks, where hunting for game was restricted. The ability to establish such areas and integrate them into state legislation and administration reflected and reinforced military proficiency and often followed victories. For example, Gregory Lubkin reports how Duke Galeazzo Maria Sforza (r. 1466–67) reserved six districts in his duchy exclusively for hunting after his successful 1467 military campaign.[41] Fifteenth-century Visconti-Sforza Lombardy was dotted with the so-called "hunts" (*cacce*), areas reserved for the lord's pleasure, which became landmarks in the local landscapes and whose administration and boundaries were spelled out in great detail in legislation.[42] At their peak, reserves amounted to about one-third of the area of Lombardy. However, these reserves were not enough: in 1476, the entire duchy of Milan was declared off-limits for non-ducal hunting.[43] In Tuscany, Cosimo I de Medici, Duke of Florence (r. 1534–69) and then Grand Duke of Tuscany (1569–74), started his family's consistent and sustained efforts to establish so-called "banned areas" (*bandite*).[44] These were very large, nonenclosed tracts of land where a number of extractive activities, including hunting, fishing, and woodcutting, were regulated for the benefit of the Florentine princes. Soon, Medici reserves formed an almost continuous string on both sides of the Arno River.

A duke's ability to control areas the size of Lombardy and Tuscany, rather than those of a town like Fucecchio, allowed him to operate in ecologically sound ways by establishing substantial preserves in which the animals and their habitats were under special protection. Large areas allowed for game to adjust to adverse local changes of any kind by simply moving.[45] Given the size of the areas, the animals could move far and still be within the confines of reserved territory. Lombard aristocratic legislation protected specific habitats, for example alongside rivers, such as the Ticino, favoring their unique animal inhabitants and food sources.[46] Large areas could contain natural corridors for the animals to move through, connecting areas of good habitat, allowing access to areas with little demographic pressure or even relatively wild areas, which could also serve as game reservoirs, such as in the Alps and Apennine Mountains. Although some of the aristocratic measures were the same as those employed by the towns when they licensed local users, prohibited access, established hunting seasons, and prohibited use of exploitative techniques, the territorial and political scale of the aristocratic actions could create much better opportunities for the game.[47]

Within certain limits, the dukes could protect game animal habitats and sources of food through an imposition of their power upon existing agro-ecosystems. The exercise of this power in the service of preserving game animal populations was a function of the immense cultural value that hunting had for aristocrats. We remain ignorant about local Italian resistance, the size of local game populations, the ability to coherently enforce legislation, excessive predation from the aristocrats themselves, and local environmental changes; however, in theory, aristocratic strategies could have produced very positive results.

In the case of boar, a prized aristocratic game animal, there is evidence for successful aristocratic management. In 1495, Duke Ludovico Sforza il Moro wrote his huntsman Baldassarre regarding boar that were ravaging (*fare gran danno*) the countryside.[48] The digging of boar can render a field a wasteland of craters. While boar may reproduce readily and in good amounts, this population's success may have been due to the protected life it led. Not only was it forbidden to hunt them, legislation also included prohibitions against cutting of stubble and allowing domestic pigs into the stubble.[49] Yet the dukes already protected woods and riverbanks from being cut for timber as well as firewood. Coppice was also protected.[50] Whether incidentally or not we don't know, but these measures would also have protected boar habitat, and in 1496, it was ordered that the collection of acorns in the woods be curtailed in order to reserve them for the boar. All of this would have promoted boar populations, which clearly were at a premium for the dukes—although in the word of the law, the collection of acorns was said not to favor the boar, but, rather, to "ruin" the woods.[51] For all this, two letters from 1490 and 1497 discussed the "paucity" of nuts.[52] Mast crops vary from year to year, which may have been the case here. A poor mast year would push more boar into agricultural areas, as seems confirmed by the complaints about boar activities from those same years. Protecting boar may have represented sacrifices for local populations who relied on mast. In the case of the "ravaging" boars, the duke asks Baldassarre to proceed with caution. First, he should try to chase the animals out of the most damaged areas, but, if everything else fails, Baldassarre and his men are allowed to kill the animals. This was not the only incident involving destructive boar, and the duke had to balance the well-being of his subjects with that of his prized boar.[53]

In Milanese ducal legislation, the natural world seems constantly under the threat of "damage" and "ruin" as a result of change in exploitation practices. The damage is always expressed in terms of a single resource, such as the hunt or woods without acorns. Preambles to legislation aimed at prohibiting non-ducal hunting speak of eliminating hunting pressure, as it would otherwise have "destroyed" the duke's pleasure.[54] The ruin never involves an entire landscape or ecosystem. In 1477, it is reiterated that "every year damage is done to the millet fields (*milii*) and various kinds of hunting done in them," and "damage" is again reported in 1478. As a consequence, the duke issues a total ban on hunting in cultivated areas.[55] Reacting to the apparent abuses of lords' excessive hunting, the duke exclaims in a document dated 1474 that if he wanted to go hunt, there would not even be a hare left.[56] The legislation expressed outrage at attacks on what represented the duke himself: his game animals. We do not know how justified this personal affront was, but posturing was part of ruling, and game was protected not for its intrinsic value but for its value as aristocratic prey. Regardless, through the prevention of a series of smaller "damages," the legislative effort as a whole at least suggests a sound assessment of what made game animals thrive, and awareness of how to protect game within a larger context (ecology). If only for political reasons, the dukes had to accommodate the claims of individuals directly damaged by their protective efforts, such as the owners of the fields devastated by boar. Even though the dukes attempted to operate outside of the "commons," they could not live outside of the layered multitude of social and political powers that coexisted in their territories.

Hunting Parks

Parks were walled hunting areas containing carefully managed landscapes, often at the doorstep of aristocratic residences. In this period, while there were none in Tuscany, three existed in Lombardy: at Cusago, near Milan; behind the Castello Sforzesco of Milan; and the largest, located behind the castle of Pavia. The rulers of Milan therefore followed suit in a long-established European tradition, and, as elsewhere, their parks contained pavilions and fishponds, housed menageries of exotic animals, had roads and irrigation, as well as shelters for the park deer. For the latter, documentation speaks of roe and red deer, but also of fallow deer (*Dama dama*).[57] Much celebration accompanied these constructions, especially surrounding the Pavia Park, flagship of Visconti-Sforza ambition.[58]

At first glance, the parks appear a contradiction in terms. How could such pampered conditions produce something reminiscent of an activity celebrated to be the best training for warriors, as Xenophon (d. 354 BCE) had proclaimed? Letters from disconsolate huntsmen at Pavia document companionate formations of ungulates that refused to separate, much less run, despite the efforts of huntsmen and dogs, rendering them worthless for the long chases on horseback that were the trademark of aristocratic prowess.[59] Walls in disrepair invited predators in for a free lunch.[60] It is important to remember that the parks were not the only hunting opportunity aristocrats held, and park hunts were not expected to be like those that took place in the open country. Duke Galeazzo Maria Sforza, for one, seemed well aware of the potential outside his parks. An accomplished and skilled hunter, the duke hunted every available game animal species all over Lombardy during his ten-year reign.[61] Hunting parks were not just for hunting, but also places of entertainment and leisure that provided hunters with large numbers of deer. In 1458, the master of the hunts, general Carlo da Cremona, sent six young fallow deer, one male, and five pregnant females, from Pavia to the duke in Milan.[62] Da Cremona explained that there were "about fifty more of the species" at the Pavia park. In another instance, a hunt at the Cusago Park involved the hunters with twenty-two deer.[63] By contrast, in 1495, huntsmen tasked with organizing a hunt for Ludovico

Sforza Il Moro in the Valsassina had been able, even with the help of dogs and local "experts," to locate only ten deer, dispersed and hidden by thick vegetation.[64] Parks also allowed for differently choreographed hunts—for example, by building corridors of canvas stretched on poles, the so-called *tele*, that guided the deer toward an open field and their likely death. The need to entertain his brother, Cardinal Ascanio Sforza, who was accustomed to the sophistications of the Vatican court, and, perhaps, venatory success, prompted Ludovico il Moro to organize just such a hunt in 1498.[65]

The historiography of hunting parks has traditionally focused on their cultural aspects. More recent research encourages us to look beyond the park as just a cultural product replete with pretty buildings and in which, with proper restocking, many deer could be taken. Parks as landscapes must be viewed in toto for both their material and cultural aspects.[66] Doing so allows us to see that, notwithstanding the glamour and conspicuous consumption associated with parks, reality involved strategies that attempted to make the park produce income, dynamically geared toward multiple uses. As Abigail P. Dowling has shown, Hesdin park woodlands, waters, and agricultural lands garnered countess Mahaut good revenue, while Sharon Farmer has pointed out that emparked areas may include very marginal lands, such as the large marshes of Hesdin, which were poorly suited for traditional agriculture but superb for falconry.[67] Finally, the choice to empark an area may have been motivated by the changing economic value of its lands due to climatic changes.[68] At the Pavia Park, there seems to have been little basis for the descriptions penned by courtly literati, who compared the densities of pheasants and other birds on the park grounds to that of ants.[69] The park also was, in Dowling's words, "intensively and self-consciously managed for internal consumable goods, for income, and for entertainment and prestige," of which both the birds and their celebration were just one aspect.[70]

Pavia Hunting Park

Pavia Park consisted of two parts, the Old Park (*Parco Vecchio*), which was attached to the castle via a garden, and the New Park (*Parco Nuovo*), which between 1378 and 1396 expanded the park's trapezoidal shape northward to the site of the newly built monastery of the Certosa di Pavia, thus ideally linking secular and religious powers through the park territory. Approximately 7.4 kilometers long and 3.7 wide, it covered an area of 27 square km circled by a 2–meter-high brick wall, pierced both by gates with draw bridges and by smaller "half-doors," to allow passage for the different watercourses.[71] A 1477, highly detailed account documenting the allotting of the park resources to a manager reveals that the New Park was used as a hunting park and reserved for the deer.[72] The Old Park was devoted to agriculture. Under the manager's care and supervision, tenants would live in the park and take its agricultural production upon themselves. Although most of the eight-page regulation did not pertain to hunting or deer, the estate still served the deer's needs, especially by producing hay that the tenants had to deliver, free of charge. The amounts consisted of one hundred sheaves of "early" (June) hay and three thousand sheaves of August hay. The Vernavola stream ran through the park, facilitating the creation of irrigated meadows that, as elsewhere in Lombardy, enabled the production of an additional harvest of hay compared to their nonirrigated counterparts, thereby allowing for a significant increase in stock breeding.[73]

For the park tenants, strictures and legislation aiming at protecting the deer population and enhancing hunting were put in place and strictly enforced. Hunting was prohibited to all except the duke and his friends. The park walls had guards tasked with upkeep and vigilance, with their salaries paid by the tenants. No wood was to be cut, and no dogs kept, except young pups destined for courtly careers as hunting dogs, which were fed and cared for by tenants until they reached the age of eight months. Upkeep of all park infrastructure, such as roads and waterways, was the responsibility of the tenants. Collection of acorns and grazing of pigs required permission, and burning of stubble was prohibited. Despite some diversion of resources away from the tenants, neither of these rules precluded sound agricultural practices and both echo measures that the dukes enforced by law in their hunts: the voracious ungulates were fenced off part of the year. For the tenants, the limitations may have outweighed the security of park living.

Park practices, although slanted toward deer keeping, are reminiscent of the kind of balances put in place by the towns on whose territories multiple resource users were to coexist. The Milanese case points to a park that could sustain deer and generate revenue, while participating in a sort of partnership with the outside world through an exchange of labor and goods. That said, parks famously subtracted resources from the world outside the wall, for example, for building the park pale and its maintenance. Wood destined for the park pale as well as coppice were the objects of a 1460 petition from disgruntled Cusago tenants to the duke. François Duceppe-Lamarre presents evidence that stone quarries and limekilns at Hesdin were part of emparkment's "major ecological impact" on a regional basis.[74] Certainly the park's walled nature, and the fact that it housed status-laden deer for the owner's exclusive enjoyment, made it a contested creation across Europe.

Was Pavia Park able to sustain its deer population? We cannot know. Much would depend on the population and the quality of available habitat, and neither can be assessed. Even under the best of circumstances, there is evidence that deer, at least beyond a certain number, do not fare well in enclosed situations. A document from 1457 illustrates how an epidemic could quickly exterminate most of a fenced-in ungulate population, making it undesirable to keep too many deer together.[75] Certainly, fencing in or emparking the animals subtracted local resources, no matter the herd's size. Sources like those cited above that record cries about dwindling food supplies for the deer (we know not how many) suggest that hay or other food had to be restocked frequently.[76] Depending on the hunts, the animals themselves may have had to be replaced. Lastly, we know little about the park's ability to generate revenue. Notwithstanding its venatory vocation, it might be useful to see the park in the context of the Lombard countryside and other ducal activities within it. Hunting parks were not the only place where the ruling elites collected an income from managing their lands. For example, Ludovico Sforza established La Sforzesca as a farming estate and a center for, among other things, the introduction of silk worms into Lombard agricultural practices.[77] While the large property included many tenants, who held the larger share, such estates represented another aspect of the aristocratic colonization of the countryside and construction of aristocratic identity. It is possible to see the park and its resource management as part of a more extensive process of landscape management, especially of the Sforzas, to increase their hegemonic power in Lombardy. While favoring deer, the Pavia Park was embedded in an agricultural society in which both intensively cultivated lands and those managed as pastures or woodlands were components of hybrid agro-ecosystems that blended anthropogenic and natural elements.[78] In this context, the park may be seen simply as an agricultural estate with deer among its resources, some of which, like Ludovico's silk worms, were unsuccessful.

Conclusion

Late medieval Italians demonstrate ecological insight tuned to territorial possibilities and a knowledgeable reading of environmental realities in their desire to sustain game animal populations. Individuals pursued game animals and birds in diverse ways, however. Whether hunted for the market or pleasure, game had a cultural value that far surpassed its caloric one. The motivation to hunt shaped the strategies needed to control and protect the game from outside threats. All parties sought to regulate access to game through legal means. Such regulation limited access to select groups at the expense of those perceived as outsiders, which for a municipality included all noncitizens, and which for aristocrats included any hunting competitor, whether elite or not. Additionally, all actors understood the ecological mechanisms that regulated game animal welfare and put in place protections based on those principles.

The towns established rules to protect their municipal hunters, consumers, and game animals from their destruction by external forces that would have broken predictable patterns of local resource use. Increasingly, lordly power created additional pressure by suppressing usage rights and linking the ability to procure game with elite status. As larger political powers, such as the Medici and the Visconti-Sforza, enhanced the authority and reach of the regional state, municipal protections were gutted. Game, all of it, was reserved for the lord. This seems to confirm Paul Warde's observation that preindustrial practices "regulated the impact of one person's actions on others and were not concerned with ecological sustainability itself."[79] This chapter speaks to the complexities—social, cultural, and economic—that underlay and shaped such self-interested behaviors.

While late medieval hunters may have tried to protect their interests, something that made sense in their primarily agricultural world, any self-interested protection of a single resource involved complex, constant negotiations with other people and institutions. Even the dukes had to arrange for the killing of the boar they protected to safeguard the interests of their citizens, while municipalities like that of Fucecchio became managers of multiple licenses. Legislative effort that centered on a single item, such as game animals, appears static because of its very nature; however, all the participants were deeply aware that medieval agricultural systems and their management were necessary to continue their existence. Thus, each control must be seen in the context of other limitations that took place simultaneously. Furthermore, it is crucial to read all legislation concerning game whether this emanated from or concerned individuals, municipalities, or elites. When read in this holistic context, it is clear that game, as a luxury natural resource, was only one, albeit important, part of complex multiuse landscapes in the late Middle Ages.

Cristina Arrigoni Martelli is Assistant Professor of History at the University of Maine at Machias. She received her PhD in Environmental History and European History at York University, Toronto. She has published articles on medieval Italian falconry and non-falconry techniques for bird capture in medieval Italy. She teaches courses in Global Environmental History, Hunting, and diseases in history. She is currently working on an article arguing for the use of positive reinforcement in medieval and late medieval hunting dog training and a project on the role of the control of water, and land, in Venetian political and social identities.

Notes

1. Cristina Arrigoni Martelli, "Nets, Lures, and Camouflage: Capturing Birds in Late Medieval North and Central Italy," *Reinardus* 27 (2015): 1–32.
2. Subsistence hunting was likely but remains invisible in the sources. Massimo Montanari, *L'alimentazione contadina nell'alto medioevo* (Naples: Liguori Editore, 1979), 268–70.
3. To see more about disputes over use-rights, see Amleto Spicciani, "Un testimoniale del 1215 sul Padule di Fucecchio," in *Atti del Convegno sull'identità geografico-storica della Valdinievole. Buggiano Castello, 24 Giugno, 1995*, ed. Comune di Buggiano Castello (Buggiano Castello, Comune di Buggiano, 1996), 183–202.
4. Archivio Comunale di Fucecchio (henceforth ACFu), *Liber Causarum* 1958, c. 42 r. (2 December 1276); Spicciani, "Un testimoniale," 183–202.
5. From 1340, when such documentation starts, waterfowling was included with fishing. The first separate *gabella* on waterfowl is in ACFu, *Rendite e Proventi*, 287, 9 April 1413.
6. Alberto Malvolti, "I Proventi dell'Incolto," in *Incolti Fiumi e Paludi. Utilizzazione delle risorse naturali nella Toscana medievale e moderna*, ed. Alberto Malvolti and Giuliano Pinto (Florence: Olschki, 2003), 254.

7. Malvolti, "I Proventi," 254.
8. For evidence of this type of thinking, see Malvolti, "I Proventi," 257.
9. Spicciani, "Un testimoniale," 198.
10. Malvolti, "I Proventi," 263.
11. Richard C. Hoffmann, *An Environmental History of Medieval Europe* (New York: Cambridge University Press, 2014), 179–80.
12. Will Richter, ed. *Petrus de Crescentiis, Ruralia Commoda* (Heidelberg: Universitätsverlag C. Winter, 1995), book X; Domenico Boccamazza, "Trattato della Caccia," in *Arte della Caccia, testi di falconeria uccellagione e altre cacce: dal Secolo XIII agli inizi del Seicento*, ed. Giuliano Innamorati (Milan: Il Polifilo, 1965), 287–5112; Arrigoni Martelli, "Nets, Lures, and Camouflage."
13. Nanni Blasi (or Blaxi) obtained the *gabella* three times between 1405 and 1422. Recurrent names are Locti or Lotti Dini, Michaelis, Guiduccini, and Mani. ACFu, *Rendite e Proventi*, 287.
14. ACFu, *Rendite e Proventi*, 287, 6 January 1409.
15. Alberto Malvolti, "Le risorse del Padule di Fucecchio nel Basso Medioevo," in *Il Padule di Fucecchio*, ed. Adriano Prosperi (Rome: Edizioni di Storia e Letteratura, 1995), 60.
16. Daniel R. Curtis, "Florence and Its Hinterlands in the Late Middle Ages: Contrasting Fortunes in the Tuscan Countryside, 1300–1500," *Journal of Medieval History* 38, no. 4 (2012): 472–99.
17. Riccardo Rao, *Comunia: Le risorse collettive nel Piemonte comunale* (Milan: LED, 2008); Riccardo Rao and Guido Alfani, eds., *La gestione delle risorse collettive: Italia settentrionale, secoli XII–XVIII* (Milan: FrancoAngeli Storia, 2011).
18. ACFu, *Statuti*, 116 (1359).
19. Bruno Casini, ed., *Statuo del comune di Santa Maria a Monte (1391)* (Florence: Olschki, 1963), 212–13. Along similar lines, see Enrico Fiumi, ed., *Statuti di Volterra. 1210/1224* (Florence: Deputazione di storia patria per la Toscana, 1951), 1: 219; Alberto M. Onori, ed., *Statuto del Comune di Pescia 1339* (Pistoia: Società pistoiese di storia patria, 2000), 125; Antonio lo Conte and Elena Vannucchi, eds., *Lo Statuto di Massa e Cozzile del 1420* (Florence: Polistampa, 2006); Corrado Buzzi, ed., *Lo statuto del comune di Viterbo del 1469* (Roma: Istituto Storico Italiano per il Medio Evo, 2004); Mario Brogi, ed., *Gli Albori del Comune di San Gimignano e lo Statuto del 1314*, Collana Documenti Di Storia, Università Di Siena (Siena: Edizioni Cantagalli, 1995); Enrico Bacchetti, ed., *Statuti di Caverzere del 1401–1402: Gherardo Ortalli. Corpus Statutario Delle Venezie* (Rome: Viella s.r.l, 2005); Hannelore Zug Tucci, "La chasse dans la législation statutaire italienne," in *La Chasse au Moyen Age: Actes du Colloque du Centre d'études médiévales de Nice*, ed. Marcelle Thiébaux (Nice: Edizioni del Galluzzo, 1979), 99–113.
20. Manila Soffici, ed., *Lo Statuto della Sambuca (1291–1340)* (Pisa: Pacini, 1996), 98–99.

21. Antonio Romiti, "Due 'societates in exercitio venandi' del secolo XIV," *Actum Luce* 2 (1973): 217–24.
22. Archivio di Stato di Firenze (henceforth ASFi), *Statuti delle Comunità Autonome e Soggette*, 337 (1353).
23. Giuseppe Liberali, *Gli Statuti del Comune di Treviso*, vol. 2: *Statuti degli anni 1231–33–1260–63* (Venice: Monumenti storici pubblicati dalla Deputazione di Storia Patria per le Venezie 1951).
24. ACFu, *Deliberazioni*, 183–84; *Statuti e Riformagioni*, 116.
25. Allen J. Grieco, "Alimentazione e classi sociali nel tardo Medioevo nel Rinascimento in Italia," in *Storia dell'alimentazione*, ed. Jean Louis Flandrin (Rome: Laterza, 1997), 371–80; Giovanna Frosini, *Il cibo e i signori: La mensa dei Priori a Firenze nel sec. XIV* (Florence: Academica della Crusca, 1993); Libreria Medicea Laurenziana, *Codice Ashburham*, 1216.
26. Claudio Benporat, *Feste e banchetti: Convivialità italiana fra Tre e Quattrocento* (Florence: Olschki, 2001).
27. Chris M. Woolgar, "Gifts of Food in Late Medieval England," *Journal of Medieval History* 37, no. 1 (2011): 6–18; Iris Origo, *Il mercante di Prato* (Milan: Bompiani, 1957); Valeria Rosati, ed., *Le lettere di Margherita Datini a Francesco di Marco (1384–1410)* (Prato: Cassa di Risparmi e Depositi, 1977), *passim*; Carolyn James and Antonio Pagliaro, trans., *Margherita Datini: Letters to Francesco Datini* (Toronto: Iter, 2012), passim; Maria Giagnacovo, "Due 'alimentazioni' del basso Medioevo: la tavola dei mercanti e la tavola dei ceti subalterni," in *Alimentazione e Nutrizione secc. XIII-XVIII*, ed. Simonetta Cavaciocchi (Rome: Le Monnier, 1996), 821–29.
28. Roberta Zazzeri, ed. *Ci desinò l'abate: Ospite e cucina del monastero di Santa Trinita, Firenze, 1360–1363* (Florence: Societá Editrice Fiorentina, 2003).
29. ACFu, *Deliberazioni*, 184 (1442; 27 January 1443; 20 September 1379/80); ASFu, *Statuti e Riformagioni*, 176, 6 September 1389.
30. Cristina Arrigoni Martelli, *Ducks and Deer, Profit and Pleasure: Hunters, Game, and the Natural Landscapes of Medieval Italy* (PhD Dissertation, York University, 2015), 215–26.
31. Jane Black, "The Limits of Ducal Authority: A Fifteenth-Century Treatise on the Visconti and Their Subject Cities," in *Florence and Italy. Renaissance Studies in Honor of Nicolai Rubinstein*, ed. Peter Denly and Caroline Elam Rubinstein (London: Committee for Medieval Studies, 1988), 153; Giorgio Chittolini, "Governo ducale e poteri ducali," in *Gli Sforza a Milano e in Lombardia e il loro rapporto con gli stati italiani ed europei (1450–1535)*, Convegno internazionale Milano, 18–21 Maggio 1981, 27–42 (Milan: Cisaplina-Goliardica, 1982).
32. Archivio di Stato di Milano (henceforth ASMi), *Potenze Sovrane*, 1461, 1468, 1469, 1483; Susanna Pietrosanti, *Le cacce dei Medici* (Florence: Vellecchi Editore, 1992), 7–24.
33. ASMi, *Panigarola Registri degli Statuti*, passim.
34. It was created in 1353 for the punishment of violent criminal incidents, and later became the policing force of first the Republic and later the Medici

Granducato. No records remain for the sixteenth century: Andrea Zagli, personal communication.

35. ASMi, *Finanza, Parte Antica, Caccia*, 485 (1470).
36. Massimo Montanari, "Vicende di un'espropriazione: Il ruolo della caccia nell'economia e nell'alimentazione dei ceti rurali," in *Campagne medievali: strutture produttive, rapporti di lavoro, sistemi alimentari*, ed. Massimo Montanari (Torino: Giulio Einaudi, 1984), 174–90.
37. Master of the hunt Carlo da Cremona was accused of stealing water to irrigate his fields: ASMi, *Potenze Sovrane*, 1483, 1467.
38. ASMi, *Panigarola Registri degli Statuti*, 11, 21 November 1494.
39. Anna Laura Trombetti Budriesi, ed., *Federico II di Svevia: De arte venandi cum avibus. L'arte di cacciare con gli uccelli. Edizione e traduzione italiana del ms. lat. 717 della Biblioteca Universitaria di Bologna collazionato con il ms. Pal. lat. 1071 della Biblioteca Apostolica Vaticana* (Milan: Laterza, 2009); Dennis P. Seniff, *Libro de la Monteria: Based on Escorial MS Y.II.19* (Madison: The Hispanic Seminary of Medieval Studies, 1983); Gunnar Tilander, ed., *Gaston Phébus, Livre de Chasse* (Karlshamn: Cynegetica, 1971).
40. Giuliano Pinto, *Città e spazi economici nel'Italia comunale* (Bologna: CLUEB, 1996); Giuliano Pinto, "I rapporti economici tra città e campagna," in *Economie urbane ed etica economica nell'Italia medievale*, ed. Roberto Greci, Giuliano Pinto, and Giacomo Todeschini (Bari: Editori Laterza, 2005), 5–73.
41. Gregory Lubkin, *A Renaissance Court: Milan Under Galeazzo Maria Sforza* (Berkeley: University of California Press, 1994), 90, note 17.
42. ASMi, *Panigarola Registri degli Statuti*, passim; Francesca Vaglienti, "Cacce e parchi ducali sul Ticino (1450–1476)", in *Vigevano e i territori circostanti alla fine del Medieoevo*, ed. Giorgio Chittolini (Vigevano: Unicopli, 1997), 256.
43. ASMi, *Panigarola Registri degli Statuti*, 9, 26 October 1476.
44. The *bandita* was an area in which *il cacciare*, *il pescare*, and *l'uccellare* (hunting, fishing, and fowling/birding) were prohibited or limited *per pubblico bando* (by public announcement): Fauzia Farneti, "La legislazione medicea sulle bandite, la caccia e la pesca," in *La legislazione medicea sull'ambiente. IV, scritti per un commento*, ed. Giovanni Cascio Pratilli and Luigi Zangheri (Florence: Olschki, 1998), 95–117; Farneti, "La legislazione," 112–17; D. Barsanti, "Tre secoli di caccia in Toscana attraverso la legislazion da 'privative' signorile sotto i Medici a 'oggetto di pubblica economia' sotto i Lorena," *Rivista di storia dell'agricoltura* 26, no. 2 (1986): 105–50.
45. A. Sforzi and S. Lovari, "Some Effects of Hunting on Wild Mammalian Populations," *Anthropozoologica* 31 (2000): 165–72.
46. ASMi, *Panigarola Registri degli Statuti*, 9, 1 July 1475.
47. ASMi, *Panigarola Registri degli Statuti*, passim. Specifically, *Registro* 1 (1393–1394) on prohibited area, prohibitions against the use of snares, dogs, and traps; *Registro* 11 (1486–1496) on prohibitions in the ducal preserves.
48. ASMi, *Finanza Parte Antica, Caccia*, 485, 9 June 1495, letter to Baldassarre.

49. ASMi, *Panigarola Registri degli Statuti*, 9 (1474–1478), 1 July 1475.
50. ASMi, *Panigarola Registri degli Statuti*, 6, f. 162; 21, f. 16; 25, 399–403; 8, 222 (coppice).
51. ASMi, *Panigarola Registri degli Statuti*, 11, 21 November 1496.
52. ASMi, *Potenze Sovrane*, 1484, 25 November 1490 and 25 November 1497.
53. ASMi, *Potenze Sovrane*, 1496, 2 and 24 October 1494.
54. ASMi, *Potenze Sovrane*, 1496, *Grida*, 19 April 1473.
55. ASMi, *Panigarola Registri degli Statuti*, 9, 9 August 1477 and 11 September 1478.
56. ASMi, *Panigarola Registri degli Statuti*, 9, 22 October 1474.
57. Naomi Sykes, "The Introduction of Fallow Deer to Britain: A Zooarchaeological Perspective," *Environmental Archaeology* 9 (2004): 75–83.
58. For contemporary literature on the Pavia park, see Luisa Erba, *Il Parco Visconteo nella letteratura:"il giardino onde svelse i gigli d'oro"* (Pavia: Tipografia Commerciale Pavese, 1999).
59. ASMi, *Finanza, Parte Antica, Cacce*, 485, 1495.
60. ASMi, *Finanza, Parte Antica, Cacce*, 485, 1491; 1495; attempts to bait the wolves, 1499.
61. Archival documentation, and Alfio Rosario Natale, ed., *I diari di Cicco Simonetta: Tomo Primo* (Milan: Giuffrè, 1962), 65, 91, 94, 118, 124, 145, 146, 155.
62. ASMi, *Potenze Sovrane*, 1457, 22 October 1458.
63. Mario Comincini, Ambrogio Palestra, and Germano Mulazzani, *Il palazzo, la chiesa, la villa: Storia ed arte a Cusago* (Cusago: Angeli, 1989), 100.
64. ASMi, *Potenze Sovrane*, 1493, Bellagio, letter of 3 August 1495 from Francesco da Cremona.
65. ASMi, *Potenze Sovrane*, 1140, letters of 3 and 4 July 1498. Higher prelates and hunting, Jeremy Kruse, "Hunting, Magnificence, and the Court of Leo X," *Renaissance Studies* 7, no. 3 (1993): 243–57.
66. Oliver H. Creighton, *Designs up on the Land: Elite Landscapes in the Middle Ages* (Woodbridge, UK: The Boydell Press, 2009).
67. Abigail P. Dowling, "Landscapes of Luxury: Mahaut d'Artois' (1302–1329) Management and Use of the Park at Hesdin," in *Rural Space in the Middle Ages and the Early Modern Age: The Spatial Turn in Premodern Studies*, ed. Albrecht Classen and Christopher Clason (Boston: de Gruyter, 2010), 367–87; Dowling, Chapter 2 in this volume.
68. William Z. MacDonald, "At the Crossroads of Culture, Economy, and Climate: Herstmonceux Park in the Late Middle Ages" (MA Thesis, University of Waterloo, 2013).
69. Erba, *Il Parco*, 73.
70. Dowling, "Landscapes," 387.
71. Margherita Azzi Visentini, "La chasse dans le duché de Milan a l'époque des Visconti et des Sforza: les parcs de Pavie et de Milan," in *Chasses Princières dans l'Europe de la Renaissance*, ed. Claude d'Athenaise and Monique

Chatenet (Paris: Actes Sud, 2007), 182, 206, notes 9 and 13; Erba, *Il Parco*, 18–20, 36–40. For more on the ideological backing from the Italian park, see A. M. Finoli and L. Grassi, eds., *Antonio Averlino detto Filarete, Trattato di architettura* (Milan: Il Polifilo, 1972); Richter, *Petrus de Crescentiis*, Book VIII: 14–15 (*Viridarium*).

72. ASMi, *Comuni*, Pavia, 331, 31 January 1477.
73. Giancarlo Andenna, "Una terra d'acqua tra due fiumi, un lago, e montagne bianche di neve," in *Una terra tra due fiumi, la provincia di Novara nella storia*, ed. Sergio Monferrini and Giancarlo Andenna (Novara: Provincia di Novara, 1998), 12–34; Enrico Roveda, "Allevamento e trasumanza nella pianura lombarda: I bergamschi nel Pavese tra '400 e '500," *Nuova Rivista Storica* 1 (1987): 49–70; Giorgio Chittolini, "Alle origini delle grandi aziende della bassa lombarda," *Quaderni Storici* 39 (1980): 828–44; Luisa Chiappa Mauri, "Riflessioni sulle campage lombarde del quattro-cinquecento," *Nuova Rivista Storica* 69 (1985): 123–30.
74. ASMi, *Archivio storico-diplomatico, Comuni* (Cusago), 29, undated petition from ca. 1460; Duceppe-Lamarre as cited in Sharon Farmer, "Aristocratic Power and the 'Natural' Landscape: The Garden Park at Hesdin, c. 1291–1302," *Speculum* 88, no. 3 (July 2013): 651.
75. ASMi, *Potenze Sovrane* 1483, letter from Cassiano on 24 October 1457.
76. ASMi, *Potenze Sovrane*, 1483, no date.
77. Mario Comincini, "Simone del Pozzo, la Sforzesca e Guglielmo da Camino," in *Metamorfosi di un borgo: Vigevano in età Visconteo-Sforzesca*, ed. Giancarlo Andenna (Milan: Angeli, 1992), 355–66.
78. See this volume's Introduction.
79. Paul Warde, "The Environmental History of Pre-industrial Agriculture in Europe," in *Nature's End: History and the Environment*, ed. Sverker Sörlin and Paul Warde (New York: Palgrave Macmillan, 2009), 72.

Bibliography

Andenna, Giancarlo. "Una terra d'acqua tra due fiumi, un lago, e montagne bianche di neve." In *Una terra tra due fiumi, la provincia di Novara nella storia*, edited by Sergio Monferrini and Giancarlo Andenna, 12–34. Novara: Provincia di Novara, 1998.

Arrigoni Martelli, Cristina. *Ducks and Deer, Profit and Pleasure. Hunters, Game and the Natural Landscapes of Medieval Italy.* PhD Dissertation, York University, 2015. https://yorkspace.library.yorku.ca/xmlui/handle/10315/30637. Accessed 1 August 2019.

_____. "Nets, Lures, and Camouflage: Capturing Birds in Late Medieval North and Central Italy." *Reinardus* 27 (2015): 1–32.

Azzi Visentini, Margherita. "La chasse dans le duché de Milan a l'époque des Visconti et des Sforza: les parcs de Pavie et de Milan." In *Chasses Princières dans l'Europe de la Renaissance*, edited by Claude d'Athenaise and Monique Chatenet, 179–215. Paris: Actes Sud, 2007.

Bacchetti, Enrico, ed. *Statuti di Caverzere del 1401–1402: Gherardo Ortalli. Corpus Statutario Delle Venezie*. Rome: Viella s.r.l., 2005.

Barsanti, D. "Tre secoli di caccia in Toscana attraverso la legislazione da 'privativa' signorile sotto i Medici a 'oggetto di pubblica economia' sotto i Lorena." *Rivista di storia dell'agricoltura* 26, no. 2 (1986): 105–50.

Benporat, Claudio. *Feste e banchetti: Convivialità italiana fra Tre e Quattrocento*. Florence: Olschki, 2001.

Black, Jane. "The Limits of Ducal Authority: A Fifteenth-Century Treatise on the Visconti and Their Subject Cities." In *Florence and Italy: Renaissance Studies in Honor of Nicolai Rubinstein*, edited by Peter Denly and Caroline Elam Rubinstein, 149–60. London: Committee for Medieval Studies, 1988.

Boccamazza, Domenico. "Trattato della Caccia." In *Arte della Caccia, Testi di falconeria uccellagione e altre cacce: Dal Secolo XIII agli inizi del Seicento*, edited by Giuliano Innamorati, 287–5112. Milan: Il Polifilo, 1965.

Brogi, Mario, ed. *Gli Albori del Comune di San Gimignano e lo Statuto del 1314*. Collana Documenti di Storia, Università di Siena. Siena: Edizioni Cantagalli, 1995.

Buzzi, Corrado, ed. *Lo statuto del comune di Viterbo del 1469*. Rome: Istituto Storico Italiano per il Medio Evo, 2004.

Casini, Bruno, ed. *Statuto del comune di Santa Maria a Monte (1391)*. Florence: Olschki, 1963.

Chiappa Mauri, Luisa. "Riflessioni sulle campage lombarde del quattrocinquecento." *Nuova Rivista Storica* 69 (1985): 123–30.

Chittolini, Giorgio. "Alle origini delle grandi aziende della bassa lombarda." *Quaderni Storici* 39 (1980): 828–44.

_____. "Governo ducale e poteri ducali." In *Gli Sforza a Milano e in Lombardia e il loro rapporti con gli stati italiani ed europei (1450–1535)*. Convegno internazionale Milano, 18–21 Maggio 1981, 27–42. Milan: Cisaplina-Goliardica, 1982.

Comincini, Mario. "Simone dal Pozzo, la Sforzesca e Guglielmo da Camino." In *Metamorfosi di un borgo: Vigevano in etá Visconteo-Sforzesca*, edited by Giancarlo Andenna, 355–66. Milan: Angeli, 1992.

Comincini, Mario, Ambrogio Palestra, and Germano Mulazzani. *Il palazzo, la chiesa, la villa: Storia ed arte a Cusago*. Cusago: Diakronia, 1989.

Creighton, Oliver H. *Designs Upon the Land: Elite Landscapes in the Middle Ages*. Woodbridge, UK: The Boydell Press, 2009.

Curtis, Daniel R. "Florence and Its Hinterlands in the Late Middle Ages: Contrasting Fortunes in the Tuscan Countryside, 1300–1500." *Journal of Medieval History* 38, no. 4 (2012): 472–99.

Dowling, Abigail P. "Landscapes of Luxury: Mahaut d'Artois' (1302–1329) Management and Use of the Park at Hesdin." In *Rural Space in the Middle Ages and the Early Modern Age: The Spatial Turn in Premodern Studies*, edited by Albrecht Classen and Christopher Clason, 367–87. Boston: de Gruyter, 2010.

Erba, Luisa. *Il Parco Visconteo nella letteratura: "il giardino onde svelse i gigli d'oro."* Pavia: Tipografia Commerciale Pavese, 1999.

Farmer, Sharon. "Aristocratic Power and the 'Natural' Landscape: The Garden Park at Hesdin, c. 1291–1302." *Speculum* 88, no. 3 (July 2013): 644–80.

Farneti, Fauzia. "La legislazione medicea sulle bandite, la caccia e la pesca." In *La legislazione medicea sull'ambiente. IV, scritti per un commento*, edited by Giovanni Cascio Pratilli and Luigi Zangheri, 95–117. Florence: Olschki, 1998.

Finoli, A. M., and L. Grassi, eds. *Antonio Averlino detto Filarete, Trattato di Architettura*. 2 vols. Milan: Il Polifilo, 1972.

Fiumi, Enrico, ed. *Statuti di Volterra. 1210/1224*. 1st volume. Florence: Deputazione di storia patria per la Toscana, 1951.

Frosini, Giovanna. *Il cibo e i signori: La mensa dei Priori a Firenze nel sec. XIV*. Florence: Accademia della Crusca, 1993.

Giagnacovo, Maria. "Due 'alimentazioni' del basso Medioevo: la tavola dei mercanti e la tavola dei ceti subalterni." In *Alimentazione e Nutrizione secc. XIII-XVIII*, edited by Simonetta Cavacciocchi, 821–29. Rome: Le Monnier, 1996.

Grieco, Allen J. "Alimentazione e classi sociali nel tardo Medioevo e nel Rinascimento in Italia." In *Storia dell'Alimentazione*, edited by Jean Louis Flandrin, 371–80. Rome: Laterza, 1997.

Hoffmann, Richard C. *An Environmental History of Medieval Europe*. New York: Cambridge University Press, 2014.

Innamorati, Giuliano, ed. *Arte della Caccia: Testi di falconeria uccellagione e altre cacce. Dal secolo XIII agli inizi del Seicento*. Milan: Il Polifilo, 1965.

James, Carolyn, and Antonio Pagliaro, trans. *Margherita Datini: Letters to Francesco Datini*. Toronto: Iter, 2012.

Kruse, Jeremy. "Hunting, Magnificence, and the Court of Leo X." *Renaissance Studies* 7, no. 3 (1993): 243–57.

Liberali, Giuseppe. *Gli Statuti del Comune di Treviso*. Vol. 2: *Statuti degli anni 1231–33–1260–63*. Venice: Monumenti storici pubblicati dalla Deputazione di Storia Patria per le Venezie, 1951.

lo Conte, Antonio, and Elena Vannucchi, eds. *Lo Statuto di Massa e Cozzile del 1420*. Florence: Polistampa, 2006.

Lubkin, Gregory. *A Renaissance Court: Milan Under Galeazzo Maria Sforza*. Berkeley: University of California Press, 1994.

MacDonald, William Z. "At the Crossroads of Culture, Economy, and Climate: Herstmonceux Park in the Late Middle Ages." MA Thesis, University of Waterloo, 2013.

Malvolti, Alberto. "I Proventi dell'Incolto." In *Incolti Fiumi e Paludi: Utilizzazione delle risorse naturali nella Toscana medievale e moderna*, edited by Alberto Malvolti and Giuliano Pinto, 247–72. Florence: Oslchki, 2003.

_____. "Le risorse del Padule di Fucecchio nel Basso Medioevo." In *Il Padule di Fucecchio*, edited by Adriano Prosperi, 35–62. Rome: Edizioni di Storia e Letteratura, 1995.

Montanari, Massimo. *L'alimentazione contadina nell'alto medioevo*. Naples: Liguori Editore, 1979.

_____. "Vicende di un'espropriazione: Il ruolo della caccia nell'economia e nell'alimentazione dei ceti rurali." In *Campagne medievali: strutture produttive, rapporti di lavoro, sistemi alimentari*, edited by Massimo Montanari, 174–90. Turin: Giulio Einaudi, 1984.

Natale, Alfio Rosario, ed. *I diari di Cicco Simonetta: Tomo Primo*. Milan: Giuffrè, 1962.

Onori, Alberto M., ed. *Statuto del Comune di Pescia 1339*. Pistoia: Società pistoiese di storia patria, 2000.

Origo, Iris. *Il mercante di Prato*. Milan: Bompiani, 1957.

Pietrosanti, Susanna. *Le cacce dei Medici*. Florence: Vallecchi Editore, 1992.

Pinto, Giuliano. *Città e Spazi Economici nell'Italia comunale*. Bologna: CLUEB, 1996.

_____. "I rapporti economici tra città e campagna." In *Economie urbane ed etica economica nell'Italia medievale*, edited by Roberto Greci, Giuliano Pinto, and Giacomo Todeschini, 5–73. Bari: Editori Laterza, 2005.

Rao, Riccardo. *Comunia: Le risorse collettive nel Piemonte comunale*. Milan: LED, 2008.

_____. *I paesaggi dell'Italia medievale*. Rome: Carocci editore, 2015.

Rao, Riccardo, and Guido Alfani, eds. *La gestione delle risorse collettive: Italia settentrionale, secoli XII–XVIII*. Milan: FrancoAngeli Storia, 2011.

Richter, Will, ed. *Petrus de Crescentiis, Ruralia Commoda*. Heidelberg: Universitätsverlag C. Winter, 1995.

Romiti, Antonio. "Due 'societates in exercitio venandi' del secolo XIV." *Actum Luce* 2 (1973): 217–24.

Rosati, Valeria, ed. *Le lettere di Margherita Datini a Franceso di Marco (1384–1410)*. Prato: Cassa di Risparmi e Depositi, 1977.

Roveda, Enrico. "Allevamento e trasumanza nella pianura lombarda: I bergamschi nel Pavese tra '400 e '500." *Nuova Rivista Storica* 1 (1987): 49–70.

Seniff, Dennis P. *Libro de la Monteria: Based on Escorial MS Y.II.19*. Madison: The Hispanic Seminary of Medieval Studies, 1983.

Sforzi, A., and S. Lovari. "Some Effects of Hunting on Wild Mammalian Populations." *Anthropozoologica* 31 (2000): 165–72.

Soffici, Manila, ed. *Lo Statuto della Sambuca (1291–1340)*. Pisa: Pacini, 1996.

Spicciani, Amleto. "Un testimoniale del 1215 sul Padule di Fucecchio." In *Atti del Convegno su l'identità geografico-storica della Valdinievole: Buggiano Castello, 24 Giugno, 1995*, edited by Comune di Buggiano, 183–202. Buggiano Castello: Comune di Buggiano, 1996.

Sykes, Naomi. "The Introduction of Fallow Deer to Britain: A Zooarchaeological Perspective." *Environmental Archaeology* 9 (2004): 75–83.

Tilander, Gunnar, ed. *Gaston Phébus, Livre de Chasse*. Karlshamn: Cynegetica, 1971.

Trombetti Budriesi, Anna Laura, ed. *Federico II di Svevia: De arte venandi cum avibus. L'arte di cacciare con gli uccelli. Edizione e traduzione italiana del ms. lat. 717 della Biblioteca Universitaria di Bologna collazionato con il ms. Pal. lat. 1071 della Biblioteca Apostolica Vaticana*. Milan: Laterza, 2009.

Vaglienti, Francesca. "Cacce e parchi ducali sul Ticino (1450–1476)." In *Vigevano e i territorio circostanti all fine del Medioevo*, edited by Giorgio Chittolini, 185–260. Vigevano: Unicopli, 1997.

Warde, Paul. "The Environmental History of Pre-industrial Agriculture in Europe." In *Nature's End: History and the Environment*, edited by Sverker Sörlin and Paul Warde. New York: Palgrave Macmillan, 2009.

Woolgar, Chris M. "Gifts of Food in Late Medieval England." *Journal of Medieval History* 37, no. 1 (2011): 6–18.

Zazzeri, Roberta, ed. *Ci desinò l'abate: Ospiti e cucina del monastero di Santa Trinita, Firenze, 1360–1363*. Florence: Societá Editrice Fiorentina, 2003.

Zug Tucci, Hannelore. "La chasse dans la législation statutaire italienne." In *La Chasse au Moyen Age: Actes du Colloque du Centre d'études médiévales de Nice*, edited by Marcelle Thiébaux, 99–113. Nice: Edizioni del Galluzzo, 1979.

CHAPTER 4

Sustaining Premodern Heathlands, 1400–1750

Collective Knowledge and Peasant Communities in the Campine, Belgium

Maïka De Keyzer

Large parts of premodern Europe were covered by heathlands. In the coversand belt, stretching from the Brecklands in England over continental Europe to Russia, a zone of quartz layers provided the fragile and infertile conditions for this particular type of landscape.[1] These heathlands were mostly open stretches of heather and sturdy grasses, interspersed with moors, scrubland, bracken, and small woodlands. The so-called marginal economies that developed in these areas were dominated by common land and collective management. Here, agriculture typically allowed only limited arable production on intensively used in-fields, while vast wastelands and valuable hay meadows were used collectively and extensively.[2] In this chapter, I will question whether in the Campine area (Belgium), the heathlands were managed sustainably, and if so, whether the premodern communities that managed them developed a form of conservationism.

From the end of the seventeenth century, the general opinion of agronomists, improvers, and governments was that these heathlands were mismanaged. Prevailing agricultural strategies in the "marginal economies" were inefficient and unsustainable. In the English Brecklands, John Evelyn (1620–1706) and William Gilpin (1724–1804) painted an almost apocalyptic picture. Gilpin described the area between Brandon and Mildenhall warrens in 1769 as a "mere African desert. In some places, this sandy waste occupied the whole scope of the eye. The whole country indeed had the appearance of a beaten sea coast."[3] John Evelyn had a similar picture in his mind when he wrote in 1677 that "travelling sands have so damaged the country, rolling from place to place, and like the sands in the deserts of Libya, quite overwhelmed some gentlemen's whole estates."[4] In the Netherlands, Johan Picardt spent most of his life pleading with the local government to allow experimental attempts to turn the unproductive heathlands of Drenthe into productive grain lands.[5] This dire image of the heathlands suited these writers' calls for reform. Throughout Europe, agronomists and reformers, such as the physiocrats, called for large-scale agricultural changes. Their views became dominant from the eighteenth century onward in most of western Europe and were adopted by governments.[6]

All these actors had the same goal: all unimproved land, such as common or private wastelands, fallow and unproductive woodlands, had to be transformed into productive, arable land or timber forests. Unproductive practices and customs had to be eradicated and replaced by modern techniques and rational use of resources.[7] The population was on the rise, and hungry mouths had to be fed.[8] Increasingly proactive governments and enlightened monarchs joined the cause of these agronomists to battle what they saw as backward, unproductive, and destructive traditional agricultural practices. In 1772 in the Austrian Low Countries, Holy Roman Empress Maria Theresa (1717–1780) issued a decree that all common wastelands had to be privatized and tilled. To ensure a successful transition toward productive agricultural practices, the government invested in extensive infrastructural works such as canal and road building to allow irrigation and the transport of crops.[9]

By the second half of the nineteenth century, public opinion had shifted. Agricultural reforms had not proven successful, and industrialization was favored over agriculture. As a backlash to industrialization, elites craved unspoiled, authentic natural landscapes. Since both dense woodlands and mountainous areas were scarce in the heavily urbanized and deforested Low Countries, heathlands became one of the last prospects for finding that ideal. Painters such as Willem Roelofs (1822–1892) in the Netherlands and Adriaan Jozef Heymans (1839–1921) in Belgium romanticized the heathland landscapes as vast and mysterious open spaces, dotted with sheep and lonely shepherds.[10] Writers too were central in shaping this romanticized image. In Belgium, the famous novelist Hendrik Conscience (1812–1883) was a champion of the Campine heathlands. He often set his stories there and ventured into remote Campine villages for inspiration as well as peace and quiet. This prompted a shift toward tourism and paved the way for a preservation movement to save these rare and shrinking landscapes.[11]

The pressure on the Kalmthout heathland in the province of Antwerp was the direct cause of the foundation in 1910 of Belgium's first nature agency, "Natural and Urban Beauty" (*Natuur en Stedeschoon*).

For many years, historians, archaeologists, and geologists have adopted the perspective introduced by eighteenth-century agronomists. The ecosystem was supposed to have been progressively degraded from the later Middle Ages onward, causing the ecologically diverse heathlands to degrade and eventually become open sand bowls.[12] The culprits were the same—namely, the village communities that managed the resources collectively and unsustainably. Because of a flawed management system and free-rider problems, large herds of sheep overgrazed the heathlands. Exacerbating these problems was plaggen fertilization, which treats the region's sandy soils by mixing animal manure with sods taken from the common heathlands, thereby destroying the remaining plant cover and leading to environmental tragedy.[13]

But was this environmental degradation a real or a perceived crisis? Did the eyewitness accounts register the environmental issues objectively, or did they try to strengthen their political agenda? Since Elinor Ostrom's groundbreaking work, *Governing the Commons*, it has become clear that in the right circumstances, common resources can be managed sustainably for centuries.[14] In spite of the traditionally pejorative image of premodern rural communities, these communities were well aware of their environmental surroundings and strived to manage their communal resources sustainably.[15] Management of the landscape was a core topic of discussion for village communities from at least the thirteenth century onward. The question remains: How do we define and perceive that management?

Several scholars have cast premodern communities as predominantly focused on the economic value of resources. Management was a way to maximize yields while sustaining resources for future generations. According to Paul Warde, premodern communities made sure "resources would be made available when and where they were essential to the functioning of the household."[16] Regulation and management of resources were not to protect an ecosystem or even agricultural system, but to protect the survival of the household.[17] Christopher Rodgers and his colleagues have a similar, but more nuanced, view: environmental sustainability is a modern concept. Resources and resource exploitation were always the underlying objectives. Rodgers et al. nonetheless argue that premodern communities managed resources sustainably because their practices considered the economy, the environment, and social equity. Premodern forms of sustainability comprised all three dimensions, which corresponds with the current idea of the "triple-bottom line" that is now often associated with sustainability.[18] The natural environment was managed to provide resources for the long term that were equitably distributed among the members of the community without degrading the ecosystem.[19]

Others, however, have stressed that premodern sustainability and even conservation are not anachronistic. Sustaining is to keep something in good condition and this state of mind was omnipresent in premodern rural communities.[20] Conservation can be linked with premodern principles, depending on the definition. Gifford Pinchot defined conservation as "Wise use of the earth and its resources for the lasting good of men. Conservation is the foresighted utilization, preservation, and/or renewal of forests, waters, lands and minerals for the greatest good of the greatest number for the longest time."[21] Similarly, this planned behavior of protecting and renewing resources for the common good has been ascribed to premodern institutions and communities in the past, especially in forest management. Monks, clerical institutions, manorial lords, and governments have been labeled excellent forest managers.[22] But woodlands were not the only resources that were looked after. Tine De Moor has shown that historical institutions for collective action in Flanders developed management plans for all types of resources that can be viewed as proto-conservation policies because the local communities adjusted the appropriation rules for communal goods based on the carrying capacity of the landscape.[23]

One of the regions that was able to develop long-term, sustainable management of the heathlands was the Campine, a region northeast of Antwerp, Belgium. This chapter analyzes the management of the Campine heathlands by local common pool institutions. After introducing the case study, the chapter examines the goals of the village communities. Were these goals limited to protecting only their economic interests and resources, or to ensuring a good neighborhood? Or can we distinguish a conservation policy and environmental awareness *avant-la-lettre*? Finally, this essay turns to the actual practices and policies that were introduced to obtain a sustainable management of the common heathlands.

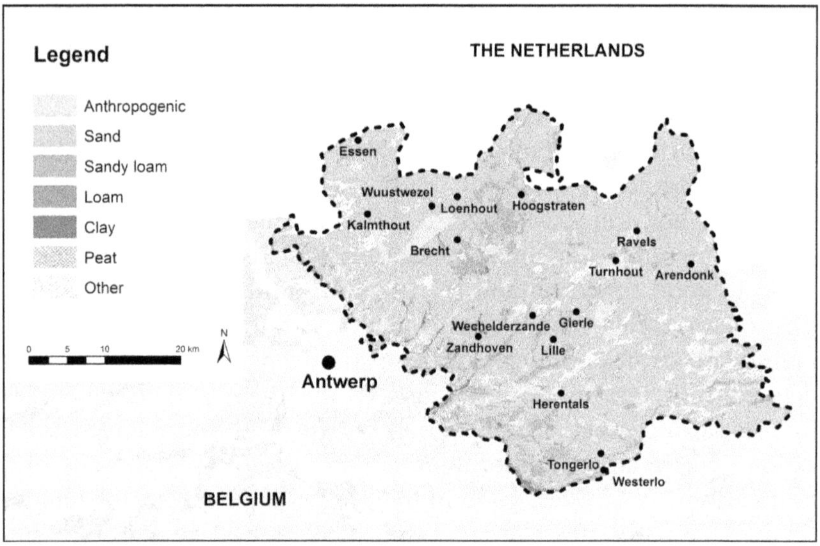

Map 4.1. A soil map of the selected case study: the Campine. Map made by Iason Jongepier. GIStorical Antwerp (UAntwerpen/*Hercules Foundation*). Published with permission.

Case Study: The Campine

The Campine is the area delimited by the Nethe River in the south and east, the border between the polder region and sandy interior in the west, and the present-day Belgian-Dutch border in the North.

As Map 4.1 shows, the Campine area was dominated by sandy soils. Because of this subsoil, the area developed a heathland environment. Originally, the climax vegetation was a mixed forest, but because of human exploitation, the wood cover gave way to an open landscape of heather and grass vegetation, together with small woodlands and peat bogs. The economy that developed was often considered marginal because it offered limited opportunities to produce grain or cash crops. Nevertheless, the resources the communities could harvest were diverse. These heathlands provided suitable grazing for sheep and occasionally cattle. In combination with hay meadows, this region was well suited to animal husbandry. In addition to fodder, the heathlands offered clippings or heather sods that could be collected or cut to mix with manure. This process provided fertilizer to allow arable farming on a limited set of fields. A third resource was peat. As in other parts of Europe where woodland was scarce, peat was an important source of fuel.[24] Apart from the area around Kalmthout, the Campine was not rich in thick peat layers but offered several smaller pits that were exploited by the peasants for their own consumption.[25] Finally, peasants derived smaller benefits from a wide variety of other resources.

This type of ecosystem was fragile. Disturbances, such as overgrazing or excessive digging for sod, could lead to a degradation of more diverse or "green heathlands," toward open sand bowls prone to sand drifts. Green heathlands include a mixture of grasses and heather vegetation, as opposed to less diverse "purple heather field" dominated by heather (*Calluna vulgaris*).[26] Between 1350 and 1750, peasant communities used the common heathlands. Elites, such as manorial lords, urban investors, or commercial farmers, were a minority in this region. The average peasant owned between one and five hectares of private land and meadow, a couple of cows, and a draft horse. A more limited group, consisting of the top 30 percent of the village, owned an additional flock of sheep. The few elite farmers had the same agricultural profile but on a larger scale.[27] They all depended on the communal resources. None of the social groups could produce enough fodder, manure, or wood on their own land, and therefore had to use communal resources to survive and maintain their economic status.[28]

The management of the Campine heathlands was not top-down. The responsibility of lords, ecclesiastical institutions, or city governments was minimal. These resources were owned and managed collectively. While in the surrounding regions, such as Flanders and Holland, most of the land was privatized from the later Middle Ages onward, in the Campine, up to 90 percent of the land was common. The largest part of this common land was labeled common wasteland, often referred to simply as "heath" (French *bruyere*; Middle Dutch *heide* or *aard*). From the thirteenth century onward, the management of the landscape was firmly in the hands of the local village communities. The arrival of the dukes of Brabant in the region ended a period of manorial dominance. In exchange for money, the dukes granted privileges and liberties to the village communities to form village governments called "aldermen's benches" (*schepenbanken*).[29] These village governments were responsible for local legislation and the management of the common natural resources, which made them common pool institutions, as defined by Ostrom.[30]

Campine Regulatory Conservation

Tracing the roots of a Campine "conservation" plan starts with the village "by-laws" (*keuren*). By-laws are a common feature in common pool institutions in northwestern Europe. As opposed to other continents and even Eastern Europe, northwestern institutions for collective action formalized their institutions and wrote their statutes and rules down in documents called by-laws.[31] As a result, we have substantial evidence of the management practices and regulatory framework of common pool institutions in the past. The village communities that were responsible for managing the communal natural resources introduced this set of (formal and informal) rules and strict policies.[32] These by-laws, though probably originating in an oral form in the thirteenth century, were only written down and preserved from the fifteenth century onward.[33] Due to the rising pressure from the Habsburg monarch Charles V (1506–58), local by-laws and regulations were increasingly homologized.

The Campine by-laws concern the organization of village life, treating criminal law, good order of the village, guild regulations, prevention measures, market rules, as well as the management of natural resources.[34] The aldermen's bench and the lord's representative created

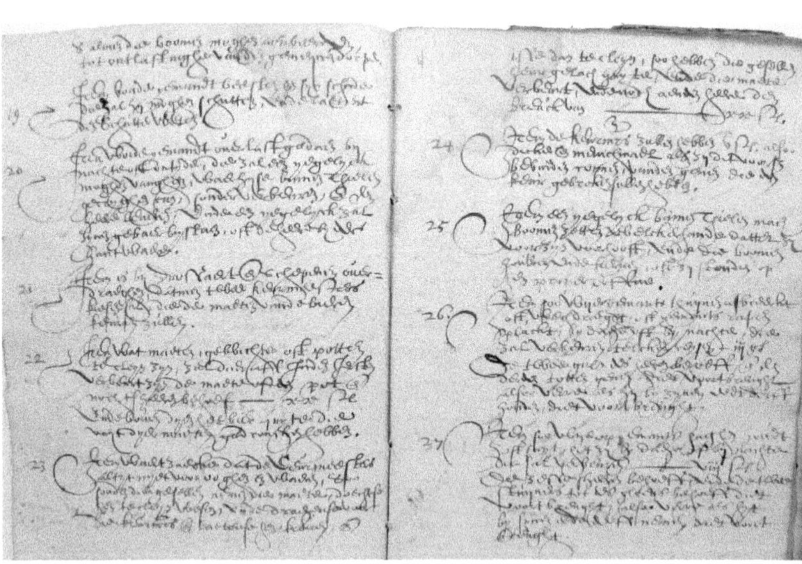

Figure 4.1. Image of the 1595 Tielen by-laws. RAA, OGA Tielen, 28 Keuren.

the rules. In theory, new rules could be proposed by all village inhabitants. In the village of Vorselaar, it was agreed that if seven people would support and suggest a new rule, it would be included in the by-laws. In practice, the village elites probably wrote most of the rules.[35] Even though the by-laws have an administrative outlook and show only normative prescriptions, with only limited references to intentions or environmental strategies, they provide a wealth of information that, in combination with archaeological reports and geological studies, allows us to reconstruct the historical management of the Campine common wastelands.

For this research, I have collected and analyzed the by-laws of twenty-four villages and hamlets.[36] Even though they represent a myriad of rules and stipulations, they dealt with very specific topics. First, it should be mentioned that I have only incorporated the rules pertaining to agricultural practices, the commons, and commercial activities in my research, leaving out guild regulations and penal codes. The remaining 1,143 rules mainly dealt with the manner in which the commons or the land in general had to be used and what punishments were to be administered if one infringed those regulations. Importantly, however, a large part of rules—22.8 percent—can be placed under the category of "management." The Campine communities were extremely strict in formulating how community members organized their usage of land. This management included the appointment of officials, delineating their powers, organizing visitations and controls, defining the sorts of punishments, and so on.

In the opening section of the by-laws, the community's main concern immediately becomes apparent. Almost all villages refer to "peace and quiet" or good neighborhood as the main objective. For example, the village of Ekeren stated: "That the statutes, rules and ordinances have been made and signed to observe the good order, to correct offences and to maintain the peace and quiet of the village subjects."[37] Not one by-law refers to sustainability or even management of the heathlands or arable fields as a concern. Instead, they take a juridical and administrative approach: a prohibition or obligation is followed by the penalty that will follow in case of a transgression. For example, a rule of the Terloo by-law states: "Every house or farmstead may according to old custom dig peat for half a day on the common of Loo, namely from the morning until 12 or from noon till sunset and not longer, on the penalty of the forfeit of the peat that will be sold to the highest bidder. The plaintiff will receive 10 *stuivers* while the remaining amount will go to the village."[38] The juridical discourse veils most of the intentions and ecological strategies, but the Campine by-laws all include a consistent and efficient policy to sustain the fragile heathlands.

The actions and strategies are planned and formulated in strict rules. Even though custom appears to be a random collection of rules that organically grew into a policy, these by-laws show that the collective action and management was carefully planned and organized:[39] access was regulated; pressure was reduced; appropriation rules were strictly delineated and limited to subsistence needs; the means of harvesting and collecting resources was described in detail; and maintenance requirements and obligations were written down. Limiting the pressure; allowing extensive grazing, mowing, and sod digging; and monitoring the vegetation cover are practices that are implemented today by preservation institutions.

One of the strongest points of these local communities is their level of commitment and organization. Contrary to the image painted by historical writers, such as Picardt, and modern scholars, such as Garrett Hardin, peasant communities were not freeriding individuals who operated according to their private interests.[40] The management tasks were a collective responsibility. Every inhabitant knew his or her obligations, and the most important management tasks were divided. Formal control over the commons was placed in the hands of a small group of officials called "commons' masters" (*aardmeesters* or *vorsters*). They were either appointed by the lord and his representative or elected, as in the case of the village of Terloo where the by-law states: "two commons' masters will be elected to serve two years like the church masters. When their term is over, they will have the opportunity of choosing other members of the community according to their understanding and judgement."[41] These commons' masters were the ones who roamed the heathlands to point out the location where resources could be collected or harvested, and they were responsible for catching trespassers. For example, the village of Ekeren stated in its by-law that the commons' masters were required to "go around the common wastelands and boundaries every 14 days together and with each other."[42] Specialists were then added to this group for issues, such as sheep scab or sand drifts.

The village government, or aldermen's bench, oversaw justice and determined the fine for infractions against the regulations. Finally, the entire community was expected to exert social control to prevent trespassing and to catch offenders in the act.[43] Many stipulations mentioned that everyone could complain about offenses, in exchange for a part of the fine. For example, in Arendonk, "everybody has to enclose their plots located next to the village meadow, so as to prevent animals from straying in the common meadow, on the penalty of 15 pence. Everybody can file a complaint about this for half of the fine."[44] As Van Weeren and de Moor argue, this type of broadly based social control is almost certainly more efficient than the imposition of high fines in preventing freeriding and trespassing.[45] This level of organization and self-control is a far stretch from the chaos and detrimental actions that are often described by modern agronomists as the behavior of premodern communities.

In the Campine area, the heathlands were managed sustainably in the long run from at least the fourteenth century and until the eighteenth century. This focus on long-term maintenance rather than short-term gain is one of the main characteristics of the Campine communities. Freeriding, overstocking, and commercial exploitations that could generate profit in the short term were forbidden so as to allow all generations to provide for their subsistence. Because of the combination of strong private claims on land and the dependence on common resources, the peasant households understood that both the current as well as multiple future generations were reliant on the sustainability of the resources. In an effort to sustain the diverse heathland vegetation for the long term, the villagers limited sand mobility, a sign of landscape degradation and reduced vegetation cover, despite high population and grazing pressure.[46] In addition, the Campine's local institutions provided equitable access to the common pool resources and adapted the management of the commons to tailor all the interest group's needs.[47]

Conservation Practices

These regulations thus secured the sustainable use of the resources, but some went even further and actively considered the ecosystem in general. This cannot be equated with a modern appreciation of wilderness or natural beauty or even ecological integrity. Nevertheless, this focus on maintaining an ecological balance was not only caused by the need to produce resources. It was caused by an awareness of their dependence on the natural environment and their influence on it. Sand drifts threatened the coversand belt, with its dominant quartz layers. During the Roman period and the Central Middle Ages, several regions, including the Campine, were struck by disastrous sand floods. During these colonization periods, when new agricultural techniques were tested and large stretches of land were cleared, the local communities were confronted with the hazardous results and learned from the experience.[48]

The communities implemented different changes and mitigation measures. First, they anticipated environmental problems, and therefore changed their agricultural system and employed prevention methods, such as windbreaks, collective planting, and limiting the uncovering of the subsoil. Even though the exact timing of these changes is unclear, somewhere between 1150 and 1300, the Campine peasants changed their agricultural model from scattered, individual farmsteads surrounded by their fields and pastures into nucleated villages.[49] In addition, between the fourteenth and sixteenth centuries, the peasants traded their open fields, which were prone to sand mobility, for a more enclosed and bocage landscape.[50]

Second, they traded isolated farmsteads and individual approaches for the development of a sense of community. Peasants organized into village governments, solidarity mechanisms arose, and communal agricultural practices developed.[51] For example, the prevention of the most common hazard, drifting sand, was handled collectively. Planting windbreaks was a communal activity that was centrally organized, and the maintenance of hedges to protect the fields from drifting sand (as well as stray animals) was collectively controlled and a communal responsibility. This is what Greg Bankoff has called "subcultures of coping." Communities who experience a recurrent threat may develop disaster subcultures. These are specific patterns geared toward the solution of problems arising from the awareness of some form of periodic threat.[52] Although in the Campine, evidence for a cultural awareness of sand disasters is only indirect, from the thirteenth century onward at the latest, local institutions had developed environmental management strategies to stabilize the landscape and prevent sand mobility. Finally, regulations were introduced that went beyond mere sustainability concerns. A strict management to maintain the balance between reforestation and degradation toward open sand dunes was the main goal. To sustain the heathlands, a set of practices was implemented by institutions for collective action.

First, the level of pressure on the landscape was regulated. The most valuable asset of the heathlands was their function as grazing land. In the Campine, the most important commercial animals were sheep.[53] They were allowed on common wastelands the entire year. Every inhabitant and member of the community could graze sheep on the commons. In most regions, the pressure was reduced by introducing stints, a numerical limitation on the number of animals per household.[54] This practice, however, was not introduced in the Campine area. The number of sheep was limited by the amount of fodder peasants and farmers could obtain from their private land; however, scholars have established that stock densities are negatively related to the level of cooperation. In communities where strong collective action exists, overstocking is less likely and total stock densities are lower.[55] The main concern of the village institutions was not the size of the flocks, but rather grazing practices. The sheep could not roam freely; they had to be monitored and kept away from fragile zones and fields. Any damage done by the grazing animals was punishable by a high fine.[56] In Retie in 1554, trespassing with animals in fragile zones was punished as follows: "Nobody can bring sheep or other animals on the common or elsewhere on the flying sand within the marks that have been installed by the sworn councilors, on the penalty of 12 *stuivers*, two shares will go to the lord and one share to the one that complains."[57] Damage was punished even more harshly. For example, in the village of Rijkevorsel, it was stipulated:

> Whoever destroyed fences, creates holes or clogged ditches during grazing in the forests, pastures, heathlands or meadows will have to pay 20 *stuivers* for every horse, cow, bovine or foal, 8 *stuivers* for every calf and 2 *stuivers* per sheep. The fine will be double if it happens during the night. All inhabitants are allowed to catch the animals on their property and deliver them to the folds. The trespassers will have to appear before the lord.[58]

The most invasive activity on the common wastelands was not grazing. These lands were also used to mow heather clippings and grasses, dig for sods and peat, and harvest building materials. These practices not only impacted ecodiversity, but also could fundamentally disturb the landscape. Intensive sod digging reduces the diversity of the flora, creating the typical purple heather fields, instead of the much more diverse "green" heathlands. In addition, the resulting pits or uncovered earth could provoke sand drifts in drier areas or swamps in poorly drained ones. As a consequence, most of the rules in the by-laws are oriented toward limiting these activities. Contrary to the image of the destructive nature of communal plaggen agriculture, the Campine communities were able to combine plaggen fertilization and a sustainable management of the heathlands until the end of the eighteenth century. Only when mixed agriculture and communal practices were transformed by the eighteenth-century "reforms" was the fragile balance disturbed and disastrous sand drifts reappeared in the region.[59] The number of sods and clippings every household could obtain was strictly limited and various precautions had to be taken. A by-law of Geel states the following:

> Nobody of the six quarters will mow more than his allowed share and between Saint John's Mass till Bamis and always on a Thursday, except when it is a holiday. Every ploughman can take a *voeder* [unit of volume] per week and the cottagers a voeder every two weeks. Nobody is allowed to sell either heather or sods from the common wastelands on the penalty of 20 *stuivers*. Every one that fails to collect its sods will have to pay 20 *stuivers*. Everybody is required to maintain the common as it has been since ancient memory.[60]

The basic premise was that common resources could be used for subsistence needs, such as fertilizing fields, repairing the farmstead, or feeding a modest flock of animals, but not more. Therefore, every household had a limited timespan per week that they could harvest heather. Often the number of people from each household that could help was also limited. The sale of resources directly obtained from the commons was strictly forbidden, as was allowing outsiders to take unclaimed shares; however, indirect products, such as wool, honey, milk, or leather, could be sold.[61]

Along with reducing the pressure on the landscape by limiting the amount of resources that could be collected, local institutions strictly defined how resources were to be appropriated. The by-laws often refer to the need "to keep the commons in good order," or "in the state as it was before."[62] There was a wide range of rules to make sure that this ideal state of the environment was maintained. First of all, the instruments or farm equipment that could be used to mow or dig were limited. This prevented a degradation of the soil and limited the amount of soil that was disturbed. Second, every activity had its appointed location. For example, peat could not be dug randomly in the common wastelands. Officials delineated a spot for the entire community or several smaller plots per household. Only when that place was depleted could a new zone be selected.[63] Finally, harvesting was only allowed when followed by maintenance. Every activity on the common wastelands was only permitted if it did not leave major scars on the landscape. For example, peat pits that were depleted had to be filled up, so as to prevent pools from appearing. Loam that was dug had to be replaced by sand.[64]

The Campine institutions of collective action invested time and effort in renewing the resources and protecting the ecological balance to prevent any disturbance. Even though they did so primarily to prevent certain hazards, these efforts show a fundamental environmental awareness and willingness to invest and develop policies to protect the heathlands as an ecosystem. Furthermore, from the later Middle Ages onward, many communities obliged all the inhabitants to restore any disturbances. The villages of Retie, Kalmthout, Ravels, Geel, and Arendonk all refer to small plantations or wooded areas, called "heather forests" (*heibossen*), that were constructed on the wastelands in order to prevent or limit drifts.[65] The community of Retie could appoint two men who were responsible for the protection of the community against the sand and for planting trees.[66] In addition, once a year, community members were required to perform communal tasks, such as maintaining woodlands planted on the common heathlands. The by-laws of Ravels and Eel, for example, stated, "The wood, needed to stop the sand, will have to be repaired by everyone on the punishment of 6 *stuivers*."[67]

Via collective, bottom-up actions, the Campine village communities therefore had developed not only sustainable policies but also effective conservation policies from the later Middle Ages onward. These policies are currently considered as the most efficient way of maintaining the Campine heathlands. Today's nature conservation agencies, such as Natuurpunt, have adopted a number of premodern practices as methods of preserving the few remaining heathlands in the Campine. Galloway cows and sheep have been reintroduced, sods are now dug regularly, and small woodlands and "heather forests" *(heibossen)* are favored over pine forests. These activities have been proven to be the most beneficial to secure ecodiversity and to protect the survival of the heathlands.

It must be acknowledged, however, that not all common pool institutions were necessarily willing or able to manage their resources sustainably or to develop a form of proto-conservationism. Premodern village communities had the tools to manage their natural resources, but they were not always able to use them to full effect. In the premodern Brecklands of eastern England, located in the same coversand belt, the heathlands were mismanaged from the seventeenth century onward. Commercial rabbit warrens and huge sheep flocks provided the opportunity to turn the arid heathlands into cash. The lords who profited from these enterprises did not wish to invest in the maintenance of the landscape, because it would reduce their profits. Here, local communities were not strong enough to resist or to protest the lack of maintenance and the resulting degradation of the landscape.[68] The implementation of conservation policies was, therefore, not uniform in either the European coversand belt or among all common-pool institutions. Sociopolitical barriers could prohibit sustainable management.

Conclusion

For a long time, the management of the heathlands by local village communities was considered detrimental to the environment. Since the eighteenth century, local practices and communal management have been attacked by agronomists and "enlightened" governments. Over the following two centuries, new conservation policies and physiocratic principles were implemented to change the drift-prone wastelands into productive fields; however, this dire picture is not supported by the evidence. In the heathlands of the European coversand belt, most resources were managed sustainably, and conservation ideas and environmental awareness were already well developed in the premodern period.

The most important incentive to manage natural resources was not nature or the environment in itself. The reliance on resources for subsistence and the need to maintain a good relationship with the members of the community provided the main motivations for sustainable practices. But more than this, Campine communities showed an environmental awareness and willingness to conserve the entire ecosystem. Because of the disastrous sand floods that had occurred during the Central Middle Ages and the constant threat of sand throughout the area, Campine communities developed a culture of coping. They realized that they depended on a fragile ecological balance and invested time and effort in conserving their ecosystem.

As a result, they developed a consistent approach that emphasized rule-making, mandated a long-term perspective, and benefited the public good. This management system was not implemented top-down by a lord, institution, or agronomist, but was developed bottom-up. In the Campine, most resources were common, and their management was a collective responsibility. From the thirteenth century onward, the heathlands were used, managed, and controlled by local common pool institutions. Communities developed practices that formed a consistent policy to limit the pressure on the landscape, prevent destruction of the fragile soils, and maintain the heathlands in good condition. These practices were written down in local regulations or by-laws. Apart from securing resources for the future generation, they actively engaged in maintaining the landscape by collective planting and maintenance works. In this way, these communities developed a premodern form of conservationism that, perhaps surprisingly, fits Pinchot's definition of conservation: they fostered "a wise use of the earth and its resources for the lasting good of men."[69] The Campine institutions showed "a foresighted utilization, preservation, and/or renewal of forests, waters, lands and minerals for the greatest good of the greatest number for the longest time."[70]

Maïka De Keyzer is a tenure track docent at the KULeuven (Belgium). She obtained her PhD in 2014 at the University of Antwerp with a dissertation on late medieval commons, sustainability, and inclusiveness. Her research deals with economic, social, and ecological history of the premodern period. Her publications include a recent book, *Inclusive Commons and the Sustainability of Peasant Communities in the Medieval Low Countries* (2018). Via an interdisciplinary and comparative approach, Maïka De Keyzer analyzes the causes and consequences of welfare, inequality, social resilience, and collective action.

Notes

1. Josef Fanta and Henk Siepel, eds., *Inland Drift Sand Landscapes* (Zeist, The Netherlands: KNNV Publishing, 2010).
2. Mark Bailey, *A Marginal Economy? East Anglian Breckland in the Later Middle Ages* (New York: Cambridge University Press, 1989).
3. William Gilpin, *Observations on Several Parts of the Counties of Cambridge, Norfolk, Suffolk, and Essex: Also on Several Parts of North Wales, Relative Chiefly to Picturesque Beauty, in Two Tours, the Former Made in the Year 1769, the Latter in the Year 1773* (New York: Cambridge University Press, 2014), 28.
4. John Evelyn, "Diary Entry for 10 September 1667," in *The Diary of John Evelyn*, ed. W. Bray (London: J. M. Dent and Sons, 1952).

5. Theo Spek, *Het Drentse esdorpenlandschap: Een historisch-geografische studie* (Utrecht: Stichting Matrijs, 2004); Michiel A. W. Gerding, *Johan Picardt (1600-1670): Drenthe's eerste geschiedschrijver*, vol. 6 (Assen: Drentse Historische Reeks, 1997).
6. The physiocrats were a group of French economists that defended the principle of economic independence and considered a modern agriculture as the true motor of economic growth and prosperity. Nadine Vivier, "Le conflict autour des biens communeaux ou la crise de la propriété collective (1760–1870)," in *Temps et éspaces des crises de l'environnement*, ed. Corinne Beck (Paris: Editions Quae, 2006), 71–82; Christine Lebeau, "La physiocratie au pouvoir? Lecture et politique à Vienne (1780–1790)," in *La diffusion internationale de la physiocratie*, ed. Bernard Delmas, Thierry Demals, and Philippe Steiner (Paris: Presses Universitaires de Grenoble, 1995), 367–79.
7. Erik Vanhaute, *De invloed van de groei van het industrieel kapitalisme en van de centrale staat op een agrarisch grensgebied: De Noorderkempen in de 19de eeuw (1750–1910)* (Brussels: Gemeentekrediet, 1990).
8. Paul M. M. Klep, "Population Estimates of Belgium, by Province (1375–1831)," in *Historiens et populations: Liber amicorum Etienne Hélin* (Louvain-la-Neuve: Academia, 1991).
9. Guy Dejongh, "De ontginningspolitiek van de overheid in de Zuidelijke Nederlanden, 1750–1830: Een maat voor niets?" *Tijdschrift van Het Gemeentekrediet* 210 (1999): 31–44.
10. Marina Van Aert and Jean Bastiaensen, *De Kalmthoutse School* (Tielt: Lannoo, 2007); Saskia De Bodt, *De Haagse School in Drenthe* (Assen: WBOOKS, 1997).
11. Kevin Absillis and Marlou de Bont, "Who's Afraid of Conscience? Naar een andere kijk op het oeuvre waarmee een volk leerde lezen," *Wetenschappelijke Tijdingen* 74, no. 1 (2015): 44–84.
12. Jan Luiten Van Zanden, "The Paradox of the Marks: The Exploitation of Commons in the Eastern Netherlands, 1250–1850," *Agricultural History Review* 47 (1999): 125–44; Ilona Castel, "Late Holocene Aeolian Drift Sands in Drenthe (the Netherlands)" (PhD Thesis, University of Utrecht, 1991).
13. On the ongoing debate about the destructiveness of plaggen fertilization, see Jan Bastiaens and Cyriel Verbruggen, "Fysische en socio-economische achtergronden van het plaggenlandbouwsysteem in de Antwerpse Kempen," *Tijdschrift voor ecologische geschiedenis* 1, no. 1 (1996): 26–32; Hans-Peter Blume and Peter Leinweber, "Plaggen Soils: Landscape History, Properties and Classification," *Plant Nutrition and Soil Science* 167 (2004): 319–27; W. van der Westeringhe, "Man-Made Soils in the Netherlands, Especially in Sandy Areas ('Plaggen Soils')," in *Man-Made Soils*, ed. W. Groenman-van Waateringe and M. Robinson (Oxford: British Archaeological Reports Limited, 1988).
14. Elinor Ostrom, *Governing the Commons: The Evolution of Institutions for Collective Action* (New York: Cambridge University Press, 1990).

15. Tine De Moor, "Avoiding Tragedies: A Flemish Common and Its Commoners under the Pressure of Social and Economic Change during the Eighteenth Century," *The Economic History Review* 62, no. 1 (2009): 1–22; C. P. Rodgers et al., *Contested Common Land: Environmental Governance Past and Present* (London: Earthscan, 2011).
16. Paul Warde, "The Environmental History of Pre-industrial Agriculture in Europe," in *Nature's End: History and the Environment*, ed. Sverker Sörlin and Paul Warde (Basingstoke: Palgrave Macmillan, 2009), 75.
17. Warde, "Environmental History," 75.
18. The term "triple-bottom line" was first coined by John Elkington, *Cannibals with Forks: The Triple Bottom Line of 21st Century Business* (Oxford: Capstone, 1997).
19. Rodgers et al., *Contested Common Land.*
20. Joachim Radkau, *Nature and Power: A Global History of the Environment* (New York: Cambridge University Press, 2008), 20–25.
21. Gifford Pinchot, *Breaking New Ground*, rev. ed., ed. Char Miller (New York: Island Press, 1995), 322–26.
22. Ellen Arnold, *Negotiating the Landscape: Environmental and Monastic Identity in the Medieval Ardennes* (Philadelphia: University of Pennsylvania Press, 2013); Péter Szabó and Radim Hédel, "Socio-Economic Demands, Ecological Conditions and the Power of Tradition: Past Woodland Management Decisions in a Central European Landscape," *Landscape Research* 38, no. 2 (2013): 243–61.
23. De Moor, "Avoiding Tragedies."
24. Iason Jongepier et al., "The Brown Gold: A Reappraisal of Medieval Peat Marshes in Northern Flanders (Belgium)," *Water History* 3, no. 2 (2011): 73–93.
25. Karel A. H. W. Leenders, *Verdwenen venen: Een onderzoek naar de ligging en exploitatie van thans verdwenen venen in het gebied tussen Antwerpen, Turnhout, Geertruidenberg en Willemstad 1250–1750* (Brussels: Gemeentekrediet, 1989).
26. Spek, *Het Drentse esdorpenlandschap.*
27. Eline van Onacker, *Village Elites and Social Structures in the Fifteenth and Sixteenth-Century Campine Area* (Turnhout: Brepols, 2017).
28. Maïka De Keyzer, *Inclusive Commons and the Sustainability of Peasant Communities in the Medieval Low Countries* (London: Routledge, 2018).
29. P. J. V. Dekkers, "Brandend zand: Hoe de hertog van Brabant zijn heerschappij op de Kempense zandgronden verwierf ten koste van de lokale en regionale adel," *Noordbrabants Historisch Jaarboek* 12 (1995): 10–40.
30. Ostrom, *Governing the Commons.*
31. Miguel Laborda Peman, "A Tale of Two Commons: Some Preliminary Hypotheses on the Long-Term Development of the Commons in Western and Eastern Europe, 11th–19th Centuries," *International Journal of the Commons* 7, no. 1 (2013): 7–33.

32. Martina De Moor, Leigh Shaw-Taylor, and Paul Warde, eds., *The Management of Common Land in North West Europe, c. 1500–1850* (Turnhout: Brepols, 2002); Rodgers et al., *Contested Common Land.*
33. Robert Stein, *De hertog en zijn staten: De eenwording van de Bourgondische Nederlanden ca. 1380–ca. 1480* (Hilversum: Verloren, 2014).
34. De Moor et. al, *Management of Common Land.*
35. De Keyzer, *Inclusive Commons.*
36. See bibliography for references.
37. "keuren en ordonnanties bij hun gemaakt en gesloten tot observatie van de goede politie, correctie van de gebreken en het onderhouden van dezelfde onderzaten in rust en vrede"; source: A. Gielens, "Keuren van Ekeren," *Oudheid en Kunst* 30, no. 1 (1939): 167–83.
38. "Elk huis of vierstede mag naar oude gewoonte turf steken op de aard van Loo voor een halve dag, te weten van 's morgens tot 12 uur op de middag of van noen tot zonsondergang en niet langer op verbeuren van de turf die men per hoogste bod zal verkopen waarvan de klager 10 stuivers krijgt en de rest voor de gemeynte"; source: J. Van Gorp, "De Aartbrief van Terloo," *Bijdragen tot de Geschiedenis* 18 (1927): 437–54.
39. Heather Falvey, "The Articulation, Transmission and Preservation of Custom in the Forest Community of Duffield (Derbyshire)," in *Custom, Improvement and the Landscape in Early Modern Britain*, ed. Richard Hoyle (Farnham: Ashgate, 2011), 65–100; Edward Palmer Thompson, *Customs in Common* (Harmondsworth: Penguin Group, 1991).
40. Garrett Hardin, "The Tragedy of the Commons," *Science* 162, no. 3859 (1968): 1243–48.
41. "Is geordonneerd dat bij gemene stemmen van de ingezetenen het gehucht van loo dat er altijd 2 aardmeesters gekozen zullen worden om 2 jaar te dienen zoals kerkmeesters. Als hun termijn beëindigd is zullen zij de keuze hebben om uit de gemeente van Loo twee andere uit te kiezen naar hun verstand en goeddunken"; source: Van Gorp, "De Aartbrief van Terloo."
42. "alle 14 dagen tezamen en met elkaar in de heide en rond de palen van Ekeren te gaan"; source: Gielens, "Keuren van Ekeren."
43. De Keyzer, *Inclusive Commons.*
44. "de erven die rondom, voor en aan het broek gelegen zullen zijn zullen hetzelfde loffelijk moeten omheinen zo dat daar over geen beesten in de gemene broeken komen op het verbeuren van 15 s. Iedereen mag hieraf klagen en de helft van de boete ontvangen"; source: Floris Prims, "Keuren der vreyheyt van Arendonk," in *Feestbundel H. J. Van De Wijer*, ed. H. Draye, vol. 1 (Leuven: Instituut voor Vlaamse topynomie, 1944).
45. René van Weeren and Tine De Moor, "Controlling the Commoners: Methods to Prevent, Detect, and Punish Free-Riding on Dutch Commons in the Early Modern Period," *Agricultural History Review* 62, no. 2 (2014): 256–77.
46. Maïka De Keyzer, "All We Are Is Dust in the Wind: The Social Causes of a 'Subculture of Coping' in the Late Medieval Coversand Belt,"

Journal for the History of Environment and Society 1, no. 1 (2016): 1–35.

47. De Keyzer, *Inclusive Commons.*
48. Cilia Derese et al., "A Medieval Settlement Caught in the Sand: Optical Dating of Sand-Drifting at Pulle (N. Belgium)," *Quaternary Geochronology* 5 (2010): 336–41; Hendrik A. Heidinga, "The Birth of a Desert: The Kootwijkerzand," in *Inland Drift Sand Landscapes*, ed. Josef Fanta and Henk Siepel (Zeist: KNNV Publishing, 2010).
49. Willy Steurs, *Naissance d'une région: Aux origines de la Mairie de Bois-Le-Duc, Recherches sur le Brabant Septentrional aux 12e et 13e siècles*, Memoire de la classe des lettres, vol. III (Brussels: Académie royale de Belgique, 1993).
50. De Keyzer, *Inclusive Commons*
51. van Onacker, *Village Elites.*
52. Greg Bankoff, *Cultures of Disaster: Society and Natural Hazards in the Philippines* (London: Routledge, 2003), 268; Greg Bankoff, "The 'English Lowlands' and the North Sea Basin System: A History of Shared Risk," *Environment and History* 19 (2013): 5.
53. Maïka De Keyzer and Eline van Onacker, "Beyond the Flock: Sheep Farming, Wool Sales and Social Differentiation in a Late Medieval Peasant Society: The Campine Area in the Low Countries," *The Agricultural History Review* 64, no. 2 (2016): 157–80.
54. Angus J. L. Winchester and Eleanor A. Straughton, "Stints and Sustainability: Managing Stock Levels on Common Land in England, c. 1600–2006," *Agricultural History Review* 58, no. 1 (2010): 30–48.
55. Nancy McCarthy, Abdul B. Kamara, and Michael Kirk, "Co-operation in Risky Environments: Evidence from Southern Ethiopia," *Journal of African Economies* 12, no. 2 (June 2003): 236–70.
56. De Keyzer, *Inclusive Commons.*
57. "Niemand mag schapen of andere beesten laten gaan op de aard of elders op het vliegend zand binnen de tekenen daarvoor gesteld of door de gezworenen gesteld zullen worden, op de boete van 12 stuivers, twee delen voor de heer en een deel voor de klager"; source: I. Helsen, "Het dorpskeurboek van Retie," *Bijdragen tot de geschiedenis* 1, no. 1 (1949): 85–107.
58. "Wie tijdens het stouwen van de voorschreven beesten in de bossen, weiden, heiden of beemden, de grachten vult, gaten maakt of tuinen, hagen en hekken van de omheining open breekt, zal overdag verbeuren 20 stuiver voor elk paard, koe, rund of veulen, 8 stuiver voor elk kalf en 2 stuiver per schaap en dubbel 's nachts bovenop de schade. (en zullen staan ter correctie van de heer en iedereen zal op zijn eigen erf de beesten mogen schutten en afleveren aan de schutterskooi)"; source: RAA, OGA Rijkevorsel, 8, By-law 20 stuiver is the equivalent of four days' wages of an Antwerp mason's laborer during the sixteenth century. Source: http://www.iisg.nl/hpw/data.php#belgium (last accessed 21 January 2020).
59. De Keyzer, "All We Are."

60. A free translation from the original: "Niemand van de zes heerdgangen zal zich vervoorderen om meer dan zijn wijkheide te maaien te weten van sint jansmis tot bamis en binnen de voorschreven tijd altijd op donderdag tenzij dit een feestdag is. Elke ploeg een voerder per week en de keuters per 14 dagen een voerder. Niemand zal noch heide, noch russen van de gemeyne aard halen en die verkopen op de pene van 20 stuivers. Wie russen steekt en deze niet haalt zal 20 stuivers verbeuren. En een ieder van de heerdgang zal zijn vrijheid houden gelijk die van oude tijden is geweest"; source: J. Ernalsteen, "Keuren van Gheel," *Oudheid en Kunst* 26, no. 2 (1935): 19–66.
61. Rijksarchief Antwerpen (RAA), Oud Gemeente-Archief (OGA) Tielen, 28.; RAA, OGA Gierle, 44, Byelaw.; RAA, OGA Herenthout, 3, Byelaw.; RAA, OGA Hoogstraten, 638, Byelaw.; RAA, OGA Rijkevorsel, 8, Byelaw.; AAT, Bundel Tongerlo I: Rules for the village of Tongerlo, Copy.; AAT, Bundel Byelaws, Veerle and Oevel, Copy.
62. All seven of the documents cited in the previous note begin with very similar expressions of this type.
63. Hilde Verboven, Kris Verheyen, and Martin Hermy, *Bos en hei in het land van Turnhout (15de-19de Eeuw): Een bijdrage tot de historische ecologie*, ed. Monumenten & Landschappen (Leuven: Ministerie van de Vlaamse gemeenschap, Monumenten & Landschappen en het Vlaams Instuut voor het Onroerend Erfgoed, 2004).
64. See bibliography for reference by-laws.
65. Ernalsteen, "Keuren van Gheel"; Helsen, "Het dorpskeurboek van Retie"; Milo Koyen, "Keuren van Ravels," *Oudheid en Kunst* 41, no. 2 (1958): 3–19; Prims, "Keuren"; P. J. Verhoeven, "Keuren van Calmpthout," *Oudheid en Kunst* (1907): 45–46.
66. Helsen, "Het dorpskeurboek van Retie," 89.
67. Koyen, "Keuren van Ravels," 7.
68. Maïka De Keyzer and Mark D. Bateman, "What Caused Late Holocene Landscape Instability in the NW European Drift Sands," *Geomorphology* 323 (2018): 123–34.
69. Pinchot, *Breaking New Ground*, 322–26.
70. Ibid., 322–26.

Bibliography

Absillis, Kevin, and Marlou de Bont. "Who's Afraid of Conscience? Naar een andere kijk op het oeuvre waarmee een volk leerde lezen." *Wetenschappelijke Tijdingen* 74, no. 1 (2015): 44–84.

Arnold, Ellen. *Negotiating the Landscape: Environmental and Monastic Identity in the Medieval Ardennes*. Philadelphia: University of Pennsylvania Press, 2013.

Bailey, Mark. *A Marginal Economy? East Anglian Breckland in the Later Middle Ages*. New York: Cambridge University Press, 1989.

Bankoff, Greg. "Cultures of Disaster, Cultures of Coping Hazard as a Frequent Life Experience in the Philippines." In *Natural Disasters, Cultural Responses: Case Studies Toward a Global Environmental History*, edited by Christof Mauch and C. Pfister. Lanham: Lexington Books, 2009.

_____. *Cultures of Disaster: Society and Natural Hazards in the Philippines.* London: Routledge, 2003.

_____. "The 'English Lowlands' and the North Sea Basin System: A History of Shared Risk." *Environment and History* 19 (2013): 3–37.

Bastiaens, Jan, and Cyriel Verbruggen. "Fysische en socio-economische achtergronden van het plaggenlandbouwsysteem in de Antwerpse Kempen." *Tijdschrift voor ecologische geschiedenis* 1, no. 1 (1996): 26–32.

Blume, Hans-Peter, and Peter Leinweber. "Plaggen Soils: Landscape History, Properties and Classification." *Plant Nutrition and Soil Science* 167 (2004): 319–27.

Castel, Ilona. "Late Holocene Aeolian Drift Sands in Drenthe (the Netherlands)." PhD Thesis, University of Utrecht, 1991.

De Bodt, Saskia. *De Haagse School in Drenthe.* Assen: WBOOKS, 1997.

Dejongh, Guy. "De ontginningspolitiek van de overheid in de Zuidelijke Nederlanden, 1750–1830: Een maat voor niets?" *Tijdschrift van het Gemeentekrediet* 210 (1999): 31–44.

Dekkers, P. J. V. "Brandend zand: Hoe de hertog van Brabant zijn heerschappij op de Kempense zandgronden verwierf ten koste van de lokale en regionale adel." *Noordbrabants Historisch Jaarboek* 12 (1995): 10–40.

De Keyzer, Maïka. "All We Are Is Dust in the Wind: The Social Causes of a 'Subculture of Coping' in the Late Medieval Coversand Belt." *Journal for the History of Environment and Society* 1, no. 1 (2016): 1–35.

_____. *Inclusive Commons and the Sustainability of Peasant Communities in the Medieval Low Countries.* London: Routledge, 2018.

De Keyzer, Maïka, and Eline van Onacker. "Beyond the Flock: Sheep Farming, Wool Sales and Social Differentiation in a Late Medieval Peasant Society: The Campine Area in the Low Countries." *The Agricultural History Review* 64, no. 2 (2016): 157–80.

De Keyzer, Maïka, and Mark D. Bateman. "What Caused Late Holocene Landscape Instability in the NW European Drift Sands?" *Geomorphology* 323 (2018): 123–34.

De Longé, G. "Coutumes de Santhoven, de Turnhout et de Rumpet." In *Coutumes du Pays et Duché de Brabant: Quartier D'anvers*, edited by G. De Longé. Brussels: Gobaerts, 1870–78.

De Molder, Th. "Keuren van Oostmalle." *Oudheid en Kunst* 26, no. 1 (1935): 3–15.

De Moor, Martina, Leigh Shaw-Taylor, and Paul Warde, eds. *The Management of Common Land in North West Europe, c. 1500–1850.* Turnhout: Brepols, 2002.

De Moor, Tine. "Avoiding Tragedies: A Flemish Common and Its Commoners under the Pressure of Social and Economic Change during the Eighteenth Century." *The Economic History Review* 62, no. 1 (2009): 1–22.

Derese, Cilia, Dimitri Vandenberghe, Nele Eggermont, Jan Bastiaens, Rica Annaert, and Peter Van den haute. "A Medieval Settlement Caught in the Sand: Optical Dating of Sand-Drifting at Pulle (N Belgium)." *Quaternary Geochronology* 5 (2010): 336–41.

Elkington, John. *Cannibals with Forks: The Triple Bottom Line of 21st Century Business.* Oxford: Capstone, 1997.

Erens, A. *De oorkonden der abdij Tongerloo.* 4 vols. Tongerlo: St.-Norbertusdr., 1948.

Ernalsteen, J. "Brecht: De keuren van 1601." *Oudheid en Kunst* 16, no. 2 (1925): 25–32.

_____. "Keuren van Gheel." *Oudheid en Kunst* 26, no. 2 (1935): 19–66.

Evelyn, John. "Diary Entry for 10 September 1667." In *The Diary of John Evelyn*, edited by W. Bray. London: J. M. Dent and Sons, 1952.

Falvey, Heather. "The Articulation, Transmission and Preservation of Custom in the Forest Community of Duffield (Derbyshire)." In *Custom, Improvement and the Landscape in Early Modern Britain*, edited by Richard Hoyle, 65–100. Farnham: Ashgate, 2011.

Fanta, Josef, and Henk Siepel, eds. *Inland Drift Sand Landscapes.* Zeist, The Netherlands: KNNV Publishing, 2010.

Gerding, Michiel A. W. *Johan Picardt (1600–1670): Drenthe's eerste geschiedschrijver.* Vol. 6. Assen: Drentse Historische Reeks, 1997.

Gielens, A. "Keuren van Ekeren." *Oudheid en Kunst* 30, no. 1 (1939): 167–83.

Gilpin, William. *Observations on Several Parts of the Counties of Cambridge, Norfolk, Suffolk, and Essex: Also on Several Parts of North Wales, Relative Chiefly to Picturesque Beauty, in Two Tours, the Former Made in the Year 1769, the Latter in the Year 1773.* New York: Cambridge University Press, 2014.

Hardin, Garrett. "The Tragedy of the Commons." *Science* 162, no. 3859 (1968): 1243–48.

Heidinga, Hendrik A. "The Birth of a Desert: The Kootwijkerzand." In *Inland Drift Sand Landscapes*, edited by Josef Fanta and Henk Siepel. Zeist: KNNV Publishing, 2010.

Helsen, I. "Het dorpskeurboek van Retie." *Bijdragen tot de geschiedenis* 1, no. 1 (1949): 85–107.

Jongepier, Iason, Tim Soens, Erik Thoen, Veerle Eetvelde, Philippe Crombé, and Machteld Bats. "The Brown Gold: A Reappraisal of Medieval Peat Marshes in Northern Flanders (Belgium)." *Water History* 3, no. 2 (2011): 73–93.

Klep, Paul M. M. "Population Estimates of Belgium, by Province (1375–1831)." In *Historiens et populations: Liber amicorum Etienne Hélin.* Louvain-la-Neuve: Academia, 1991.

Koyen, Milo. "Keuren van Ravels." *Oudheid en Kunst* 41, no. 2 (1958): 3–19.

Lauwerys, J. "Keuren van Westerloo." *Oudheid en Kunst* 28, no. 4 (1937): 95–120.

Laborda Peman, Miguel. "A Tale of Two Commons: Some Preliminary Hypotheses on the Long-Term Development of the Commons in Western and Eastern

Europe, 11th–19th Centuries." *International Journal of the Commons* 7, no. 1 (2013): 7–33.

Lebeau, Christine. "La physiocratie au pouvoir? Lecture et politique à Vienne (1780–1790)." In *La diffusion internationale de la physiocratie*, edited by Bernard Delmas, Thierry Demals, and Philippe Steiner, 367–79. Paris: Presses Universitaires de Grenoble, 1995.

Leenders, Karel A. H. W. *Verdwenen venen: Een onderzoek naar de ligging en exploitatie van thans verdwenen venen in het gebied tussen Antwerpen, Turnhout, Geertruidenberg en Willemstad 1250–1750*. Brussels: Gemeentekrediet, 1989.

McCarthy, Nancy, Abdul B. Kamara, and Michael Kirk. "Co-operation in Risky Environments: Evidence from Southern Ethiopia." *Journal of African Economies* 12, no. 2 (June 2003): 236–70.

Meeusen, Gerard. "Keuren van Esschen, Calmpthout en Huybergen." *Oudheid en Kunst* 23 (1932): 112–24.

Michielsen, J. "Keuren van Brecht." *Oudheid en Kunst* (1907): 71–81.

Ostrom, Elinor. *Governing the Commons: The Evolution of Institutions for Collective Action*. New York: Cambridge University Press, 1990.

Peeters, K. C. "De Wuustwezelsche dorpskeuren (XVe–XVIIe eeuw)." *Verslagen en mededelingen van de koninklijke Vlaamsche academie voor taal en letterkunde* (1932): 595–709.

_____. 'De Wuustwezelsche dorpskeuren (XVe–XVIIe eeuw)." *Wesalia, Tijdschrift voor plaatselijke Geschiedenis en Folklore* 8, no. 1–2 (1933): 2–48.

Peeters, Raymond. "De keuren van Turnhout (1550)." *Taxandria* 29, no. 1–4 (1957): 61–122.

Pinchot, Gifford. *Breaking New Ground*. Rev. ed., edited by Char Miller. New York: Island Press, 1995.

Prims, Floris. "Keuren der vreyheyt van Arendonk." In *Feestbundel H. J. Van De Wijer*, edited by H. Draye, vol. 1. Leuven: Instituut voor Vlaamse topynomie, 1944.

Radkau, Joachim. *Nature and Power: A Global History of the Environment*. New York: Cambridge University Press, 2008.

Rodgers, Christopher P., Eleanor A. Straughton, Angus J. L. Winchester, and Margherita Pieraccini. *Contested Common Land: Environmental Governance Past and Present*. London: Earthscan, 2011.

Spek, Theo. *Het Drentse esdorpenlandschap: Een historisch-geografische studie*. Utrecht: Stichting Matrijs, 2004.

Stein, Robert. *De hertog en zijn staten: De eenwording van de Bourgondische Nederlanden ca. 1380–ca. 1480*. Hilversum: Verloren, 2014.

Steurs, Willy. *Naissance d'une région: Aux origines de la Mairie de Bois-le-Duc, Recherches sur le Brabant Septentrional aux 12e et 13e siècles*. Memoire de la Classe des Lettres. Vol. III. Brussels: Académie royale de Belgique, 1993.

Szabó, Péter, and Radim Hédel. "Socio-Economic Demands, Ecological Conditions and the Power of Tradition: Past Woodland Management Decisions in a Central European Landscape." *Landscape Research* 38, no. 2 (2013): 243–61.

Thompson, Edward Palmer. *Customs in Common.* Harmondsworth: Penguin Group, 1991.

Van Aert, Marina, and Jean Bastiaensen. *De Kalmthoutse School.* Tielt: Lannoo, 2007.

van der Westeringhe, W. "Man-Made Soils in the Netherlands, Especially in Sandy Areas ('Plaggen Soils')." In *Man-Made Soils*, edited by W. Groenman-van Waateringe and M. Robinson. Oxford: British Archaeological Reports Limited, 1988.

Van Gorp, J. "De Aartbrief van Terloo." *Bijdragen tot de Geschiedenis* 18 (1927): 437–54.

_____. "Het keurboek van Casterlee." *Bijdragen tot de geschiedenis* 18 (1927): 182–96.

Vanhaute, Erik. *De invloed van de groei van het industrieel kapitalisme en van de centrale staat op een agrarisch grensgebied: De Noorderkempen in de 19de eeuw (1750–1910).* Brussels: Gemeentekrediet, 1990.

Van Olmen, E. H. A. "De keuren van Vorselaar." *Taxandria* 7 (1910): 35–49.

van Onacker, Eline. *Village Elites and Social Structures in the Fifteenth and Sixteenth-Century Campine Area.* Turnhout: Brepols, 2017.

Van Weeren, René, and Tine De Moor. "Controlling the Commoners: Methods to Prevent, Detect, and Punish Free-Riding on Dutch Commons in the Early Modern Period." *Agricultural History Review* 62, no. 2 (2014): 256–77.

Van Zanden, Jan Luiten. "The Paradox of the Marks: The Exploitation of Commons in the Eastern Netherlands, 1250–1850." *Agricultural History Review* 47 (1999): 125–44.

Verbist, F. *Costuymen van de hoofdrechtbank van Zandhoven, Uitgave 1664: Keuren en breuken, uitgave 1665.* Zandhoven: Gemeentebestuur Zandhoven, 2007.

Verboven, Hilde, Kris Verheyen, and Martin Hermy. *Bos en hei in het Land van Turnhout (15de-19de eeuw): Een bijdrage tot de historische ecologie*, edited by Monumenten & Landschappen. Leuven: Ministerie van de Vlaamse gemeenschap, Monumenten & Landschappen en het Vlaams Instuut voor het Onroerend Erfgoed, 2004.

Verellen, J. R. "De keuren van Herentals (1410–1567)." *Taxandria* XVI, no. 1 (1950): 20–107.

Verhoeven, P. J. "Keuren van Calmpthout." *Oudheid en Kunst* (1907): 45–46.

Vivier, Nadine. "Le conflict autour des biens communeaux ou la crise de la propriété collective (1760–1870)." In *Temps et espaces des crises de l'environnement*, edited by Corinne Beck, 71–82. Paris: Editions Quae, 2006.

Warde, Paul. "The Environmental History of Pre-industrial Agriculture in Europe." In *Nature's End: History and the Environment*, edited by Sverker Sörlin and Paul Warde. Basingstoke: Palgrave Macmillan, 2009.

Winchester, Angus J. L., and Eleanor A. Straughton. "Stints and Sustainability: Managing Stock Levels on Common Land in England, c. 1600–2006." *Agricultural History Review* 58, no. 1 (2010): 30–48.

PART II

The Governance of Aquatic Resources

Fishing and Flowing Freshwater

CHAPTER 5

Fisheries Regulations in Late Medieval Europe

Authorities, Concerns, Measures

Richard C. Hoffmann

Present-day fishers, governments, scientists, and environmentalists place great store on governmental regulation of fisheries. Laws or the lack of laws get praised or blamed for protection, overfishing, consumption, thoughtless waste of individual species, and destruction of entire aquatic communities. At least in the setting of western European Christendom, such intervention by rulers in the exploitation of fish stocks gained some regularity after 1200 CE. From then on, state regulations are a common but hitherto rarely analyzed source of information about human relations with aquatic systems. They report the fish varieties that drew official attention and the social and technical arrangements whereby medieval societies turned fish into a resource. If read with critical care as normative prescriptive texts, medieval fish laws can help contextualize other written sources as well as archaeological fish remains and assess the sustainability of human exploitation of European fishes. This brief initial survey of thirteenth- through sixteenth-century fisheries legislation on both sides of the Alps from Iberia to the eastern Baltic describes the sources and authority for regulation, the grounds articulated by legislators, the characteristically selective attention to certain fish varieties, and something of the bewildering diversity of substantive provisions people then thought worth enacting. The aim is not only to demonstrate medieval cultural concern for what some now call "conservation," but also to provide a context for closer individual study of evolving local or regional regulatory regimes.[1]

Regulatory Authority

For purposes of this paper, fisheries regulations (laws, ordinances, statutes, etc.) are defined as authoritative rules governing the capture of wild fish, as distinct from tolls, tax schedules, or other rules regulating their commerce (although some limits on fishing seasons and catches did result from restrictions on legal sales). Under typical medieval conditions of extensive private (seigneurial) rights over fisheries and other natural resources, the basis for regulating the activities of fishers could range from individual or collective private ownership (a lord or commune setting conditions of use for subjects or licensees) through corporate self-regulation to the assertion of sovereign authority over "public" resources on the part of municipalities, princes, parliaments, or interstate treaties.

Seigneurial regulations resemble instructions from a property manager to employees or contractors. Hence, the seigneur of Villeneuve-les-Maguelonne in 1173 forbade a new technique for stunning fish in his Languedoc lagoon, and the count of St. Pol banned the trammel net from his waters in 1225.[2] In the fifteenth century, private owners of fishing rights on the Rio Arlanzon near Burgos, Castile, set strict rules for fishing licenses, seasons, sites, and gear. The line between property management and governance blurred when the latter lords or their contemporary counterpart at Auenheim on the Rhine ceased to consume or sell the catch domestically but rather were licensing subject fishers to market it elsewhere.[3]

Collective self-regulation is ostensibly evident when organized groups or formal fishers' guilds agree on the rules their membership will follow—though in many instances, the legitimacy and enforcement of those rules were assured, even imposed, by municipal or territorial authorities. Brescia approved or certified a "statute of fishers" (*statuta piscatorum*) for Lake Garda in 1029, 1313/38, and 1429.[4] The self-governing assemblies (*Fischermaien*) and sworn regulations (*Fischer-Einung*) of fishers on the Zürichsee were held and supervised at the order of the Zürich city council, with other municipalities playing a secondary role.[5] Master fishers of the lower Traun River agreed on a set of rules in 1418 with the implied consent of private holders of fishing rights who employed them and had this confirmed by the duke's governor of Upper Austria.[6]

Public fisheries regulation asserted an authority over those waters and fishes not in private ownership or even to override acknowledged private rights. (Both the sovereign's ownership of "unowned" resources and right to constrain use of private property go back in law or by imitation to Roman imperial authority.) The town of Perugia rested its control over Lake Trasimeno on an 817 imperial grant to Pope Paschal I and subsequently that of Innocent III to the free commune in 1212.[7] Alfonso VIII of Castile set rules for fishing the waters of Cuenca as part of the royal municipal code he granted shortly after conquering the city in 1177.[8] Like Spanish monarchs, kings of France claimed imperial authority within their own domains, so Philip IV in 1289 and all his successors asserted the right to legislate for public (i.e., large, navigable) waters and gradually extended this also over private uses of other natural water bodies (artificial ponds were not included).[9] The oldest known Scottish regulations came as a precedent-setting judgment of King William I (1165–1214) sitting as the country's top feudal lord, but subsequent rules took the form of parliamentary statutes.[10]

The institutional evolution over time of fisheries regulations depended to some extent on regional governmental/administrative practice, including incremental development of literate record keeping. This process was under way earlier in French, British, and some Hispanic monarchies, as well as Italian city-states, than in the territorial principalities of central Europe or Scandinavia. But on Western Europe's largest lake, Lake Constance (the *Bodensee*), in 1481, five independent authorities—the cities of Constance and Überlingen, the abbey of Salem, the Mainau commandery of the Teutonic Order, and the Count of Heiligenberg—together established the first interstate fisheries protection treaty, an arrangement which would grow and continue deep into the nineteenth century. Michael Zeheter's contribution to this collection summarizes his important book-length case study.[11] The long-term trend for fisheries was part of a larger expansion of public, princely, or state intervention in all sorts of environmental and resource issues.[12]

Grounds for Intervention

Authorities asserted that legislative intervention would serve their communities by resolving evident practical problems in their fisheries. In managing fisheries, they appealed to the same rationale as was increasingly applied to woodlands, flood control, and pollution: collective benefit to the "common good" (*bien publique*, *gemeinen Nutz*, *ad commune beneficium*, etc.) was superior to mere private profits. French King Philip IV in 1289 and generations of his successors asserted that excessive and improper fishing caused "no small loss to the rich and poor of our realm" (*dampnum non modicum tam divitum quam pauperum regni nostri*).[13] Venetian regulation of fishing in Lake Garda was "for common benefit and abundance of fishes" (*ad comune beneficium et copiam piscium*), and the 1537 imperial ordinance for Upper Austria was for "common usefulness" (*gemainen nutz*).[14]

And what problems called for intervention? Concern for public safety led London and Austrian authorities alike to ban fish weirs in the Thames and the Danube as hazards to navigation; Florentine councilors did the same for the weirs' role in the flood disaster of 1333.[15] Continually from 1287, Venice tried to limit the seasons and locations of "fish fences" (*valli de pesce*) because they blocked tidal flow and contributed to siltation in the lagoon.[16] Guild statutes and municipal ordinances in particular—at Zürich, Madrid, and along the upper Rhine, for instance—directed attention to conflict and disorder in their fisheries and provided rules and institutional mechanisms to avoid or settle them.[17]

But the most commonly articulated reason for regulation was the fear and perception of fisheries depleted especially from overfishing. As early as 1278, the county assizes of Cumberland shut down weirs to preserve the salmon, and just more than a decade later, Philip IV blamed fishers' practice of taking fish that were too small for the fact that some French rivers were "yield[ing] nothing."[18] Duke Albrecht V of Austria in 1412 found "the waters everywhere in this land have become almost desolate and fishless" because no fishes therein could ever reach their proper growth. In the middle of the fifteenth century, city governments of Strasbourg and Florence shared that opinion, blaming on overfishing the remarkable declines of fish in the Rhine and in all Florentine territories. Venice, too, feared depletion of the lagoon fisheries, although corrective measures conflicted with worries over siltation.[19]

Now the conundrum: While plainly the appeal to the general good was meant to resonate positively with some kind of "political public," how much were the regulations actually a response to fisheries issues, depletion or conflict, and how much assertions of power on the part of authorities? How much did the *fish* matter? Or the fishers? A look at legislative content may help.

Regulatory Measures

Fish taxa are a good place to begin. If the legal form of fish laws shows public claims spreading fitfully across Europe, this aspect of their substance varied regionally. Some authorities worried about particular species, others did not. While quite a few ordinances survive from the upper Rhine—where other sources reveal diverse local fisheries—scarcely any of these wide-ranging laws name fishes other than salmon.[20] Elsewhere, we can differentiate between rules for single species and more comprehensive codes. Scottish royal and parliamentary legislation worried only about salmon, and English King Edward I instituted seasonal closure of the Channel Island fishery "for the protection of the congers" (*pro salvacione congrorum*).[21] Florentine statutes of the 1450s identified trout as needing protection against use of piscicides. Other kings, fishing communities, and municipalities issued laws naming many species. An exemplary selection of the latter (see Table 5.1) includes some from Danubian, Rhine, and Atlantic watersheds, as well as Iberian and Italian cases. (We may speculate, perhaps, why I have failed to turn up such extensive codes from Britain[22] or the Baltic.) Whether individually or collectively, the fishes named in such legislation may be thought of some serious interest to those who drafted the rules. What features are now visible?

Laws arguably indicate the probable (or at least then-traditional) presence of named taxa in local waters. Note that with exception of the invasive carp, all the fishes mentioned were indigenous to the watersheds and habitats involved. Legislators knew what varieties were locally available, whether for consumption at home or export. Huchen but not eel are present in the Austrian Danube and the reverse is true for the Rhine, Seine, central Spain, and central Italy. Coregonid whitefishes occur in Alpine lakes. This perhaps obvious point deserves emphasis because it suggests real familiarity with the resource, not mere compilation from learned texts. Of course, the converse is not true: important local species such as sturgeon in the middle Danube and salmonids in the Seine system simply do not appear in the regulatory texts.

So the fisheries regulations do not reflect local aquatic ecosystems alone, but also some kind of "political" significance. The named species had to matter to some influential group or individual. Each had acquired some sort of cultural, commercial, or economic value in a

Table 5.1. Fish taxa present in select medieval fisheries regulations.

Fish taxa present in select medieval fisheries regulations	Cuenca, Castile, ca.1185/90	Lago Trasimeno & area, 1276–1342	France (Seine basin), 1268–1326
	Kg. Alfonso VIII, Fuero de Cuenca, cap. XLII: 7–9	Perugia communal statuti	Kgs. Louis IX, Philip IV, Philip V
PETROMY-ZONIDAE			
Lamprey			•
ANGUILLIDAE			
A. anguilla **Eel**	•	•	•
SALMONIDAE			
Salmo trutta **(Brown) trout, Sea trout**	•	•	
Hucho hucho **Huchen**			
COREGONIDAE			
Whitefish			
Coregonus lavaretus **Laveret**			
THYMALLIDAE			
Thymallus thymallus **Eur. grayling**			
ESOCIDAE			
Esox lucius **Pike**		•	•
CYPRINIDAE			
Cyprinus carpio **(Common) carp**			•
Alburnus alburnus **Bleak**			
Aspius aspius **Asp**			
Barbus barbus **Barbel**			•
Barbus spp. **Barbel sp.**	•		
Protochondro-stoma genei **Lasca**		•	
Leuciscus leuciscus **Dace**			•
Leuciscus (Squalius) cephalus **Chub**		•	•
Rutilus rutilus **Roach**			•
Rutilus rubilio **Rovella, Triotto, Italian roach**		•	

Zürichsee, 1386	Traun R, 1418	Chiemsee, 1448	Austrian Danube, 1506	Bodensee, 1536
Compulsory sworn agreement among fishers	Agreement among master fishers	Dk. Henry VI, Bavaria-Landshut	Emperor Maximilian, Patent for Upper and Lower Austria	Fisheries agreement between Konstanz & Überlingen
				•
	•		•	•
	•		•	
		•		
•				•
	•			
•	•	•	•	•
			•	•
•				•
		•		
		•	•	•
				•
•		•		
•				

Table 5.1. (continued)

Fish taxa present in select medieval fisheries regulations	Cuenca, Castile, ca.1185/90	Lago Trasimeno & area, 1276–1342	France (Seine basin), 1268–1326
	Kg. Alfonso VIII, Fuero de Cuenca, cap. XLII: 7–9	Perugia communal statuti	Kgs. Louis IX, Philip IV, Philip V
Scardinius erythropthalmus **Rudd**		•	
Tinca tinca **Tench**		•	•
SILURIDAE			
Silurus glanis **(European) Catfish, Wels**			
GADIDAE			
Lota lota **Burbot**			
PERCIDAE			
Perca fluviatilis **Perch**			
Zingel spp. **Apron, Streber, Zingel**			
NUMBER OF TAX NAMED*	3	8	9
*Exclusive of distinctive names for different size or year classes	Powers 2000, Ureña y Smenjaud 1935	Biganti 1995, Scialoja 1910, Vincenti 2002	Rouillard 2004, Duples-Agier 1852, Lespinasse 1886

regional medieval society and so now deserve at least special awareness on the part of historians and zooarchaeologists alike. Legislative attention to Scottish salmon has directed attention to their export value and subsequently to the surprisingly extensive fiscal records of their arrival in the late medieval Low Countries.[23] Perugia's diligent attention to preserving and enhancing diverse species in Lake Trasimeno for the sake of sales to larger Italian towns has not. The ubiquity of pike in ordinances across its range surely bears some relation to this fish's potentially impressive size, aspect, and mythic reputation[24]—though that might shrink in view of minimum legal sizes set as low as about 30 cm. Put simply, if medieval authorities thought certain fish varieties worth

Zürichsee, 1386	Traun R, 1418	Chiemsee, 1448	Austrian Danube, 1506	Bodensee, 1536
Compulsory sworn agreement among fishers	Agreement among master fishers	Dk. Henry VI, Bavaria-Landshut	Emperor Maximilian, Patent for Upper and Lower Austria	Fisheries agreement between Konstanz & Überlingen
		•		
				•
		•	•	
		•	•	
•		•		•
			• ?	
6	4	9	8	10
Amacher 1996, 387–90	Scheiber 1930, 152	Höfling 1987, 115–18	Hoffmann & Sonnlechner 2011	Zeheter 2014, 122–38

special treatment in their laws, we should find commensurate presence in other written and archaeological sources—or be pursuing the reasons for their absence. (Ambiguous nomenclature? Taphonomy?[25] Prestige values greater than economic? Etc.)

Whether driven by ecological, sociopolitical, or other variables, the particulars of regulations reveal great local and regional diversity, but also, at the same time, common close familiarity with the immediate local situation and, as a close look at actual provisions reveals, considerably parallel problems and solutions.

Fisheries techniques must be the most common material element in the extant regulations, but also the most generally obscure. Extensive use of local vernacular dialects and minute attention to such detail as precise size of mesh imply access to technical knowledge on the part of the legislators, but also grave difficulty now in assessing the motives and implications, both social and environmental, of the various provisions. Only in some cases and measures are these reasonably clear to a modern reader. As already remarked, these included rules meant to reduce the hazards and inconveniences fisheries could pose to other users of waterways.

But at some indeterminate point, considerations of safety blend into often-complex regulations of gear and practices of ambiguous intent, possibly combining conflicts internal to the fishery, a wish to balance competition, and some protection of the resource. As a case in point, take the ordinances of the fisher's guild approved by the town council of Tulln in 1469 and renewed in 1535. These men had the privilege of fishing public water, including a reach of the Danube later known as a spawning site for the big beluga sturgeon (*Huso huso*) that migrated up from the Black Sea. Besides controlling membership in the guild and limiting each master fisher to a single assistant, the rules obliged each master to declare every Christmas which one of the three permitted methods—seine, trap, or [set]line—he would use in the coming year. Certain kinds of trap nets were forbidden during the summer months, and special provisions governed the placement of devices for fishing through the winter ice.[26] Madrid's thirteenth-century royal charter (*Fuero*) closed fishing in summer months, permitted certain named types of nets (*mandil*, *manga*, and *asadega*), and itemized the reaches where townspeople were allowed to fish.[27]

Capture techniques could be regulated in several ways. A minimum legal size of mesh gave an easily verified index of selectivity for small fish. The standard in early French legislation was the size of a coin, the *gros tournois* of about 2.5 cm diameter, and later the smaller Paris *denier.* London's 1279/80 ordinance demanded two inches for the "peternet" used in the Thames estuary.[28] A Scottish statute of 1318 used the same means to protect "little salmon or smolts or fry of other kinds of fishes of the sea or fresh water," and such provisions became the norm in fourteenth- and fifteenth-century regulations from German-speaking lands.[29] Officials along the north slope of the Alps carried a standard gauge called *prittlmas* to measure the openings (Figure 5.1). The Duke of Brittany, who claimed royal jurisdiction up to thirty leagues offshore, made his chief fiscal official (*receveur*) in Nantes responsible for inspecting mesh.[30]

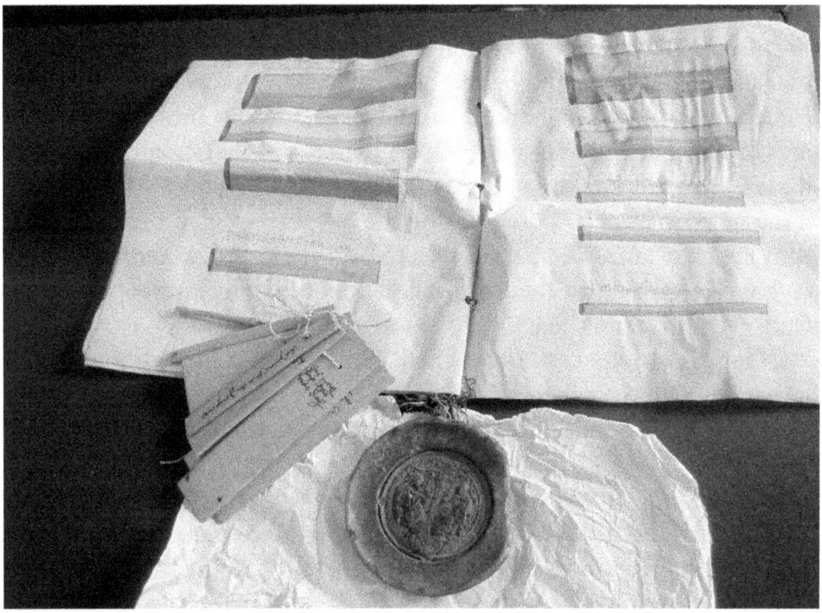

Figure 5.1. *Prittlmas* (gauge to measure mesh of fishing nets) on the Mondsee. Upper Austria, under Mondsee abbey fishing regulations of 1544. Photograph by Richard C. Hoffmann with permission of Museum Mondsee.

Other measures targeted named forms of weirs or traps, forbidding them entirely or sharply restricting them to certain seasons, waters, or privileged users. King Alfonso's founding charter for Cuenca allowed no seines or trammel nets within nine paces of a mill or in a channel, and in 1444, the queen regent of Aragon ordered Tortosa's municipal council to control certain nets that were harming the run of shad in the Ebro.[31] The town of Cannes refused to license more than four boats to use the *lamparo* technique of netting by torchlight in coastal waters. In the Rhone's estuarine marshes, mid-channel traps had to be removed in the spring when fishes were running into the salt ponds to spawn.[32] The large *niewod* seine of the Baltic littoral was often banned outright—as to town-based fishers in fourteenth-century Poland—or used only by special license of the sovereign.[33] Governments across Mediterranean Europe outlawed piscicides, led by Frederick II's Sicilian legislation of 1231, Madrid's municipal charter a generation thereafter, and ontinually reiterated enactments at Florence and other places.[34]

A different legislative strategy specified the particular gear allowed, as the "pridenet" permitted on London waters in 1279/80 during eight days only of the autumn season between Michaelmas (Sept. 29) and Martinmas (Nov. 11), the angling or simple basket traps conceded to townsfolk at Toul, or the baited hook, rod, and horsehair line used at Vercelli. On the other hand, Count-Palatine Philip prohibited all set lines and most angling in his 1489 ordinance for the Neckar.[35] But why in particular did the *Fuero* of Cuenca restrict fishing on only one reach of the Jucar to hook and line during the low waters of summer and fall, from Pentecost until All Saints' Day (Nov. 1)? This document elsewhere envisaged legal sale on the town market of both trout and barbel measuring even less than the length of a finger.[36]

Other provisions certainly aimed at the protection and wise use of fish stocks with the expectation of long-term exploitation. Much of this is evident in seasonal and size restrictions oriented to spawning and juvenile fish.

Seasonal regulations visibly worked the interlaced calendars of fishes' natural reproduction and vulnerability and of human fish consumption. French ordinances closed fishing for roach during their April and May spawning time.[37] At Ivrea and on Lake Garda, fishing for (fall-spawning) trout was closed during October and November, while a 1472 law on the Swiss Bielersee protected miller's thumb (*bundollen*) after Easter, when demand for fish fell off and this species was spawning.[38] Fishing for conger off the Cotentin peninsula was closed from Easter to John the Baptist's Day (June 24), and the Teutonic Order established seasonal fish sanctuaries off the mouths of Prussian rivers with important runs of migratory fishes and, since 1426, applied a general rule of closure during spawning.[39] None were to fish inland waters of Sardinia during the summer dry season up to September 29.[40] A variant strategy of annually rotated closures of lakes is reported in Brandenburg since 1311 and reflected in seasonal alternation between Sicily and Provence of fishing for migratory tuna.[41]

Size limits might protect juveniles—"fry," called "*inrudines*" in the Florentine statutes, and "*alevins*" in Lorraine's *Loi de Beaumont*[42]—or measure actual size. The French ordinance of 1289 set the minimum in terms of market price—eel at a quarter denier, barbel and carp at a half, and pike at two denier—while later redactions called for dace, chub, and roach to be at least five inches (*pouces*). Brescia eventually required a minimum weight for pike, tench, and trout. At Nice, grayling and mullet were legal at a "half foot" (*semissa*) and trout at Burgos keepers at a "palm."[43] A Douai ordinance of 1391 set pike at ten "inches."[44] To avoid ambiguity, Bavarian and Austrian rulers since the mid-fifteenth century used graphic means. The former had special posters first hand-written, then printed, with inked bars showing the actual smallest legal size for each variety from the Danube and its navigable tributaries: the pike standard measured 39 cm, carp 36 cm, huchen 42 cm, barbel 32 cm, ide and bream 18 cm.[45] Emperor Maximilian I tried in 1506 to facilitate his Austrian officials' negotiations with rights holders by providing accurate full-color pictures of eight key species.[46] Of bag limits or quotas directly limiting kill, however, medieval sources seem mute. It is important to acknowledge, moreover, that all the medieval size limits so far known fall well below the norms of modern regulations for the named species and so offered little protection for fish not yet up to their spawning size.

Sometimes protective measures were extended to fish habitat. Since 1279, Perugia obliged shoreline residents every third October to "refresh" (*reinfrascare*) the lake with bundles of reeds and branches to shelter spawning and overwintering fish. Madrid banned private diversion channels and the assembled fishers of the lower Traun River their drainage before September 8, both to ensure adequate flows for juveniles. The Count of Holland, Philip of Burgundy, ordered full protection of riparian vegetation in the Haarlemmermeer and other waters north of Leiden, hoping this cover would harbor enough forage fish to keep pike in those waters and sustain the value of Leiden's fishery.[47] Habitat and forage mattered to the city of Narbonne, too, when in 1227 it banned a bottom trawl or seine, the *bourgin*, from its lagoon:

> because such a net takes away from the whole pond [and] lays waste all types of nourishment for fishes both large and small, and thus the pond would be emptied out and desolate of all fishes, and not any fishes would be found there and it would be sterile and yield nothing.[48]

Not in evidence in the medieval sources, however, are restrictions on the quantity of fish taken (bag limits, quotas) and only most rarely the designation of protected areas where all fishing was forbidden.[49] It is also the case that some gear restrictions applied only outside the season of Lent, when Christian religious practice forbade most other protein-rich foods and those who could afford fish ate them eagerly.

Even pending the necessary systematic local studies, there are grounds to acknowledge active and successful enforcement efforts in well-governed communities. London authorities in 1320 seized and burned sixteen illegal traps from locations along the tidal Thames, and in 1444, jailed a local brewer for using illegal gear.[50] Charged to enforce French ordinances were royal officers or, as on the Seine around Paris, subordinates of those who held royal rights in fief. Hence, in 1422, royal officials arrested Pierre le Nourrissier and crewmen Jean Fournier, Colin Jobart, and Liebaut de Gandelus for fishing in the city moat with illegally small mesh and catching the fry of protected varieties, which the officers then confiscated, precipitating a judicial wrangle over boundaries of public waters.[51] Florentine agents in 1460 caught Buono di Francesco Beroni poisoning fish, while in 1490, Perugia even placed an armed galley on Lake Trasimeno to patrol that fishery.[52] Princes along the upper Rhine appointed "Counts of the Rhine" (*Rheingrafen*) to investigate and judge violations and disputes.[53]

Conservation for What?

Plainly, then, numbers of medieval law-makers and their advisors were well-aware of threats to their fish stocks and knew some measures to reduce or mitigate those threats, whether affecting particular species or stocks in general. In that sense, we might say they sought what is now called sustainability. But taken as a whole, the "common good" pursued by fisheries regulation was primarily economic local supply of a culturally desired commodity, and secondarily, the maintenance of a certain social order among fishers and between the fishing industry and a larger society. Fish stocks came further down the list of priorities. Their wise use was, however, a genuine concern for would-be rulers over inland and inshore fisheries, though not often on the open sea.

In summary, then, what is to be learned from medieval fisheries regulations depends to some degree on the questions asked. A long-term trend since about 1200 increased state intervention "for the common good," often but not exclusively in response to perceived depletion. Local and regional regulations pay special attention to certain valued indigenous fish varieties; the biological and cultural status of these deserves study from other perspectives, both textual and from zooarchaeology. Seasonal and technical particulars suggest drafters of the regulations had access to considerable expertise about local fishes and methods of capture. Research needs to consider the likely ecological and institutional impacts of the laws. Would each serve the apparent purpose? Local and regional diversity in particulars should not obscure legislators' recognition of similar problems and their selection of remedies from a common regulatory repertoire. Fisheries protection was *one* of several priorities for legislative action. Medieval governing elites were not ignorant of the need for preservation and sustainable use, but other considerations might have greater importance.

Richard C. Hoffmann, FRSC, is Professor Emeritus and Senior Scholar in the Department of History at York University, Toronto, Canada. After training in History at the University of Wisconsin (1965) and in Medieval Studies at Yale (1970), he practiced the craft of premodern economic and environmental history at York until formal retirement in 2008. His award-winning publications have included *Land, Liberties, and Lordship in a Late Medieval Countryside* (1989); "Economic Development and Aquatic Ecosystems in Medieval Europe" (1996); *Fishers' Craft and Lettered Art* (1997); and *An Environmental History of Medieval Europe* (2014). An environmental history of medieval fisheries is forthcoming.

Notes

1. An earlier version of this chapter was presented to the Fish Remains Working Group, International Congress of Archaeozoology, meeting in Lisbon, Portugal, 28 September–3 October 2015. Text completed 2017 November; minor revisions 2019 June.
2. Monique Bourin-Derruau, Daniel Le Blévec, Claude Raynaud, and Laurent Schneider, "Le littoral languedocien au Moyen Âge," in *Castrum 7: Zones côtières littorales dans le monde méditerranéen au Moyen Âge*, ed. Jean-Marie Martin (Rome-Madrid: Ecole française de Rome, Casa de Velázquez, 2001), 384–85; Robert Fossier, *La terre et les hommes en Picardie jusqu'à la fin du XIIIe siècle* (Paris: Béatrice-Nauwelaerts, 1968), 397.
3. Hilario Casado Alonso, *Señores, mercaderes y campesinos: La comarca de Burgos a fines de la edad media* (Valladolid: Junta de Castilla y León, Consejería de Cultura y Bienestar Social, 1987), 207–209; Franz J. Mone, "Ueber die Flußfischerei und den Vogelfang vom 14. bis 16. Jahrhundert," *Zeitschrift für die Geschichte des Oberrheins* 4 (1853): 79–81.
4. Mattia Butturini, "La pesca sul lago di Garda," *Archivio Storico Lombarda* 8 (1881): 175.
5. Urs Amacher, *Zürcher Fischerei im Spätmittelalter*. Mitteilungen der Antiquarischen Gesellschaft in Zürich, Bd. 63 (Zürich: Verlag Hans Rohr, 1996): 167–73 and 387–91.

6. Artur Maria Scheiber, *Zur Geschichte der Fischerei in Oberösterreich, insbesondere der Traunfischerei:* Sonderabdruck aus den "*Heimatgauen*": *Zeitschrift für oberösterreichische Geschichte, Landes- und Volkskunde* (Linz: Verlag R. Pirngruber, 1930), 30–35 and 152.
7. Valentina Vincenti, "La tutela ambientale del Lago Trasimeno in età medioevale," in *Thinking about the Environment: Our Debt to the Classical and Medieval Past*, ed. Thomas M. Robinson and Laura Westra (Lanham, MD: Lexington Books, 2002), 133; Tiziana Biganti, "La pesca nel lago Trasimeno: Sfruttamento et tutela delle risorse ittiche del lago di Perugia (seccoli XIII-XV)," in *Gli archivi per la storia dell'alimentazione*, ed. Paolo Carucci and Massimo Buttazzo (Roma: Ministero per i Beni Culturali e Ambientali, Ufficio Centrale per i Beni Archivistici, 1995), 772–99.
8. James F. Powers, ed. *The Code of Cuenca. Municipal Law on the Twelfth-Century Castilian Frontier* (Philadelphia: University of Pennsylvania Press, 2000). In Portugal, ultimate royal authority over seas and rivers was acknowledged from the twelfth century, but its exercise by regulating activity in public and private waters seems to date only to the 1540s; see Inês Amorim, "The Evolution of Portuguese Fisheries in the Medieval and Early Modern Period: A Fiscal Approach," in *Beyond the Catch: Fisheries of the North Atlantic, the North Sea and the Baltic, 900–1850*, ed. Louis Sicking and Darlene Abreu-Ferreira (Boston: Brill, 2009), 245–71.
9. Joséphine Rouillard, "La législation royale de la pêche en eau douce du XIIIe au début du XVe siècle," in *Pêche et pisciculture en eau douce: la rivière et l'étang au Moyen Age: actes des 1res Rencontres internationales de Liessies: 27, 28, 29 avril 1998*, ed. Paul Benoît, Frédéric Loridant, and Olivier Mattéoni [unpaginated booklet and CD-Rom] (Lille: Conseil Général du Nord, 2004.; with texts in Henri Duplés-Agier, "Ordonnances inédites de Philippe-le-Bel et de Philippe-le-Long sur la police de la péche fluviale," *Bibliothèque de l'Ecole des Chartes* 14 (1852): 43–55; René de Lespinasse, ed., *Les métiers et corporations de la ville de Paris*, vol. 1: *XIV-XVIIIe siècle: Ordonnances générales, métiers de l'alimentation* (Paris: Imprimerie Nationale, 1886); Paul Guilhiermoz, ed., "Ordonnance inédite de Philippe le Bel sur la police de la pêche fluviale (17 Mai 1293)," *Bibliothèque de l'Ecole des Chartes* 63 (1902): 331–37; and Paul Jeulin, "L'élaboration par la monarchie de l'ancienne réglementation de la pêche fluviale," *Revue historique du droit français et étranger* 4, no. 40 (1962): 127–28, and 42 (1964): 728–30.
10. Richard C. Hoffmann, "*Salmo salar* in Late Medieval Scotland: Competition and Conservation for a Riverine Resource," *Aquatic Sciences* 77 (2015): 355–66 and sources there cited.
11. Michael Zeheter, *Die Ordnung der Fischer: Nachhaltigkeit und Fischerei am Bodensee (1350–1900)*, Umwelthistorische Forschungen (Köln-Weimar-Wien: Böhlau, 2014).
12. Richard C. Hoffmann, *An Environmental History of Medieval Europe* (Cambridge, UK: Cambridge University Press, 2014), 263–78.

13. Duplès-Agier, "Ordonnances inédites," 49. For this concept, see: Diana Wood, *Medieval Economic Thought* (Cambridge, UK: Cambridge University Press, 2002), 21–40; and Matthew S. Kempshall, *The Common Good in Late Medieval Political Thought* (Oxford: Clarendon Press, 1999).
14. Butturini, "La pesca," *Archivio Storico Lombarda* 6 (1879): 142–49; and for Venice in general, see Fabien Faugeron, *Nourrir la ville: ravitaillement, marchés et métiers de l'alimentation à Venise dans les derniers siècles du Moyen Age*, Bibliothèque des Ecoles françaises d'Athènes et de Rome, 362 (Rome: Ecole française de Rome, 2014), 281–92; Richard C. Hoffmann and Christoph Sonnlechner, "Vom Archivobjekt zum Umweltschutz: Maximilians Patent von 1506," *Studien zur Wiener Geschichte: Jahrbuch des Vereins für Geschichte der Stadt Wien* 62/63 (2006/2007) [2011]: 120.
15. Derek Keene, "Issues of Water in Medieval London to c.1300," *Urban History* 28 (2001): 161–79; Hoffmann and Sonnlechner, "Vom Archivobjekt," 131; Gerrit J. Schenk, "'... prima ci fu la cagione de la mala provedenza de'Fiorentini ...': Disaster and Life World Reactions in the Commune of Florence to the Flood of November 1333," *Medieval History Journal* 10 (2007): 355–86.
16. Faugeron, *Nourrir la ville*, 181–84.
17. Amacher, *Zürcher Fischerei*, 187–94; Tomás Puñal Fernández, *El Mercado en Madrid en la baja edad media: Estructura y sistemas de abastecimiento de un concejo medieval castellano (s. xv)* (Madrid: Caja de Madrid, 1992), 175–76; Julián Sánchez Quiñones, *Pesca y Comercio en el Reino de Castilla Durante la Edad Media: Los valles del Guadiana, Júcar y Tajo (siglos xii y xvi)*, Serie Histórica 6 (Madrid: Ediciones de La Ergástula, 2014), 110–20; Mone, "Ueber die Flußfischerei," 79–81, 82–84, and 89–94.
18. Angus J. L. Winchester, *Landscape and Society in Medieval Cumbria* (Edinburgh: John Donald, 1987), 110; Duplès-Agier, "Ordonnances inédites," 49.
19. Hoffmann and Sonnlechner, "Vom Archivobjekt," 117; Mone, "Ueber die Flußfischerei," 82; Richard Trexler, "Measures Against Water Pollution in Fifteenth Century Florence," *Viator* 5 (1974): 465–66. For more general discussions of late medieval regulations as responses to regional overfishing, see Jan Materné, "Exploitatiemetoden, uitrustung en reglementering in de vlaamse binnenvisserij. Een regionale analyze: De Wase durmestreek van de late middeleeuwen tot op het einde van het ancien régime," *Annalen van de koninklijke oudheidkundige kring van het land van waas* 91 (1988): 205–66; and Pierre Plouchard, "La Scarpe et les gens de rivière aux XIVe et XVe siècles," *Revue du Nord* 79, no. 322 (1997): 849–53. For the complex Venetian situation, see Elisabeth Crouzet-Pavan, "Mythes et réalités de la crise d'un écosystème littoral: Venise et sa lagune à la fin du Moyen Âge," in *Castrum 7: Zones côtières littorales dans le monde méditerranéen au Moyen Âge*, ed. Jean-Marie Martin (Rome-Madrid: Ecole française de Rome, Casa de Velázquez, 2001), 95–106, and the ensuing discussion there reported, with treatment more centered on the fishery itself in Faugeron, *Nourrir la ville*, 180–84.

20. Mone "Flußfischerei," 67–97; Henry J. Cohn, *The Government of the Rhine Palatinate in the Fifteenth Century* (Oxford: Oxford University Press, 1965), 240; Arnold Nauwerck, "Der Lachsfang in der Kinzig: In biologischer Beleuchtung. Nach Akten des Fürstlich Fürstenbergischen Rentamtes Wolfach aus dem 18. Jahrhundert," *Ortenau* 66 (1986): 499–518.
21. Jean-Claude Hocquet, "Les pêcheries médiévales," in *Histoire des pêches maritimes en France*, ed. Michel Mollat (Toulouse: Privat, 1987), 76.
22. See Michael Bartrip, "Food for the Body and Food for the Mind: The Regulation of Freshwater Fisheries in the 1870s," *Victorian Studies* 28 (1985): 285–304.
23. Scottish records of salmon exports are discussed in Martin Rorke, "English and Scottish Overseas Trade, 1300–1600," *Economic History Review* 59, no. 2 (2006): 278, and tabulated in Martin Rorke, "Scottish Overseas Trade, 1275/86–1597," 2 vols. (PhD dissertation, University of Edinburgh, 2001), Tables A2–6, 564–571, and Graphs A3/I-6, 662–669, accessed 12 November 2017, https://www.bris.ac.uk/Depts/History/Maritime/Sources/2001phdRorke1.html. Their arrival in the Low Countries is documented in Willem S. Unger, *De tol van Iersekeroord: Documenten en rekeningen 1321–1572* (The Hague: Martinus Nijhoff, 1939), 294–316 and 504–516, accessed 12 November 2017, http://resources.huygens.knaw.nl/toliersekeroord.
24. Richard C. Hoffmann, "The Protohistory of Pike in Western Culture," in *An Annotated Bibliography of the Pike Esox lucius*, ed. Edwin J. Crossman and John Casselman (Toronto: Royal Ontario Museum, 1987), vii–xvi; reprinted in *The Medieval World of Nature: A Book of Essays*, ed. Joyce E. Salisbury (New York: Garland, 1993), 61–76.
25. A collective term for the processes undergone by a past object or organism from deposition or death until it has been recovered by an archaeologist or paleontologist. By extension, it ought to apply as well to preservation, identification, and recovery of textual objects. Different remains survive differently or are more or less likely destroyed under diverse conditions.
26. Silvia Petrin, "Das Archiv der Tullner Fischerzeche," *Mitteilungen aus dem Niederösterreichischen Landesarchiv* 3 (1979): 29–34.
27. Puñal Fernández, *El Mercado*, 175–80.
28. London Letterbook A, fols. 89v–91v, cited in Laura Wright, *Sources of London English: Medieval Thames Vocabulary* (Oxford: Clarendon Press, 1996), 68–69, who also provides later rules on mesh size. More elaborate regulations of net size, form, mesh, and other features are published in Henry T. Riley, ed. and trans., *Liber Albus: The White Book of the City of London compiled A.D. 1419 by John Carpenter, Common Clerk, and Richard Whitington, Mayor* (London: R. Griffin, 1861), 331–34.
29. Keith M. Brown, Gillian H. MacIntosh, Alastair J. Mann, Pamela E. Ritchie, and Roland J. Tanner, eds., *The Records of the Parliaments of Scotland to 1707* (Saint Andrews: University of Saint Andrews, 2007–2017), online ed., accessed 2011 September, http://www.rps.ac.uk/, 1318/13 [19]: "*salmunculi vel smolti seu*

fria alterius generis piscium maris vel aque dulcis"; Ernst Cahn, *Das Recht der Binnenfischerei im deutschen Kulturgebiet von den Anfängen bis zum Ausgang des 18. Jahrhunderts*, ed. Ekkehard Kaufmann (Frankfurt a.M: Klostermann, 1956), 137; Mone, "Flußfischerei," 78, has the 1434 rules from Strasbourg.

30. Zeheter, *Ordnung der Fischer*, 43 et passim; Amacher. *Zürcher Fischerei*, 35–40; Hoffmann and Sonnlechner, "Vom Archivobjekt," 81, 126, and 133; Hocquet, "Pêcheries médiévales," 73–75 and 105–6.
31. Cuenca *fuero*, xxxv:15, Rafael de Ureña y Smenjaud, ed., *Fuero de Cuenca* (Madrid: Tipografia de Archivos, 1935), 728–31; Powers, *Code of Cuenca*, 191; F. Pastor y Lluís, "La pesca de la saboga en el Ebro (Notas históricas)," *La Zuda* [Tortosa] 40, año. 4 (1916): 108–9.
32. Roger Aubenas, *Le droit de pêche de l'abbaye de Lérins*, Equip des Historiens Cannois, fascic. no. 2. (Cannes: Equipe des historiens cannois, 1953), 10–12; Paul Amargier, "La pêche en Petite Camargue au XIVe siècle," in *Bulletin Philologique et Historique (jusqu'à 1610) du Comité des Travaux Historiques et Scientifiques*, 99eme Congrès des Sociétés savantes, Tours, 1968, vol. 1: *Les Problèmes de l'alimentation* (Paris: Bibliothèque National, 1971), 338 ; and Amargier, "Notes sur l'ichthyophagie des serviteurs de dieu en Provence au Moyen Age," in *Manger et boire au Moyen Age*, ed. Denis Menjot (Paris: Belles lettres, 1984), 2: 318–19.
33. Stefan Górzyski, *Zarys historii rybołówstwa w dawnej Polsce* (Warszawa: PWRiL, 1964), 37–38; Guido Kisch, *Das Fischereirecht im Deutschordensgebiete: Beiträge zu seiner Geschichte*, 2nd ed., Forschungen und Quellen zur Rechts- und Sozialgeschichte des Deutschordenlandes, vol. 3 (Sigmaringen: Thorbecke, 1978), 173–82; Arthur Seligo, "Zur Geschichte der Fischerei in Westpreussen," *Mitteilungen des westpreussischen Fischereivereins* 14 (1902): 17–19; Jürgen Sarnowsky, *Die Wirtschaftsführung des Deutschen Ordens in Preußen (1382–1454)*, Veröffentlichungen aus den Archiven Preussischer Kulturbesitz, Bd. 34 (Köln-Weimar-Wien: Böhlau, 1993), 208–10; Berthold Benecke, "Beiträge zur Geschichte der Fischerei in Ost- und Westpreussen," *Altpreussisch Monatsschrift* 17 (1880): 302–5. The river-spanning seine called *regfaut* was banned from the Saône in Franche-Comté; see Jean-Pierre Kempf, "L'économie et la société aux XIIe et XIIIe siècles d'apres le cartulaire de l'abbaye cistercienne de Cherlieu en comté de Bourgogne," *Bulletin de la Société d'agriculture, lettres, sciences et arts du Département de la Haute-Saône* 9 (1973): 44.
34. *The Liber Augustalis or Constitutions of Melfi Promulgated by the Emperor Frederick II for the Kingdom of Sicily in 1231*, trans. James M. Powell (Syracuse, NY: Syracuse University Press, 1971), 144; Puñal Fernández, *El Mercado*, 175–76; Trexler, "Measures," 460–61: "Nullus audeat vel presumat pisces in flumine, vivariis, puteis, vel aliis quibuscunque aquis tossicare"; from 1322 and later, laws ever more explicitly directed against the use of lime as a fish poison. Likewise, see Barabra Fois, "Annotazioni sull'alimentazione nella Sardegna del trecento: i prodotti, le vivande, prezzi e salari," in *Manger et boire au Moyen Age*, ed. Denis Menjot (Paris: Belles lettres, 1984), 1: 189 (Sardinia,

expressly both freshwater and marine); Anna Maria Nada Patrone, *Il cibo del ricco ed il cibo del povero: Contributo alla storia qualitativa dell'alimentazione. L'area pedemontana negli ultimi secoli del Medio Evo* (Torino: Centro studi piemontesi, 1981), 325 (Piedmonte); Casado Alonso, *Señores, mercaderes y campesinos*, 210 (Old Castile); and a Pistoiese ordinance of 1296; see Lodovico Zdekauer, ed., *Statutum potestatis comunis Pistorii anni MCCLXXXXVI* (Milan: Apud Ulricum Hoepli, 1888), 131 and 139 (a reference generously provided by John Muendel).

35. London Letterbook A, fol. 89v–91v, as quoted in Wright, *Sources*, 71; Guy Cabourdin, *Terre et hommes en Lorraine (1550–1635): Toulois et Comté de Vaudémont*. Annales de l'Est Mémoire no. 55 (Nancy: Université de Nancy, 1977), 657; Nada Patrone, *Il cibo*, 325; Wilhelm Koch, "Die Geschichte der Binnenfischerei von Mitteleuropa, " in *Handbuch der Binnenfischerei Mitteleuropas*, 4 vols., ed. Reinhard Demoll and Hermann N. Maier (Stuttgart: Schweizerbart, 1925), 4: 27, and works there cited.
36. Cap. XLIII: §§7 and 13–14, in Powers, ed., *Code of Cuenca*, 215–16.
37. Duplès-Agier, "Ordonnances inédites," 49–53.
38. Nada Patrone, *Il cibo*, 326; Butturini, "La pesca," (1881): 158, and compare 173 note; Cahn, *Recht de Binnenfischerei*, 137. Rules set for the whole Zürichsee by treaty in 1386 protected small perch, roach/rudd (the taxonomy is ambiguous), and bleak from mid-April to the end of May (Amacher, *Zürcher Fischerei*, 387). Giuseppi Mira, *La pesca nel medioevo nelle acque interne italiane* (Milano: Dott. A. Giuffré, 1937), 46, mentions more Italian cases.
39. Hocquet, "pêcheries médiévales," 75; Benecke, "Beiträge zur Geschichte," 312; Seligo, "Fischerei in Westpreussen," 25–26; Kisch, *Das Fischereirecht*, 191. On the north German Ratzeburger See what is likely a seine or trawl (*iagnette*) was since the thirteenth century forbidden when the smelt or bream were spawning; see Angelika Lampen, *Fischerei und Fischhandel im Mittelalter: Wirtschafts- und sozialgeschichtliche Untersuchungen nach urkundlichen und archäologischen Quellen des 6. bis 14. Jahrhunderts im Gebiet des Deutschen Reichs*. Historische Studien, Bd. 461 (Husum: Matthiesen Verlag, 2000), 118. Douai similarly protected spawning pike and roach; see Jean-Pierre Leguay, *L'eau dans la ville au Moyen Age* (Rennes: Presses Universitaires de Rennes, 2002), 291–92. See more French parallels in Thérèse Sclafert, *Le Haut-Dauphiné au Moyen Age* (Paris: Société anonyme du Recueil Sirey, 1926), 145–47; and Roger Grand and Raymond Delatouche, *L'Agriculture au Moyen Age de la fin de l'empire romain au XVIe siècle*, L'Agriculture à travers les âges, vol. 3 (Paris: E. de Boccard, 1950), 544. Scotland's closed season on salmon is repeatedly mentioned in parliamentary statutes: Brown et al., eds, *Records of the Parliaments*, 1318/13 [19], 1425/10, 1425/3/13, 1450/1/2, 1458/3/34, 1504/3/117, and 1535/25.
40. Fois, "Annotazioni," 189
41. "*parcere in piscando*," Friedrich Bestehorn, "Die geschichtliche Entwicklung des märkischen Fischereiwesens, ein Beitrag zur Kultur—und

Wirtschaftsgeschichte der Mark Brandenburg," *Archiv für Fischereigeschichte* 1 (1913): 140, as understood by Cahn, *Recht der Binnenfischerei*, 136; Hocquet, "Pêcheries médiévales," 64; Henri Bresc, "Pêche et habitat en méditerranée occidentale," in *Castrum 7: Zones côtières littorales dans le monde méditerranéen au Moyen Âge*, ed. Jean-Marie Martin (Rome-Madrid: Ecole française de Rome, Casa de Velázquez, 2001), 526–28.

42. Trexler, "Measures," 461; Hubert Collin, "Les ressources alimentaires en Lorraine pendant la première partie du XIVeme siècle," in *Bulletin Philologique et Historique (jusqu'à 1610) du Comité des Travaux Historiques et Scientifiques*, 99eme Congrès des Sociétés savantes, Tours, 1968, vol. 1: *Les Problèmes de l'alimentation* (Paris: Bibliotheque National, 1971), 43.
43. Duplès-Agier, "Ordonnances inédites," 49–53; Butturini, "La pesca," (1881): 175; Nada Patrone, *Il cibo*, 326; Casado Alonso, *Señores, mercaderes y campesinos*, 210.
44. Plouchard, "La Scarpe," 20.
45. Manuscript ordinances from as early as 1489 and 1490 and a printed placard *Vischordnung* from 1528 survive in Munich, Bayerische Hauptstaatsarchiv, Ältere Bestände, Staatsverwaltung, Bd. I, Nrs. rot 1608 and 1612b, and Generalregister Fasc. 444, Fischwesen Nr. 1, 16–17, 18–19, and 389–98.
46. For a facsimile and analysis of the Austrian *Patent*, see Hoffmann and Sonnlechner, "Vom Archivobjekt," and older literature there cited.
47. Biganti, "La pesca," 792–95; Vincenti, "tutela ambientale," 134–35; Puñal Fernández, *El mercado*, 180; Scheiber, *Zur Geschichte*, 152; Petra J. E. M. van Dam, "Feestvissen en vastenvissen: Culturele, ecologische en economische aspecten van de visconsumptie in de Nederlanden in de Late Middeleeuwen," *Tijdschrift voor Sociale Geschiedenis* 29, no. 4 [Themanummer Eetregimes in de Middeleeuwen] (2003): 489; van Dam, "Fish for Feast and Fast: Fish Consumption in the Netherlands in the Late Middle Ages," in *Beyond the Catch. Fisheries of the North Atlantic, the North Sea and the Baltic, 900–1850*, ed. Louis Sicking and Darlene Abreu-Ferreira (Leiden: Brill, 2009), 326.
48. Bourin-Derruau et al., "Le littoral languedocien," 384: *Nam tali rete brugina tali modo piscari ut multorum didicimus veridico testimonio totum exhaurat stagnum eremat omni genere piscium nutrimento tam magnorum quam parvorum et ita esset stagnum exhaustum et heremum omnibus piscibus, non invenirentur ibi aliqui pisces et esset sterile et infructuosum.* The Bishop of Agde in 1321 likewise ruled against the same gear in the Etang de Thau; see Carole Puig, "Les ressources de l'étang et de la mer, dans la partie occidentale du Golfe du Lion (du XIIe au XIVe siècle)," in *L'homme et l'animal dans les sociétés méditerranéennes*, ed. Marie-Claude Marandet (Perpignan: Presses Universitaires de Perpignan, 2000), 118–19.
49. A possibly mythic exception is Salzburg's Pinzgauer Zellersee, where fifteenth-century sources report that fourteenth-century depletion of the whitefish from overfishing had induced the fishing community to rest the lake for three years

and then ban fishing in the area where the whitefish were known to spawn; see Hans Freudlsperger, "Kurze Fischereigeschichte des Erzstiftes Salzburg," *Mitteilungen der Gesellschaft für Salzburger Landeskunde* 76 (1936): 100–2.
50. Wright, *Sources*, 58.
51. André Bossuat, "La pêche en Seine au XVe siècle," *Bulletin de la Société de l'histoire de Paris* 90 (1963 [1965]) : 69–74.
52. Trexler, "Measures," 465; Vincenti, "Tutela ambientale," 135.
53. Mone, "Ueber der Flußfischerei," 70.

Bibliography

Amacher, Urs. *Zürcher Fischerei im Spätmittelalter*. Mitteilungen der Antiquarischen Gesellschaft in Zürich, Bd. 63. Zürich: Verlag Hans Rohr, 1996.

Amargier, Paul. "La pêche en Petite Camargue au XIVe siècle." In *Bulletin Philologique et Historique (jusqu'à 1610) du Comité des Travaux Historiques et Scientifiques*, 99eme Congrès des Sociétés Savantes, Tours, 1968. Vol. 1: *Les problèmes de l'alimentation*, 331–46. Paris: Bibliothèque Nationale, 1971.

_____. "Notes sur l'ichthyophagie des serviteurs de dieu en Provence au Moyen Age." In *Manger et boire au Moyen Age*, edited by Denis Menjot, 2: 313–23. Paris: Belles Lettres, 1984.

Amorim, Inês. "The Evolution of Portuguese Fisheries in the Medieval and Early Modern Period: A Fiscal Approach." In *Beyond the Catch: Fisheries of the North Atlantic, the North Sea and the Baltic, 900–1850*, edited by Louis Sicking and Darlene Abreu-Ferreira, 245–71. Boston: Brill, 2009.

Aubenas, Roger. *Le droit de pêche de l'abbaye de Lérins*. Equip des Historiens Cannois, fascic. no. 2. Cannes: Equipe des historiens cannois, 1953.

Bartrip, Michael. "Food for the Body and Food for the Mind: The Regulation of Freshwater Fisheries in the 1870s." *Victorian Studies* 28 (1985): 285–304.

Benecke, Berthold. "Beiträge zur Geschichte der Fischerei in Ost- und Westpreussen." *Altpreussisch Monatsschrift* 17 (1880): 300–22 and 385–416.

Benoît, Paul, Frédéric Loridant, and Olivier Mattéoni, eds. *Pêche et pisciculture en eau douce: la rivière et l'étang au Moyen Age. Actes des 1res Rencontres internationales de Liessies: 27, 28, 29 avril 1998*. [Unpaginated booklet and CD-Rom.] Lille: Conseil Général du Nord, 2004.

Bestehorn, Friedrich. "Die geschichtliche Entwicklung des märkischen Fischereiwesens, ein Beitrag zur Kultur—und Wirtschaftsgeschichte der Mark Brandenburg." *Archiv für Fischereigeschichte* 1 (1913): 1–199.

Biganti, Tiziana. "La pesca nel lago Trasimeno: Sfruttamento et tutela delle risorse ittiche del lago di Perugia (seccoli XIII-XV)." In *Gli archivi per la storia dell'alimentazione*, edited by Paolo Carucci and Massimo Buttazzo, 772–99.

Atti del convegno, Potenza-Matera, 5–8 settembre 1988. Pubblicazioni degli Archivi di Stato, Saggi, 34. Rome: Ministero per i Beni Culturali e Ambientali, Ufficio Centrale per i Beni Archivistici, 1995.

Bossuat, André. "La pêche en Seine au XVe siècle." *Bulletin de la Société de l'histoire de Paris* 90 (1963 [1965]): 60–81.

Bourin-Derruau, Monique, Daniel Le Blévec, Claude Raynaud, and Laurent Schneider. "Le littoral languedocien au Moyen Âge." In *Castrum 7: Zones côtières littorales dans le monde méditerranéen au Moyen Âge*, edited by Jean-Marie Martin, 345–423. Rome-Madrid: Ecole française de Rome, Casa de Velázquez, 2001.

Bresc, Henri. "Pêche et habitat en méditerranée occidentale." In *Castrum 7: Zones côtières littorales dans le monde méditerranéen au Moyen Âge*, edited by Jean-Marie Martin, 525–39. Rome-Madrid: Ecole française de Rome, Casa de Velázquez, 2001.

Brown, Keith M., Gillian H. MacIntosh, Alastair J. Mann, Pamela E. Ritchie, and Roland J. Tanner, eds. *The Records of the Parliaments of Scotland to 1707*. Saint Andrews: University of St. Andrews, 2007–2017. Online ed. Retrieved September 2011. http://www.rps.ac.uk/.

Butturini, Mattia. "La pesca sul lago di Garda." *Archivio Storico Lombarda* 6 (1879): 142–49; 7 (1880): 73–91; and 8 (1881): 157–95.

Cabourdin, Guy. *Terre et hommes en Lorraine (1550–1635): Toulois et Comté de Vaudémont*. Annales de l'Est Mémoire no. 55, 2 vols. paginated continuously. Nancy: Université de Nancy, 1977.

Cahn, Ernst. *Das Recht der Binnenfischerei im deutschen Kulturgebiet von den Anfängen bis zum Ausgang des 18. Jahrhunderts*, edited by Ekkehard Kaufmann. Frankfurt a.M: Klostermann, 1956.

Casado Alonso, Hilario. *Señores, mercaderes y campesinos: La comarca de Burgos a fines de la edad media*. Valladolid: Junta de Castilla y León, Consejería de Cultura y Bienestar Social, 1987.

Cohn, Henry J. *The Government of the Rhine Palatinate in the Fifteenth Century*. Oxford: Oxford University Press, 1965.

Collin, Hubert. "Les ressources alimentaires en Lorraine pendant la première partie du XIVeme siècle." In *Bulletin Philologique et Historique (jusqu'à 1610) du Comité des Travaux Historiques et Scientifiques*. 99eme Congrès des Sociétés savantes, Tours, 1968. Vol. 1: *Les Problèmes de l'alimentation*, 37–75. Paris: Bibliothèque National, 1971.

Crouzet-Pavan, Élisabeth. "Mythes et réalités de la crise d'un écosystème littoral: Venise et sa lagune à la fin du Moyen Âge." In *Castrum 7: Zones côtières littorales dans le monde méditerranéen au Moyen Âge*, edited by Jean-Marie Martin, 95–106. Rome-Madrid: Ecole française de Rome, Casa de Velázquez, 2001.

de Lespinasse, René, ed. *Les métiers et corporations de la ville de Paris*. Vol. 1: *XIV-XVIIIe siècle: Ordonnances générales, métiers de l'alimentation*. Paris: Imprimerie Nationale, 1886.

de Ureña y Smenjaud, Rafael, ed. *Fuero de Cuenca*. Madrid: Tipografía de Archivos, 1935.

Duplés-Agier, Henri. "Ordonnances inédites de Philippe-le-Bel et de Philippe-le-Long sur la police de la pêche fluviale." *Bibliothèque de l'Ecole des Chartes*, 14 (1852): 43–55.

Faugeron, Fabien. *Nourrir la ville: ravitaillement, marchés et métiers de l'alimentation à Venise dans les derniers siècles du Moyen Age*. Bibliothèque des Ecoles françaises d'Athènes et de Rome, 362. Rome: Ecole française de Rome, 2014.

Fois, Barbara. "Annotazioni sull'alimentazione nella Sardegna del trecento: i prodotti, le vivande, prezzi e salari." In *Manger et boire au Moyen Age*, edited by Denis Menjot, 1: 183–97. Paris: Belles lettres, 1984.

Fossier, Robert. *La terre et les hommes en Picardie jusqu'à la fin du XIIIe siècle*. Publications de la Faculté des Lettres et Sciences Humaines de Paris-Sorbonne, Série Recherches. Vols. 48–49. Paris: Béatrice-Nauwelaerts, 1968.

Freudlsperger, Hans. "Kurze Fischereigeschichte des Erzstiftes Salzburg." *Mitteilungen der Gesellschaft für Salzburger Landeskunde* 76 (1936): 81–128 and 77 (1937): 145–175.

Górzyński, Stefan. *Zarys historii rybołówstwa w dawnej Polsce*. Warszawa: PWRiL, 1964.

Grand, Roger, and Raymond Delatouche. *L'Agriculture au Moyen Age de la fin de l'empire romain au XVIe siècle*. L'Agriculture à travers les âges. Vol. 3. Paris: E. de Boccard, 1950.

Guilhiermoz, Paul, ed. "Ordonnance inédite de Philippe le Bel sur la police de la pêche fluviale (17 Mai 1293)." *Bibliothèque de l'Ecole des Chartes* 63 (1902): 331–37.

Hocquet, Jean-Claude. "Les pêcheries médiévales." In *Histoire des pêches maritimes en France*, edited by Michel Mollat, 36–129. Toulouse: Privat, 1987.

Hoffmann, Richard C. *An Environmental History of Medieval Europe*. Cambridge, UK: Cambridge University Press, 2014.

_____. "The Protohistory of Pike in Western Culture." In *An Annotated Bibliography of the Pike Esox lucius*, edited by Edwin J. Crossman and John Casselman, vii–xvi. Life Sciences Miscellaneous Publications. Toronto: Royal Ontario Museum, 1987. Reprinted in *The Medieval World of Nature: A Book of Essays*, edited by Joyce E. Salisbury, 61–76. New York: Garland, 1993.

_____. "*Salmo salar* in Late Medieval Scotland: Competition and Conservation for a Riverine Resource." *Aquatic Sciences* 77 (2015): 355–66. DOI: 10.1007/s00027-015-0397-4.

Hoffmann, Richard C., and Christoph Sonnlechner. "Vom Archivobjekt zum Umweltschutz: Maximilians Patent von 1506." *Studien zur Wiener Geschichte: Jahrbuch des Vereins für Geschichte der Stadt Wien* 62/63 (2006/2007) [2011]: 79–133.

Höfling, Paul, in collaboration with Torsten Gebhard. *Die Chiemsee-Fischerei: Beiträge zu ihrer Geschichte*. Munich: Institut für Volkskunde, 1987.

Jeulin, Paul. "L'élaboration par la monarchie de l'ancienne réglementation de la pêche fluviale." *Revue historique du droit français et étranger* 4, no. 40 (1962): 127–28; and 42 (1964): 728–30.

Keene, Derek. "Issues of Water in Medieval London to c.1300." *Urban History* 28 (2001): 161–79.

Kempf, Jean-Pierre. "L'économie et la société aux XIIe et XIIIe siècles d'après le cartulaire de l'abbaye cistercienne de Cherlieu en comté de Bourgogne." *Bulletin de la Société d'agriculture, lettres, sciences et arts du Département de la Haute-Saône* 9 (1973): 1–55 and 10 (1975): 55–121.

Kempshall, Matthew S. *The Common Good in Late Medieval Political Thought*. Oxford: Clarendon Press, 1999.

Kisch, Guido. *Das Fischereirecht im Deutschordensgebiete: Beiträge zu seiner Geschichte*. 2nd ed. Forschungen und Quellen zur Rechts- und Sozialgeschichte des Deutschordenlandes. Vol. 3. Sigmaringen: Thorbecke, 1978.

Koch, Wilhelm. "Die Geschichte der Binnenfischerei von Mitteleuropa." In *Handbuch der Binnenfischerei Mitteleuropas*, edited by Reinhard Demoll and Hermann N. Maier, vol. 4: 1–52d. Stuttgart: Schweizerbart, 1925.

Lampen, Angelika. *Fischerei und Fischhandel im Mittelalter: Wirtschafts- und sozialgeschichtliche Untersuchungen nach urkundlichen und archäologischen Quellen des 6. bis 14. Jahrhunderts im Gebiet des Deutschen Reichs*. Historische Studien, Bd. 461. Husum: Matthiesen Verlag, 2000.

Leguay, Jean-Pierre. *L'eau dans la ville au Moyen Age*. Rennes: Presses Universitaires de Rennes, 2002.

The Liber Augustalis or Constitutions of Melfi Promulgated by the Emperor Frederick II for the Kingdom of Sicily in 1231. Translated by James M. Powell. Syracuse, NY: Syracuse University Press, 1971.

Martin, Jean-Marie, ed. *Castrum 7: Zones côtières littorales dans le monde méditerranéen au Moyen Age: défense, peuplement, mise en valeur*. Actes du colloque international organisé par l'Ecole française de Rome et la Casa de Velázquez, en collaboration avec le Collège de France et le Centre interuniversitaire d'histoire et d'archéologie médiévales (UMR 5648—Université Lyon II—C.N.R.S.—E.H.E.S.S.). Rome, 23–26 October 1996. Rome-Madrid: Ecole française de Rome, Casa de Velázquez, 2001.

Materné, Jan. "Exploitatiemetoden, uitrustung en reglementering in de vlaamse binnenvisserij. Een regionale analyze: De Wase durmestreek van de late middeleeuwen tot op het einde van het ancien régime." *Annalen van de koninklijke oudheidkundige kring van het land van waas* 91 (1988): 205–66.

Menjot, Denis, ed. *Manger et boire au Moyen Age*. 2 vols. Colloque de Nice, 1982. Publications de la Faculté des lettres et sciences humaines de Nice, no. 27–28, 1ére serie. Paris: Belles Lettres, 1984.

Mira, Giuseppi. *La pesca nel medioevo nelle acque interne italiane*. Milano: Dott. A. Giuffré, 1937.

Mone, Franz J. "Ueber die Flußfischerei und den Vogelfang vom 14. bis 16. Jahrhundert." *Zeitschrift für die Geschichte des Oberrheins* 4 (1853): 67–97.

Nada Patrone, Anna Maria. *Il cibo del ricco ed il cibo del povero: Contributo alla storia qualitativa dell'alimentazione. L'area pedemontana negli ultimi secoli del Medio Evo*. Torino: Centro studi piemontesi, 1981.

Nauwerck, Arnold. "Der Lachsfang in der Kinzig: In biologischer Beleuchtung. Nach Akten des Fürstlich Fürstenbergischen Rentamtes Wolfach aus dem 18. Jahrhundert." *Ortenau* 66 (1986): 499–525.

Pastor y Lluís, F. "La pesca de la saboga en el Ebro (Notas históricas)." *La Zuda* [Tortosa] 40, no. 4 (1916): 108–9.

Petrin, Silvia. "Das Archiv der Tullner Fischerzeche." *Mitteilungen aus dem Niederösterreichischen Landesarchiv* 3 (1979): 29–34.

Plouchard, Pierre. "La Scarpe et les gens de rivière aux XIVe et XVe siècles." *Revue du Nord* 79, no. 322 (1997): 841–58.

Powers, James F., ed. *The Code of Cuenca: Municipal Law on the Twelfth-Century Castilian Frontier*. Philadelphia: University of Pennsylvania Press, 2000.

Puig, Carole. "Les ressources de l'étang et de la mer, dans la partie occidentale du Golfe du Lion (du XIIe au XIVe siècle)." In *L'homme et l'animal dans les sociétés méditerranéennes*, edited by Marie-Claude Marandet, 93–121. Perpignan: Presses Universitaires de Perpignan, 2000.

Puñal Fernández, Tomás. *El Mercado en Madrid en la baja edad media: Estructura y sistemas de abastecimiento de un concejo medieval castellano (s. xv)*. Madrid: Caja de Madrid, 1992.

Riley, Henry T., ed. and trans. *Liber Albus: The White Book of the City of London compiled A.D. 1419 by John Carpenter, Common Clerk, and Richard Whitington, Mayor*. London: R. Griffin, 1861.

Rorke, Martin. "English and Scottish Overseas Trade, 1300–1600." *Economic History Review* 59, no. 2 (2006): 265–88.

_____. "Scottish Overseas Trade, 1275/86–1597." 2 vols. PhD dissertation, University of Edinburgh, 2001. Retrieved 12 November 2017. https://www.bris.ac.uk/Depts/History/Maritime/Sources/2001phdRorke1.html.

Rouillard, Joséphine. "La législation royale de la pêche en eau douce du XIIIe au début du XVe siècle." In *Pêche et pisciculture en eau douce: la rivière et l'étang au Moyen Age: actes des 1res Rencontres internationales de Liessies: 27, 28, 29 avril 1998*, edited by Paul Benoît, Frédéric Loridant, and Olivier Mattéoni [unpaginated booklet and CD-Rom]. Lille: Conseil Général du Nord, 2004.

Sánchez Quiñones, Julián. *Pesca y Comercio en el Reino de Castilla Durante la Edad Media: Los valles del Guadiana, Júcar y Tajo (siglos xii y xvi)*. Serie Histórica 6. Madrid: Ediciones de La Ergástula, 2014.

Sarnowsky, Jürgen. *Die Wirtschaftsführung des Deutschen Ordens in Preußen (1382–1454)*. Veröffentlichungen aus den Archiven Preussischer Kulturbesitz, Bd. 34. Köln-Weimar-Wien: Böhlau, 1993.

Scheiber, Artur Maria. *Zur Geschichte der Fischerei in Oberösterreich, insbesondere der Traunfischerei:* Sonderabdruck aus den *"Heimatgauen": Zeitschrift für oberösterreichische Geschichte, Landes- und Volkskunde*. Linz: Verlag R. Pirngruber, 1930.

Schenk, Gerrit J. "'. . . prima ci fu la cagione de la mala provedenza de'Fiorentini . . .': Disaster and Life World Reactions in the Commune of Florence to the Flood of November 1333." *Medieval History Journal* 10 (2007): 355–86.

Scialoja, Antonio. "Statuta et ordinamento artis piscium civitatis Perusii (1296–1376)." *Bollettino della Regia Deputazione di Storia patria per l'Umbria* (1910): 813–73.

Sclafert, Thérèse. *Le Haut-Dauphiné au Moyen Age*. Paris: Société anonyme du Recueil Sirey, 1926.

Seligo, Arthur. "Zur Geschichte der Fischerei in Westpreussen." *Mitteilungen des westpreussischen Fischereivereins* 14 (1902): 2–72.

Trexler, Richard. "Measures Against Water Pollution in Fifteenth-Century Florence." *Viator* 5 (1974): 455–67.

Unger, Willem S. *De tol van Iersekeroord: Documenten en rekeningen 1321-1572*. The Hague: Martinus Nijhoff, 1939. Retrieved 12 November 2017. http://resources.huygens.knaw.nl/toliersekeroord.

van Dam, Petra J. E. M. "Feestvissen en vastenvissen: Culturele, ecologische en economische aspecten van de visconsumptie in de Nederlanden in de Late Middeleeuwen." *Tijdschrift voor Sociale Geschiedenis* 29, no. 4 [Themanummer Eetregimes in de Middeleeuwen] (2003): 467–96.

_____. "Fish for Feast and Fast: Fish Consumption in the Netherlands in the Late Middle Ages." In *Beyond the Catch: Fisheries of the North Atlantic, the North Sea and the Baltic, 900–1850*, edited by Louis Sicking and Darlene Abreu-Ferreira, 309–36. Leiden: Brill, 2009.

Vincenti, Valentina. "La tutela ambientale del Lago Trasimeno in età medioevale." In *Thinking about the Environment: Our Debt to the Classical and Medieval Past*, edited by Thomas M. Robinson and Laura Westra, 130–40. Lanham, MD: Lexington Books, 2002.

Winchester, Angus J. L. *Landscape and Society in Medieval Cumbria*. Edinburgh: John Donald, 1987.

Wood, Diana. *Medieval Economic Thought*. Cambridge, UK: Cambridge University Press, 2002.

Wright, Laura. *Sources of London English: Medieval Thames Vocabulary*. Oxford: Clarendon Press, 1996.

Zdekauer, Lodovico, ed. *Statutum potestatis comunis Pistorii anni MCCLXXXXVI*. Milan: Apud Ulricum Hoepli, 1888.

Zeheter, Michael. *Die Ordnung der Fischer: Nachhaltigkeit und Fischerei am Bodensee (1350–1900)*. Umwelthistorische Forschungen. Köln-Weimar-Wien: Böhlau, 2014.

CHAPTER 6

Managing the Lake Constance Fisheries, ca. 1350–1800

Michael Zeheter

The commonly used woodlands and pastures of what today is southwest Germany have for some time attracted the interest of historians. Coming in all forms, shapes, and sizes, villages and towns relied on commons as a backbone of their local economy from the Middle Ages to the beginning of the nineteenth century when many, though by no means all, commons were transformed into private properties.[1] Especially the poor depended on the resources they could obtain free of charge from the commons, thus being able to supply themselves with animal protein and wood products. Given the importance of the commons to the village economy and society, the allocation of the resources they provided was a continuous source of conflict but also of cooperation between villagers and their lords, who were often the legal owners of the land in question. Since these conflicts left a paper trail, historians have used them to explore rural life during the centuries before about 1800, the workings of the village communes, as well as the authorities' attempts to gain and exert control over their subjects and property.[2]

Yet, this story leaves out probably the largest commons in the south of the German-speaking lands: Lake Constance, which did not conform to this narrative due to its peculiar characteristics. Located between present-day Austria, Germany, and Switzerland, Lake Constance is one of the largest lakes in Central Europe. In the period under consideration here, the overwhelming majority of its surface was not private property, but open to usage for all inhabitants living on its shores. Everyone could cast a line or use a small net to catch fish for his or her own consumption; although in practice, fishing was mostly restricted to a relatively small group of professional fishermen who plied a demanding, dangerous, and highly regulated trade. From around 1350 until the first years of the nineteenth century, the fishermen were subjected to a considerable number of rules that specified their gear, restricted the times of fishing, and prescribed closed seasons for certain species of fish.

At first sight, the free access to the resources of the lake, including fish, for shoreline inhabitants seems to contrast sharply with the strict regulatory regime for the professional fishermen. The difference between these two forms of resource use lay in their scale and potential impact. While the region's inhabitants might have caught fish to sustain themselves, the professional fishermen sought to sell their catch on the local markets to mostly urban customers or nearby monasteries. This access to and dependency on the local markets made the fisheries susceptible to the lure of short-term profits and overexploitation, thus potentially starting a series of events similar to those described centuries later by Garrett Hardin in his influential article on the "Tragedy of the Commons."[3] In order to avoid such an outcome, the local authorities subjected the fishermen to a set of rules that would ensure the quality of the product and a steady supply of a culturally significant resource. Since fish was a Lenten food, it was the task of a good lord to ensure that his subjects had access to it at a reasonable price. Therefore, the authorities had an interest in maintaining the long-term viability of the fisheries in a manner that we today would call "sustainable."

This chapter explores this question: How did the regional authorities and the fishermen try to regulate and manage the Lake Constance fisheries over the course of 450 years in a way that would ensure both a constant supply of fish and preserve the resource for future generations? This was by no means an easy feat. There were geographical, ecological, and technical challenges, as well as political and social demands they had to overcome in order to develop a functioning regulatory framework that would allow them to sustainably manage the fish stocks. To answer this question, this chapter will introduce the Lake Constance region with its physical properties and political actors. It will then explore the practices of the lake's premodern fisheries, describe their regulatory regimes, and finally come to a conclusion about their failures and successes in achieving a sustainable form of resource use despite the complexities of the situation.

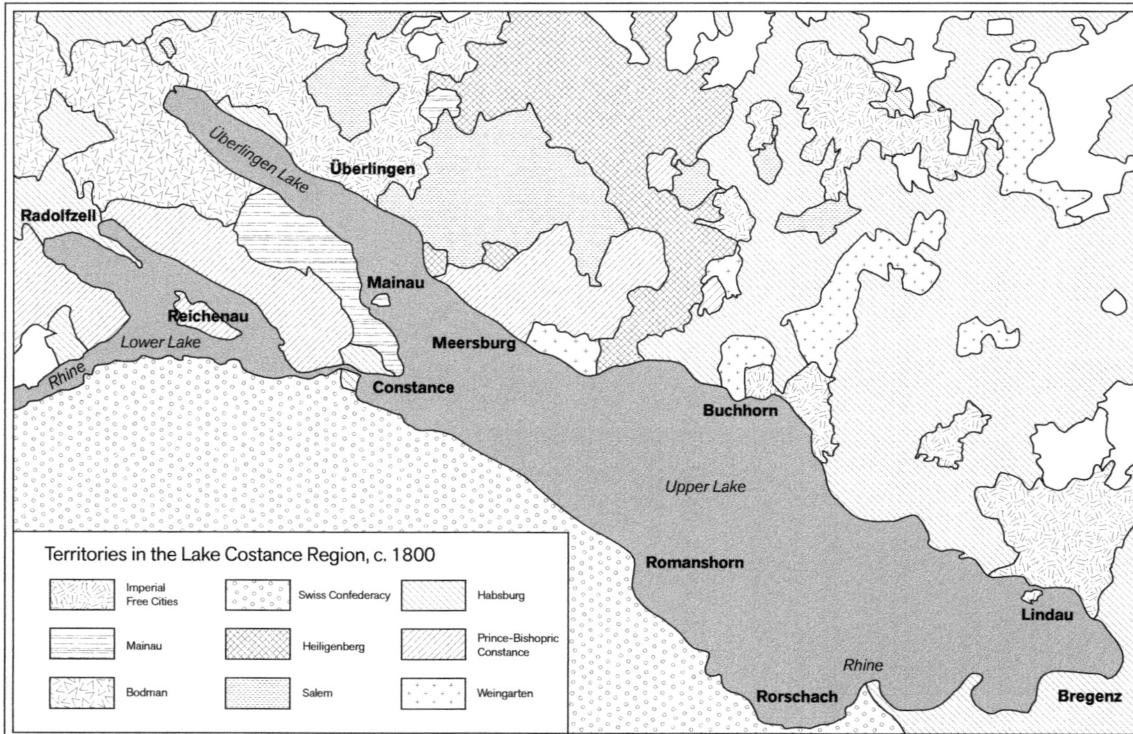

Map 6.1. The Lake Constance region, ca. 1800. Based on Erwin Hölzle, *Der Deutsche Südwesten am Ende des Alten Reiches* (Stuttgart: Württembergisches Statistisches Landesamt, 1938).

The Lake Constance Region

When viewed on a map, Lake Constance, or *Bodensee* in German, appears to be not one lake but two. In fact, it consists of three parts with very distinct characteristics that translate to a whole range of different ecosystems that were home to different species of fish. The largest part of Lake Constance is the Upper Lake (*Obersee*), which covers 473 square kilometers. With a maximum depth of 252 meters, its waters are cold, clear, and rich in oxygen, making it nutrient poor, or oligotrophic, and a perfect habitat for *salmonidae* species, such as lake trout (*Salmo trutta f. lacustris*), lake char (*Salvelinus alpinus*), and, most importantly for the fisheries, whitefish (*Coregonus lavaretus*). The Lower Lake (*Untersee*) is very different. It is much smaller (63 square kilometers), shallower (46 meters maximum depth), and its waters are relatively warm and rich in nutrients, or eutrophic. Its densely vegetated shallow parts are the preferred habitat of *cyprinidae* such as roach (*Rutilus rutilus*), rudd (*Scardinius erythrophtalmus*), tench (*Tinca tinca*), and carp (*Cyprinus carpio*, which was introduced to Lake Constance during the Middle Ages), but also pike (*Esox lucius*), perch (*Perca fluviatilis*), and eel (*Anguilla anguilla*). Connecting the Upper and the Lower Lakes is the Lake Rhine (*Seerhein*). The Rhine is Lake Constance's most important tributary, carrying alpine waters into the Upper Lake, and its only outlet, leaving the Lower Lake at Stein. The Lake Rhine, which begins at Constance and discharges into the Lower Lake after only a little more than four kilometers, provides for a slow current and a habitat for grayling (*Thymallus thymallus*), chub (*Leuciscus cephalus*), and barbel (*Barbus barbus*).[4]

This general description of the lake and its piscine inhabitants, however, fails to do justice to the complexity of its topography. Both the Upper and Lower Lakes are themselves subdivided into numerous bays and feature several islands and peninsulas. Their littoral is at certain places shallow, in others it falls steeply to the deep parts of the lake, the pelagic. All this affects the local flora and fauna, creating many different aquatic habitats within a relatively small space. Adding the effects of the seasons, the vagaries of the weather, and the migratory habits of many species of fish, navigating and exploiting Lake Constance was a highly complex task that required practical knowledge of the topography and ecology.

Complexity was not only a central feature of Lake Constance itself, but also of the political situation on its shores. A number of principalities had territorial possessions in the region. The oldest and most prestigious were the prince-bishopric of Constance and the Benedictine imperial abbeys of Reichenau and St. Gallen. Founded in the early Middle Ages, they quickly became local centers of worship and learning and acquired extensive landholdings around the lake and beyond. Thus Reichenau, named after the island on which it stands in the Lower Lake, came to own most of the surrounding land, while St. Gallen dominated the southern shore of the Upper Lake. The bishopric of Constance owned not only the city but also the land around Meersburg castle on the northern shore of the Upper Lake, along the Lake Rhine, and from 1540 on, Reichenau Abbey.[5]

The dominance of these venerable religious principalities began to decline in the twelfth and thirteenth centuries. In order to take advantage of Lake Constance's role as an intersection of two important trade routes—north–south across the Alps and east–west between France and the alpine upland—the principalities around the lake had bestowed privileges on their market towns, hoping to develop them into hubs of long-distance trade. Not all were equally successful, but the most advantageously located market towns quickly gained enough power and prosperity to shake off the control of their overlords. Constance was foremost among them, but Lindau, Überlingen, and Buchhorn also acquired the prestigious status of free imperial city that gave them independence and control over their political affairs.[6]

The free imperial cities' positions, however, were threatened by the emergence of two predominant powers in the region: the Habsburgs and the Swiss Confederacy. The former lost their possessions on the southern shore of the lake due to the latter's northward expansion during the fourteenth century. The Habsburgs sought to compensate these losses by territorial gains to the north of Lake Constance, and by asserting an informal dominance over smaller principalities such as the commandery of the Teutonic Order on the island of Mainau, the Cistercian Abbey of Salem, and the County of Heiligenberg, as well as the free imperial cities. Most of these smaller principalities managed to survive by aligning themselves with one of the two powers; however, the city of Constance, which tried to maintain an independent position between the Habsburgs and Swiss, was conquered by the former in 1548 and lost its status as a free imperial city. Thus, in the early- to mid-sixteenth century, the Lake Constance region was split between two political spheres of influence. The politically still-heterogeneous north of the lake was dominated by the Habsburgs while the south was

Swiss.[7] This constellation remained stable until the early nineteenth century when Napoleonic secularization and mediatization put an end to most of the principalities around the lake by integrating them into the new modern states of Baden, Bavaria, Württemberg, Austria, and Switzerland.[8]

The Lake Constance Fisheries

In all these political developments, the fisheries played no role. Catching fish to supplement the diet was as old as human settlement on the lake's shores. Archeological excavations of stilt houses dating back 6000 years have discovered the remains of fishing gear, such as hooks and net sinkers.[9] We can safely assume that the lake's rich fish stocks have been exploited ever since. Although the specific designs have changed, the fishermen appear to have continuously employed four different kinds of gear. Large seines were cast into the water and then hauled into the boat or at land. They came in various sizes and shapes and were used particularly during the spawning season when fish congregated in large numbers and were easy prey.[10] Gill nets were used to target specific species. With floaters attached to the top and sinkers at the bottom, these nets stood in the water like a wall. The fishermen adjusted the mesh size to the intended catch and hoped that fish would swim into the net and get stuck with their gills.[11] In addition to these two forms of nets, fishermen also used fish traps made of willow twigs or netting[12] and fishing lines to which several fishing hooks were attached.[13] All this gear was produced and repaired by the fishermen and their families who were involved in almost every aspect of the craft.

If the fishing gear was a necessary precondition of the fisheries, so was access to fish. In early medieval times, the right to fish began to be restricted by and associated with the owner of the neighboring land. For large bodies of water such as Lake Constance, however, such control by shoreline landowners remained incomplete. Instead, a legal difference was introduced between the deep lake, which was considered ownerless, and the littoral, which was placed under the jurisdiction of the owner of the shore, but nominally was still accessible to all. In practice, the most productive parts of the littoral were privately owned, since property rights of fishing grounds independent from the owner of the shore were possible there.[14] This distinction, however, was only relevant for the Upper Lake.[15] All of the shallow Lower Lake was under the jurisdiction of Reichenau Abbey, if only a small part was the Abbey's private property. The Lake Rhine was a special case, since almost all of it was in private hands. Usually, the owners of fishing rights did not fish themselves, but instead leased their rights to professionals who paid them either a share of the catch or a fixed annual sum of money.[16]

Compared to other kinds of commons like woods or pastures, exploiting an aquatic resource like Lake Constance's fish stocks was dangerous and required specialized equipment, knowledge, experience, and skill. Consequently, fishermen competed mostly among themselves, although the legal basis for their craft varied. Those who lived in villages around the lake were subjects of a lord from whom they received the privilege to fish and the right to exploit the lord's private fishing rights as a fief. In return, they ceded part of their catch.[17] The situation of the fishermen in the free imperial cities was quite different. There, the guilds—the fishermen's among them—had taken over the city councils from the patrician families in the mid-fourteenth century and established a political order that allowed them to regulate their own trades and crafts by issuing ordinances and excluding all but their members from practicing them.[18]

Foremost, the fishermen's ordinances were intended to set common rules that were legally binding for all members of the guild. On the water, the rules applied only in those areas under the jurisdiction of the issuing city; however, the cities demanded that all fish sold on their fish market were caught in compliance with the local ordinance, so their fishermen would not face disadvantages compared to outsiders. This gave the cities of Constance and Lindau, which featured the most important fish markets, regulatory powers that extended far beyond their jurisdiction and over fishermen who were subjects of other principalities.

Exercising this regulatory authority mattered because fish was a culturally important food. Catholics had to forego meat for around 140 days per year, and many sought to substitute it with fish.[19] The fish markets were the only place that could satisfy the demand and especially that of the many monasteries around Lake Constance, which followed even stricter rules and had considerably higher fish consumption than their fishermen subjects could catch.[20] It was the task of the city councils to ensure that everyone—primarily their own subjects, then outsiders—had access to fish so they would not endanger salvation. Since the fishermen's guilds were represented in the city councils, they also considered supplying the markets with sufficient amounts of affordable fish their social responsibility, as was clearly stated in the preamble to a fisheries treaty between Constance and Mainau in 1566.[21]

The responsibility for the markets also extended to the quality of the fish on offer. Fish is a highly perishable foodstuff that, before the invention of refrigeration technology, had to be sold and consumed quickly. Therefore, to ensure that customers were not taken advantage of, medieval and early modern fish markets were highly regulated. All fish had to be brought directly to the market on the day of the catch.[22] It was forbidden to mix dead with living fish, and slaughtered fish had to be salted and clearly marked if sold the next day.[23] The authorities also set maximum prices for certain species of fish to keep them affordable even in periods of scarcity or high demand. Prices were either included in the fish market ordinances or in separate fish price lists.[24] Fish inspectors named by the council ensured compliance with the regulations and prices and were responsible for destroying rotten fish.[25] Supervision was necessary because some fishermen were willing to defraud their customers, as is demonstrated by the many reprimands issued by the council.[26]

The regulation of the fisheries and the fish markets had one dominant purpose: to fulfill the urban authorities' social obligation to supply the cities' inhabitants with sufficient fish of good quality; however, there was a secondary benefit. The central role of the markets for the fishermen's business allowed the city councils to extend their authority to parts of the lake beyond their jurisdiction. If the fishermen wanted to sell their catch, they had to adhere to the local market's rules—regardless of the rules of their own lord or the location of the catch. Thus, the cities gained considerable power. The fishermen also profited from the regulation. First, the ordinances limited competition, thus ensuring that urban fishermen organized in guilds had no disadvantage compared to rural fishermen. Second, the ordinances gave them reliable access to the markets they depended on to make a living. The fisheries were not based on subsistence, but on the quick sale of their perishable commodity. Regulated markets made their unstable and dangerous craft more calculable and brought some sense of security.

Regulating the Fisheries

But how did the fishermen's ordinances try to accomplish these goals, and were they effective? These ordinances covered six areas of the craft. First, they defined the design and the characteristics of the gear, including size, form, as well as mesh size, and prohibited innovations. A second group of rules restricted the use of the gear at certain times or places. Third were protections for specific species of fish, either by proscribing closed seasons, by general prohibition as in the case of certain young fish, or by defining minimum lengths for fish to be caught legally. The conduct among the fishermen on the water was regulated by a fourth set of rules, while a fifth, supplementing the fish market ordinances, concerned the sale of the catch on the local market. Finally, the ordinances set the punishments for misconduct, which ranged from fines to confiscation of gear and imprisonment. Thus, the fishermen's ordinances covered virtually all aspects of their work on the lake that were directly linked to maintaining the fish stocks and supplying the markets.

The oldest known fishermen's ordinance for Lake Constance was issued in Lindau in 1349,[27] although it remains unclear if it built on oral traditions or lost predecessors. Around the same time, the city of Constance followed suit.[28] Over the next century, the ordinances, as well as the rules included in them, grew in length and complexity, making them highly specific. This development cannot have been driven in a top-down fashion by the authorities who lacked the knowledge of the lake, its fish, and the fishermen's equipment and practices. Instead, the fishermen themselves were closely involved in the creation of these detailed rules. At the annual fishermen's assemblies, the ordinances were discussed in the presence of the city's clerk, who noted complaints and proposals. At the end of the assembly, the fishermen had to swear an oath to observe the ordinance during the following year. The ordinances were the result of a continuous process of negotiation between the cities' authorities and the fishermen and represent a means to balance diverging interests.[29] Their success also expressed itself through emulation, as Reichenau and St. Gallen, whose fishermen were not guild members but subjects, issued their own ordinances in 1455 and 1534 respectively.[30]

The introduction of ordinances beyond the confines of the free imperial cities was part of a trend that led to four similar yet distinct regulatory regimes for the Lake Constance fisheries. First, from the mid-fifteenth century, the fishermen exploiting the Lower Lake were subject to the ordinances issued by Reichenau Abbey. Second, the city of Lindau continued to issue ordinances, but it extended their reach by integrating into its fishermen's guild the rural fishermen east of an imagined line between the mouths of the rivers Argen and Rhine. Thus all fishermen of the most eastern part of the Upper Lake, regardless of their local lord, had to obey these ordinances, but also had a say in their development.[31]

That this arrangement was prone to conflict within the guild should not surprise, as the guild masters in Lindau had to rely on other authorities to punish guild members for wrongdoings.[32] These conflicts were underpinned by a different understanding of what it meant to be a fisherman. The urban fishermen considered themselves craftsmen and felt threatened by rural guild members, whom they considered to be "half-fishermen" at best, since they could rely on their land and livestock for subsistence and were less exposed to the vagaries of the market.[33] A third regulatory regime developed from the late fifteenth century onward for the western Upper Lake where Constance began to negotiate fisheries treaties with neighboring principalities, mainly Überlingen, Salem, Heiligenberg, and Mainau.[34] These treaties were similar in substance to the ordinances and were negotiated with the guild masters and other senior fishermen present.

Compared to the ordinances issued by single principalities, the treaties had distinct advantages for the authorities and the fishermen. For the former, the treaties improved the surveillance of the fishermen on the lake. It is probably safe to assume that there was some sense of solidarity among the members of a guild, and that they were reluctant to report each other to the authorities, despite a reward of one third of the fine. With fishermen from elsewhere, there was no such esprit de corps. They were certainly more than willing to denounce outsiders, as numerous letters to other principalities demanding punishment attest.[35] For the fishermen, the treaties created more security and options. They had only one set of rules to follow, which reduced the risk of making a mistake, and they could sell their catch on a market of their choice without prior consideration of diverging local rules. The fishermen's treaties also found an imitator in St. Gallen, which negotiated a treaty with the Swiss condominiums of Thurgau and Rheintal in 1544 that covered all of the Upper Lake's southern shore and remained in place until the end of the eighteenth century, thus forming the fourth and final regulatory regime.[36]

The emergence of these four regulatory regimes covering Lake Constance cannot be explained, however, solely by the authorities' interest in extending their reach into the deep lake beyond their jurisdiction, or the fishermen's interest in more predictability and security for their dangerous occupation. That these ordinances had far more than nominal value is shown not only by their detailed, pragmatic character, but also by the fact that the parties continued to fine-tune many particular rules over the course of several centuries. The regulatory contents of the ordinances and treaties themselves had a clear purpose, which was expressed in a letter sent on 11 August 1531 by the guild master of Constance in reply to the request of his colleague from Überlingen to transmit the most recent fishermen's ordinance.[37] In this letter, the guild master of Constance affirmed the value of such an ordinance: if all the fishermen adhered to these rules and protected the young fish, then the almost emptied lake would soon again have an abundance of fish. Thus, the guild master advocated a practice that today we call "sustainable."

The concept of sustainability was not yet in existence in the early sixteenth century,[38] and we cannot ascribe a modern ecological understanding of sustainability to the fishermen and their authorities; however, they grasped that the maintenance of Lake Constance's fish stocks was in their hands. That the guild master singled out the protection of young fish is particularly telling. Young perch and whitefish were considered a delicacy by the inhabitants of the Lake Constance region, and they were willing to pay for the pleasure. Meeting this demand at a considerable price must have been tempting for the fishermen who could earn a nice sum by specializing on young fish. The damaging effects of this practice were apparently clear not only to us today but also to the premodern fishermen. By catching young fish that had yet to procreate, there would be increasing pressure on the fish stocks of those two species. Additionally, using very small-meshed nets for the young fish would produce a lot of by-catch of other species, hurting them too. The long-term consequences of the short-term profits would be dwindling fish stocks and declining income for the fishermen that would ultimately endanger the craft's existence.

The motivation behind the regulation was, however, not environmental or ecological, but economic and social, as is demonstrated by the aforementioned preamble of the fisheries treaty agreed between Constance and Mainau in 1566.[39] It called for the protection of spawn and young fish to safeguard the interests of "the poor man," who should not suffer from a scarcity of fish. The fishermen and the authorities were aware of their social responsibility to supply the inhabitants of the region with affordable fish, as noted above. The protection of the fish stocks was a precondition to living up to this responsibility—and to maintaining their own livelihood. The fishermen served the community of which they were a part, and this informed the rules they agreed to obey as well as their conduct on the lake.

This approach required a long-term commitment to adaptable regulation. Therefore, only a serial analysis of fishermen's ordinances and fisheries treaties issued over the course of several centuries can reveal how the authorities and fishermen tried to achieve the protection of piscine resources. By comparing changes in the rules chronologically, one can discern if they conformed to any pattern. It is, indeed, possible to identify four strategies in the ordinances and treaties: continuation, tightening, differentiation, and loosening. I will introduce each strategy very briefly through a few clear-cut examples from the Lower Lake ordinances.

Continuation means the unchanged perpetuation of a rule. A mesh size of 37 "mesh per Constance twisted yarn cubit"[40] for gill nets was introduced by the first ordinance for the Lower Lake in 1455 and remained unaltered through to 1774.[41] This remarkable continuity could be interpreted as a sign of neglect; but given the numerous changes in the ordinances, this seems implausible. I rather interpret the continuation of the rule over the course of more than three centuries as evidence that it stood the test of time. There was simply no need to revise it.

That such stability was rare is demonstrated by the rule defining the mesh size of seines used in the Lower Lake. A fishermen's ordinance for the Lower Lake issued in 1522 introduced a mesh size of 24 mesh per cubit, which was confirmed several times until it was changed to 21 mesh in 1594. This stipulated a larger mesh, making the seines less effective and letting somewhat bigger fish escape than was the case with the narrower mesh size. Although this tightened rule, which allowed more fish to remain in the lake and thus was a more restrictive standard from the fishermen's perspective, was confirmed in 1613, a special rule was introduced for one variety of seine, the *Fürnwatt*, which was used for catching roach. This rule set the mesh size for the *Fürnwatt* at 24 mesh per cubit, the mesh size for all seines before the tightening of 1594. While the mesh size of the *Fürnwatt* remained at 24 mesh per cubit until 1774, the mesh size for the other seines was further increased to 20 mesh per cubit in 1717, then set back to 21 in 1774, only to return to 20 again in 1790.[42] The example of the seines allows us to observe two different strategies. First, the use of seines was restricted twice, by introducing a mesh size and by increasing it markedly in 1594. Second, this stricter rule was then differentiated by the introduction of a specific rule that applied only to one variety of seine. I interpret this change as an adaptation of the rules to the complexities of the craft on the lake. Apparently, after restricting the mesh size to 21 mesh per cubit, one unitary rule proved to be too general. Its differentiation resulted in a partial loosening of the rule.

But there are also outright examples of loosening when a tightening of the rules proved to be overly harsh. For example, a limit for employing lines with hooks was introduced in the Lower Lake in 1542 and set at four hundred caught fish. This rule was confirmed in the following ordinance in 1550, but halved to two hundred caught fish. The ordinance gives us no clue as to why this drastic step was taken, but apparently it was an overreaction. Although it was confirmed in 1594, it was partially rescinded in 1613 when the limit was set to three hundred caught fish, a compromise that remained in place until 1774.[43]

These examples give a good idea of the scope of changes to the rules contained in the fishermen's ordinances and fisheries treaties. Since these legal texts do not contain explicit comments on the motivation and considerations behind these specific changes, the exact reasons for each alteration remain obscure. In their sheer number, they suggest a deliberate and elaborate approach to the regulation of the Lake Constance fisheries. Changes were not random or arbitrary. They were based on observation and the practical experiences of the fishermen, but they also demonstrate a long institutional memory that was aware of prior regulation and could refer to rules contained in ordinances and treaties issued decades or even centuries before. This is not to say that the fishermen and their authorities followed some grand scheme, or that they had a quasi-scientific understanding of the lake and its fish, which would have allowed them to devise precise regulations that would deliver an intended outcome. Instead, the many changes to the rules reveal an approach of trial and error, which sometimes resulted in a stable solution that would work over the course of several centuries. In most instances, the rules required regular tweaking and continuous attention to detail. The result was a flexible regulatory regime that took into account the availability of the resource, the interests of the fishermen and of their authorities, and the demands of the region's population.

Sustainability and the "Tragedy of the Commons"

If the purpose of these regulations was the sustainable management of the Lake Constance fish stocks, it is pertinent to ask if they succeeded. Since statistics came into existence only in the late nineteenth century, it is impossible to give a direct answer to this question. We simply cannot know the amounts of fish extracted from Lake Constance and how the catch was distributed among the different species. Therefore, it is nigh impossible to exactly assess the human impact on the lake and the difference the regulation of the fisheries might have had. Using nets made from flax and relying solely on their muscles, the premodern fishermen's possible impact on Lake Constance was certainly much smaller than it would be today with nylon nets and motorboats. It would be easy to dismiss the potential impact of the fishermen on the fish stocks. Consequently, the regulation of the fisheries would lose significance. If it had had no discernable effect on the lake, the attempts to regulate the fisheries would have been a social practice of limited relevance to the lake itself. Yet we know from the correspondence between principalities and from the preambles to fisheries treaties that the fishermen perceived themselves as having an impact on the lake and its fish stocks despite their simple tools. If the ordinances were based on this perception and a reaction to them, those normative texts in combination with other premodern sources and recent historical research on the social conditions of the Lake Constance region during the period under consideration might give us a better idea of the ordinances' and treaties' effectiveness as a means of resource management.

The causes of some of the changes in the lake were clearly beyond the scope of the premodern fisheries. In two instances, epizootics ravaged the fish stocks. In 1536, Überlingen warned Constance about a disease among roach in the eastern part of the Upper Lake where, consequently, trade in the species was prohibited.[44] The abbot of Reichenau took different measures when pike in the Lower Lake suffered from a plague in 1790. He extended the minimum length of pike by three inches (7.6 cm) to protect the decimated stocks.[45] In both cases, the measures were temporary and did not find their way into the local ordinances. Nonetheless, they demonstrate that authorities were responsive to sudden changes in the lake's fish population.

An area of human activity that had direct and easily observable consequences for the fish stocks was the building of mills. From the late medieval period onward, numerous rivers and creeks were dammed in order to harness the water's power. The resulting ponds were an ideal habitat for species such as carp, tench, and pike, while the mills' dams inhibited movement of such migratory fish as eel, salmon, trout, and sturgeon.[46] In the Lake Constance region, mills certainly had an impact on the many small tributaries that lake trout ascended to spawn, although we have no sources that would confirm this. In the case of the mill built in 1580 on the bridge crossing the Rhine at Constance, however, we have ample evidence regarding its negative impact. Since the Rhine has only a weak current at this place, a funnel made of willow fences directed the water onto the wheel. From 1582 on, there were numerous complaints about the negative effects of the new mill. The city of Zurich complained that it hindered transportation by boat.[47]

The funnel was not only obstructing the waterway for boats, but also for a variety of whitefish, the *Gangfisch*, which spawned in the Lake Rhine in great numbers. This made the Lake Rhine the most profitable and productive part of the lake and property rights that allowed for the catch of whitefish during spawn commanded a high price. For this privilege, the fishermen paid a fixed number of whitefish to the owner. Although the fishermen were already behind their dues in 1581, their debts began to spiral out of control in the following years. In 1604, the fishermen asked for the reduction of their debts and dues, which was granted and explicitly justified by reference to the new mill.[48] The owners tried to recoup at least some of their losses by demanding compensation from the proprietor of the mill, the city of Constance.[49] This episode is telling, even if we consider other factors such as established trends away from paying dues in kind to payment in money and debts being a frequent occurrence centuries before the construction of the new mill.[50] Although these factors might explain the willingness of the owners of fishing rights to accommodate the fishermen, they cannot explain the compensation paid to them by the city of Constance. An even more striking piece of evidence is the market price of a share of the large seines that were used to exploit the *Gangfisch* during spawning season. In the years between 1594 and 1604, the price declined from 46 to 13 gulden and it recovered only marginally to 14 gulden in 1615. Such a development at a time of inflation can only be explained by a marked decrease in the value of the share, the most plausible reason for which must be a decline in the fish stocks due to the mill.[51]

The case of the mill on the Rhine bridge clearly demonstrates the potential harm that human actions could do to the fish stocks if they hit a vulnerable point in the life cycle, such as spawning. However, the most direct and potentially most destructive human intervention were the fisheries. Even the simple late medieval and early modern gear could have devastating effects on the fish stocks, as the example of the Zeller Lake in Salzburg shows. In the middle of the fourteenth century, the local fishermen promised to pay twenty-seven thousand whitefish and eighteen lake trout per year to the archbishop for the right to exploit the lake. After one human generation, the stocks of whitefish collapsed and the attempt to compensate by introducing pike into the lake only hurt the lake trout. In the end, the fishermen had to accept a complete ban of fishing for three years and the strict regulation of their craft.[52]

As far as we know, Lake Constance never suffered such a disaster. It is far larger as well as more heterogeneous, but the example of the Zeller Lake is nonetheless instructive, as it clearly demonstrates the devastating potential of premodern gear. Continued and targeted fishing of certain species could apparently lead to overfishing if there were no regulations in place. Particularly perilous was the catch of young fish that had not yet procreated, as the fishermen themselves had clearly recognized. Here, the market pressure to increase supply was evidently high, and the fishermen had an obvious short-term economic incentive to satisfy this demand. Unfettered competition between the fishermen without regulatory restrictions would have made a tragedy of the commons an easily conceivable outcome. The most vulnerable part of the ecosystem being also highly profitable was a potentially devastating combination.

Although they refrained from realizing these available profits, they also never succeeded in completely banning the practice of catching young fish. In 1790, the city of Constance tried to negotiate a common fisheries treaty that would cover all of the Upper Lake. The reason for this diplomatic endeavor was a necessity to curb the catch of young perch that apparently had gone too far recently, as the number of fishermen had increased beyond sustainable levels. At a conference where most principalities of the Lake Constance region were represented, the delegates agreed not only to a ban of catching young perch for ten years, but also to a host of measures that were clearly intended to ease the pressure on the fish stocks, including a reduction in the number of fishermen. In the end, the treaty collapsed because the Swiss authorities would not join and fishermen would not accept to be treated differently compared to their colleagues from the next village, but the effort showed the awareness of the dangers of overfishing.[53]

This was not a singular event. There were two periods of increased regulatory activity concerning the Lake Constance fisheries, both in terms of ordinances and treaties issued and the tightening of the rules contained in them. In the fourteenth and fifteenth centuries, the instruments of regulation were introduced and developed, and in the eighteenth century, there was a revival of regulation after an intermission of more than seventy years. Both these periods were also times of rapid population growth when food production in the region was pushed to its limits.[54] One could expect that higher demand would lead to a loosening of the rules in order to supply a hungry population, thereby setting off a series of events that could eventually result in a tragedy of the commons. In fact, the opposite was true. Through the introduction of new rules and the tightening of existing ones, the fishermen and their authorities limited the supply at a time of need. This must have been a political decision that put the long-term interests first and would be difficult to imagine without an implicit understanding of sustainability.

If we try to explain why a tragedy of the commons could be avoided in the case of the Lake Constance fish stocks, it is helpful to look at the eight "design principles" that Elinor Ostrom suggested might secure the long-term viability of a commonly used resource, and apply them to the Lake Constance fish stocks. It is clear that seven of these eight principles did apply: the boundaries of the common pool resource (fish) were clearly defined, if not the number of those using it; the rules regulating the use of the resource were conceived locally and reacted to local conditions; the users collectively were involved in the process of devising and modifying those rules; the users themselves monitored adherence to the rules; sanctions for violating the rules were graduated according to their gravity; there were mechanisms to resolve conflicts either within the group of users or by the authorities; and the organization of the users and its rights to devise its own rules were recognized by the authorities.[55]

Thus, the regulation of the Lake Constance fisheries provided for a tool of exclusion and integration that prevented the abuse of and encouraged adherence to the rules. It is hard to overestimate the importance of the guilds for putting this understanding into practice. In the cities, they provided the institutional framework for the development of a regulatory regime. Through them, the fishermen were represented in the city councils and they made sure that the authorities heard the fishermen's interests.[56] Once they had established fishermen's ordinances and fisheries treaties, other principalities could adapt them for their own purposes. Neither St. Gallen nor Reichenau were required to introduce this regulatory means nor to take the considerations of their fishermen into account, yet they did both. The result was a comprehensive regime of resource management that succeeded in preserving the Lake Constance fish stocks over the course of 450 years to the benefit of the fishermen, the authorities, and the inhabitants of the region.

Michael Zeheter is a historian, lecturer, and research associate at the University of Trier, Germany. He studied history, philosophy, art history, and media studies at the University of Konstanz, Germany, where he also obtained his PhD. He is currently studying the cultural history of mineral water consumption in Europe from the early nineteenth century to the present. His previous research explored the sustainability of the Lake Constance fisheries from a *longue durée* perspective and the environmental history of cholera epidemics in the colonial cities of Madras and Quebec during the nineteenth century.

Notes

1. Harmut Zückert, *Allmende und Allmendaufhebung: Vergleichende Studien zum Spätmittelalter bis zu den Agrarreformen des 18./19. Jahrhunderts* (Stuttgart: Lucius and Lucius, 2003), 14–73; Werner Konold, "Allmenden in Baden-Württemberg zwischen Veränderungsdruck und Gemeinschaftssinn," *Berichte zur deutschen Landeskunde* 81 (2007): 367–89; Werner Rösener, "Der Wald als Wirtschaftsfaktor und Konfliktfeld in der Gesellschaft des Hoch- und Spätmittelalters," *Zeitschrift für Agrargeschichte und Agrarsoziologie* 55 (2007): 14–31.
2. Paul Warde, *Ecology, Economy and State Formation in Early Modern Germany* (New York: Cambridge University Press, 2006).
3. Garrett Hardin, "The Tragedy of the Commons," *Science* 162 (1968): 1243–48.
4. Internationale Gewässerschutzkommission für den Bodensee, ed., *Der Bodensee: Zustand—Fakten—Perspektiven*, 2nd ed. (Bregenz: Internationale Gewässerschutzkommission für den Bodensee, 2004), 9; Rainer Berg, "Über die Fische im Bodensee," in *Bodenseefischerei: Geschichte—Biologie und Ökologie—Bewirtschaftung: Zum 100jährigen Jubiläum der Internationalen Bevollmächtigtenkonferenz für die Bodenseefischerei*, ed. Benno Wagner et al. (Sigmaringen: Thorbecke, 1993), 92–101.
5. Otto Feger, *Geschichte des Bodenseeraumes*, vol. 1: *Anfänge und frühe Größe*, 4th ed. (Sigmaringen: Thorbecke, 1975).

6. Peter Eitel, "Die Städte des Bodenseeraums: Historische Gemeinsamkeiten und Wechselbeziehungen," *Schriften des Vereins für die Geschichte des Bodensees und seiner Umgebung* 99/100 (1981/82): 577–96; Bernhard Kirchgässer, "Strukturfragen von Handel und Verkehr des Bodenseeraums im Mittelalter," in *Bernhard Kirchgässer: Wirtschaft—Finanzen—Gesellschaft: Ausgewählte Schriften zu seinem 65. Geburtstag*, ed. Josef Wysocki (Sigmaringen: Thorbecke, 1988), 235–59; Stefan Sonderegger, "Politik, Kommunikation und Wirtschaft über den See," *Heimatkundliche Blätter für den Kreis Biberach* 31 (2008): 34–45.
7. Wolfgang Dobras, "Konstanz zur Zeit der Reformation," in *Konstanz in der Frühen Neuzeit: Reformation, Verlust der Reichsfreiheit, Österreichische Zeit*, ed. Martin Burkhardt, Wolfgang Dobras, and Wolfgang Zimmermann (Constance: Stadler, 1991), 130–46; Alois Niederstätter, "Die ersten Regierungsjahre Kaiser Friedrichs III. und der Südwesten des Reiches," in *Die Eidgenossen und ihre Nachbarn im Deutschen Reich des Mittelalters*, ed. Peter Rück (Marburg: Basilisken-Presse, 1991), 111–29; Volker Press, "Vorderösterreich in der Habsburgischen Reichspolitik des späten Mittelalters und der frühen Neuzeit," in *Vorderösterreich in der frühen Neuzeit*, ed. Hans Maier and Volker Press (Sigmaringen: Thorbecke, 1989), 1–41; Dieter Stievermann, "Österreichische Vorlande," in *Der Südwesten*, vol. 5 of *Die Territorien des Reichs im Zeitalter der Reformation und Konfessionalisierung: Land und Konfession 1500–1650*, ed. Anton Schindling and Walter Ziegler (Münster: Aschendorff, 1993), 257–77.
8. Georg Wieland, "Die Integration der Städte in die neuen Staaten," in *Das Ende reichsstädtischer Freiheit 1802: Zum Übergang schwäbischer Reichsstädte vom Kaiser zum Landesherrn, Begleitband zur Ausstellung "Kronenwechsel: Das Ende reichsstädtischer Freiheit,"* ed. Daniel Hohrath, Gebhard Weig, and Michael Wettengel (Ulm: Kohlhammer, 2002), 56–110.
9. Helmut Schlichterle, *Siedlungsarchäologie im Alpenvorland I: Die Sondagen 1973–1978 in den Ufersiedlungen Hornstaad-Hörnle I: Befunde und Funde zum frühen Jungneolithikum am westlichen Bodensee* (Stuttgart: Konrad Theiß, 1990), 121–23, 132.
10. Carl Klunzinger, *Bodenseefische, deren Pflege und Fang* (Stuttgart: Enke, 1892), 178–207.
11. Klunzinger, *Bodenseefische*, 159–78.
12. Klunzinger, *Bodenseefische*, 210–22.
13. Klunzinger, *Bodenseefische*, 119–48.
14. Ursula Lampen, *Fischerei und Fischhandel im Mittelalter: Wirtschafts- und sozialgeschichtliche Untersuchungen nach urkundlichen und archäologischen Quellen des 6. bis 14. Jahrhunderts im Gebiet des Deutschen Reiches* (Husum: Matthiesen, 1997), 83–85; Felix Stoffel, *Die Fischereiverhältnisse des Bodensees unter besonderer Berücksichtigung der an ihm bestehenden Hoheitsrechte: Historisch-dogmatische Studie* (Berne: Stämpfli, 1906), 3–5.
15. There were slightly different customs in different parts of the Upper Lake: Stoffel, *Fischereiverhältnisse*, 221.

16. Richard Kunz, "Fischereirechte im Untersee und Seerhein: Eine rechtshistorische Untersuchung über die Entstehung, Ausbildung und Weiterentwicklung von Fischereirechten" (PhD diss., Université de Fribourg, 1994).
17. For the Lower Lake, see Kunz, *Fischereirechte*, 44–49, 58–60.
18. Gerda Leipold-Schneider, "Die Lindauer Schiffer- und Fischerzunft von den Anfängen bis ins 18. Jahrhundert," in *Das Recht im kulturgeschichtlichen Wandel: Festschrift für Karl Heinz Burmeister zur Emeritierung*, ed. Bernd Marquart and Alois Niederstätter (Constance: Universitätsverlag Konstanz (UVK), 2002), 59–85.
19. Richard C. Hoffmann, "Footprint Metaphor and Metabolic Realities: Environmental Impacts of Medieval European Cities," in *Natures Past: The Environment and Human History*, ed. Paolo Squatriti (Ann Arbor: University of Michigan Press, 2007), 301; Lampen, *Fischerei*, 41.
20. Anton Strigel, "Die Fischereipolitik der Bodenseeorte in älterer Zeit mit besonderer Rücksicht auf Überlingen" (PhD diss., Universität Freiburg, 1910).
21. Fisheries treaty Constance and Mainau 1566, Stadtarchiv Konstanz (hereafter StA KN) DI Fasc. 35.
22. Fishermen's ordinance Constance 1527, StA KN DI Fasc. 44.
23. Fish market ordinance Constance 1433, StA KN DI Fasc. 44; Strigel, *Fischereipolitik*, 51.
24. Many fish price lists issued have survived in the archives. For Constance for the years 1612, 1660, 1665, 1680, 1752, 1787, and 1802, see StA KN DI Fasc. 28, 36, 44 and DI Band 43; for Lindau for the year 1705, see Stadtarchiv Lindau (hereafter StA LI) A III 55, 2; and for Überlingen for the years 1647 and 1688, see Stadtarchiv Überlingen (hereafter StA ÜL) C 988.
25. Fish market ordinance for Constance 1433; Fishermen's ordinance for Constance 1491, StA KN DI Fasc. 44.
26. Klaus D. Bechtold, *Zunftbürgerschaft und Patriziat: Studien zur Sozialgeschichte der Stadt Konstanz im 14. und 15. Jahrhundert* (Sigmaringen: Thorbecke, 1981), 56.
27. Leipold-Schneider, "Lindauer Schiffer- und Fischerzunft," 65–66.
28. Undated [fourteenth century?] fishermen's ordinance, StA LI A III 101, 8.
29. Fishermen's assemblies were a common practice: Urs Amacher, "Die Fischermaien: Die Gerichtstage der Fischer zwischen eigener Reglementierung und herrschaftlicher Machtausübung," in *Wirtschaft und Herrschaft: Beiträge zur ländlichen Gesellschaft in der östlichen Schweiz (1200–1800)*, ed. Thomas Meier and Roger Sablonier (Zurich: Chronos, 1999), 279–94.
30. Fishermen's ordinance Lower Lake 1455, StA KN DI Fasc. 44; fishermen's ordinance St. Gallen 1534, printed in Stoffel, *Fischereiverhältnisse*, 21.
31. Stoffel, *Fischereiverhältnisse*, 152–54.
32. Correspondence Lindau to bailiff of Fussach 23 Nov. 1723, StA LI A III 100, 7.
33. Memorandum fishermen's guild to Lindau city council 22 July 1663, StA LI A III 100, 9.

34. A first fisheries treaty between Constance, Überlingen, Salem, Mainau, and Heiligenberg was negotiated in 1481, copies in StA KN DI Fasc. 44 and StA ÜL C 976/1.
35. Report by Constance to Überlingen on the arrest of a fisherman from Überlingen in 1513, StA ÜL C976/1; Complaint by Lindau to Constance because of the fishermen of Staad in 1716, StA LI A III 110, 16.
36. Fisheries treaty St. Gallen 1544, StA KN DI Band 37.
37. Letter, guild master of Constance to guild master of Überlingen, 11 August 1531, StA ÜL C 976/1.
38. On the history of practices and the concept of sustainability: Ulrich Grober, *Sustainability: A Cultural History* (Totnes: Green Books, 2012); Jeremy L. Caradonna, *Sustainability: A History* (New York: Oxford University Press, 2014).
39. Fisheries treaty Constance and Mainau, 1566.
40. The length of the specific unit used in the ordinances, the "Konstanzer Zwillichelle" (hereafter simply "cubit"), is unclear, as there were two different cubits in use at the time, a long and a short one. It seems likely that the fishermen used the long one, but this cannot be ascertained.
41. Fishermen's ordinances Lower Lake, 1455, 1613, StA KN DI Fasc. 44; 1522, StA KN DI Fasc. 37; 1542, 1550, both StA KN DI Band 37; 1594, StA KN DI Fasc. 26; 1707, StA KN DI Band 39; 1717, StA KN DI Band 40; 1774, StA KN DI Band 42.
42. Fishermen's ordinance Lower Lake, 1532, StA KN DI Band 37; 1522, 1542, 1550, 1594, 1613, 1707, 1717, 1774.
43. Fishermen's ordinance Lower Lake, 1470, 1583, both StA KN Di Fasc. 44; 1542, 1550, 1594, 1613, 1707, 1717, 1774.
44. Correspondence Constance to Überlingen, 11 Nov. 1536, StA ÜL C 976/1.
45. Correspondence Reichenau to Constance, 8 May 1790, StA KN DI Fasc. 29.
46. Richard C. Hoffmann, "Economic Development and Aquatic Ecosystems in Medieval Europe," *American Historical Review* 101 (1996): 640–43.
47. Correspondence Zurich to Constance, 26. Aug. 1852, StA KN DIII Fasc. 33.
48. Memorandum 1604, Generallandesarchiv Karlsruhe (hereafter GLA KA) 209/306; Treaty between the provost of the Constance cathedral chapter and fishermen 1591, GLA KA 5/291.
49. Report of 1581, GLA KA 95/44.
50. Several treaties, GLA KA 5/398.
51. Lease treaties of 1594, 1595, 1603, 1616, StA KN DI Fasc. 29. Hans-Jürgen Gerhard, "Ursachen und Folgen der Wandlungen im Währungssystem des Deutschen Reiches 1500–1625: Eine Studie zu den Hintergründen der sogenannten Preisrevolution," in *Geld und Währung vom 16. Jahrhundert bis zur Gegenwart. Referate der 14. Arbeitstagung der Gesellschaft für Wirtschafts- und Sozialgeschichte vom 9. bis 13. April 1991 in Dortmund*, ed. Eckart Schremmer (Stuttgart: Steiner, 1993), 69–84.
52. Hoffmann, "Economic Development," 648.

53. Correspondence between principalities of the Lake Constance region 1789 and 1790, StA KN DI Fasc. 36, StA LI A III 55, 2; Minutes fisheries conference 1790, StA KN DI Fasc. 45; Correspondence between Lindau and Constance, 24 July 1790, StA LI A III 55, 2; Correspondence between Constance and St. Gallen, 30 July 1790, Correspondence Münsterlingen to Constance 3 Jan. 1791, StA KN DI Fasc. 36.
54. Alexander Kessler, *Die Bevölkerung der Stadt Radolfzell am Bodensee im 17. und 18. Jahrhundert: Demographische Strukturen einer "Ackerbürgerstadt" vor Beginn der Industrialisierung* (Constance: Hartung-Gorre, 1992), 13–21.
55. Elinor Ostrom, *Governing the Commons: The Evolution of Institutions for Collective Action* (New York: Cambridge University Press, 1990), 88–102.
56. Tine de Moor, "The Silent Revolution: A New Perspective on the Emergence of Commons, Guilds, and Other Forms of Corporate Collective Action in Western Europe," *International Review of Social History* 58 (2008): 179–212.

Bibliography

Amacher, Urs. "Die Fischermaien: Die Gerichtstage der Fischer zwischen eigener Reglementierung und herrschaftlicher Machtausübung." In *Wirtschaft und Herrschaft: Beiträge zur ländlichen Gesellschaft in der östlichen Schweiz (1200–1800)*, edited by Thomas Meier and Roger Sablonier, 279–94. Zurich: Chronos, 1999.

Bechtold, Klaus D. *Zunftbürgerschaft und Patriziat: Studien zur Sozialgeschichte der Stadt Konstanz im 14. und 15. Jahrhundert*. Sigmaringen: Thorbecke, 1981.

Berg, Rainer. "Über die Fische im Bodensee." In *Bodenseefischerei: Geschichte—Biologie und Ökologie—Bewirtschaftung: Zum 100jährigen Jubiläum der Internationalen Bevollmächtigtenkonferenz für die Bodenseefischerei*, edited by Benno Wagner, Herbert Löffler, Theo Kindle, Manfred Klein, and Erich Staub, 92–101. Sigmaringen: Thorbecke, 1993.

Caradonna, Jeremy L. *Sustainability: A History*. New York: Oxford University Press, 2014.

de Moor, Tine. "The Silent Revolution: A New Perspective on the Emergence of Commons, Guilds, and Other Forms of Corporate Collective Action in Western Europe." *International Review of Social History* 58 (2008): 179–212.

Dobras, Wolfgang. "Konstanz zur Zeit der Reformation." In *Konstanz in der Frühen Neuzeit: Reformation, Verlust der Reichsfreiheit, Österreichische Zeit*, edited by Martin Burkhardt, Wolfgang Dobras, and Wolfgang Zimmermann, 130–46. Constance: Stadler, 1991.

Eitel, Peter. "Die Städte des Bodenseeraums: Historische Gemeinsamkeiten und Wechselbeziehungen." *Schriften des Vereins für die Geschichte des Bodensees und seiner Umgebung* 99/100 (1981/82): 577–96.

Feger, Otto. *Geschichte des Bodenseeraumes*. Vol. 1: *Anfänge und frühe Größe*, 4th ed. Sigmaringen: Thorbecke, 1975.

Gerhard, Hans-Jürgen. "Ursachen und Folgen der Wandlungen im Währungssystem des Deutschen Reiches 1500–1625: Eine Studie zu den Hintergründen der sogenannten Preisrevolution." In *Geld und Währung vom 16. Jahrhundert bis zur Gegenwart. Referate der 14. Arbeitstagung der Gesellschaft für Wirtschafts- und Sozialgeschichte vom 9. bis 13. April 1991 in Dortmund*, edited by Eckart Schremmer, 69–84. Stuttgart: Steiner, 1993.

Grober, Ulrich. *Sustainability: A Cultural History*. Totnes: Green Books, 2012.

Hardin, Garrett. "The Tragedy of the Commons." *Science* 162 (1968): 1243–48.

Hoffmann, Richard C. "Economic Development and Aquatic Ecosystems in Medieval Europe." *American Historical Review* 101 (1996): 630–69.

_____. "Footprint Metaphor and Metabolic Realities: Environmental Impacts of Medieval European Cities." In *Natures Past: The Environment and Human History*, edited by Paolo Squatriti, 288–325. Ann Arbor: University of Michigan Press, 2007.

Internationale Gewässerschutzkommission für den Bodensee, ed. *Der Bodensee: Zustand—Fakten—Perspektiven*, 2nd ed. Bregenz: Internationale Gewässerschutzkommission für den Bodensee, 2004.

Kessler, Alexander. *Die Bevölkerung der Stadt Radolfzell am Bodensee im 17. und 18. Jahrhundert: Demographische Strukturen einer "Ackerbürgerstadt" vor Beginn der Industrialisierung*. Constance: Hartung-Gorre, 1992.

Kirchgässer, Bernhard. "Strukturfragen von Handel und Verkehr des Bodenseeraums im Mittelalter." In *Bernhard Kirchgässer: Wirtschaft—Finanzen—Gesellschaft: Ausgewählte Schriften zu seinem 65. Geburtstag*, edited by Josef Wysocki, 235–59. Sigmaringen: Thorbecke, 1988.

Klunzinger, Carl. *Bodenseefische, deren Pflege und Fang*. Stuttgart: Enke, 1892.

Konold, Werner. "Allmenden in Baden-Württemberg zwischen Veränderungsdruck und Gemeinschaftssinn." *Berichte zur deutschen Landeskunde* 81 (2007): 367–89.

Kunz, Richard. "Fischereirechte im Untersee und Seerhein: Eine rechtshistorische Untersuchung über die Entstehung, Ausbildung und Weiterentwicklung von Fischereirechten." PhD diss., Université de Fribourg, 1994.

Lampen, Ursula. *Fischerei und Fischhandel im Mittelalter: Wirtschafts- und sozialgeschichtliche Untersuchungen nach urkundlichen und archäologischen Quellen des 6. bis 14. Jahrhunderts im Gebiet des Deutschen Reiches*. Husum: Matthiesen, 1997.

Leipold-Schneider, Gerda. "Die Lindauer Schiffer- und Fischerzunft von den Anfängen bis ins 18. Jahrhundert." In *Das Recht im kulturgeschichtlichen Wandel: Festschrift für Karl Heinz Burmeister zur Emeritierung*, edited by Bernd Marquart and Alois Niederstätter, 59–85. Constance: Universitätsverlag Konstanz (UVK), 2002.

Niederstätter, Alois. "Die ersten Regierungsjahre Kaiser Friedrichs III. und der Südwesten des Reiches." In *Die Eidgenossen und ihre Nachbarn im Deutschen Reich des Mittelalters*, edited by Peter Rück, 111–29. Marburg: Basilisken-Presse, 1991.

Ostrom, Elinor. *Governing the Commons: The Evolution of Institutions for Collective Action*. New York: Cambridge University Press, 1990.
Press, Volker. "Vorderösterreich in der Habsburgischen Reichspolitik des späten Mittelalters und der frühen Neuzeit." In *Vorderösterreich in der frühen Neuzeit*, edited by Hans Maier and Volker Press, 1–41. Sigmaringen: Thorbecke, 1989.
Rösener, Werner. "Der Wald als Wirtschaftsfaktor und Konfliktfeld in der Gesellschaft des Hoch- und Spätmittelalters." *Zeitschrift für Agrargeschichte und Agrarsoziologie* 55 (2007): 14–31.
Schlichterle, Helmut. *Siedlungsarchäologie im Alpenvorland I: Die Sondagen 1973–1978 in den Ufersiedlungen Hornstaad-Hörnle I: Befunde und Funde zum frühen Jungneolithikum am westlichen Bodensee*. Stuttgart: Konrad Theiß, 1990.
Sonderegger, Stefan. "Politik, Kommunikation und Wirtschaft über den See." *Heimatkundliche Blätter für den Kreis Biberach* 31 (2008): 34–45.
Stievermann, Dieter. "Österreichische Vorlande." In *Der Südwesten*. Vol. 5 of *Die Territorien des Reichs im Zeitalter der Reformation und Konfessionalisierung: Land und Konfession 1500–1650*, edited by Anton Schindling and Walter Ziegler, 257–77. Münster: Aschendorff, 1993.
Stoffel, Felix. *Die Fischereiverhältnisse des Bodensees unter besonderer Berücksichtigung der an ihm bestehenden Hoheitsrechte: Historisch-dogmatische Studie*. Berne: Stämpfli, 1906.
Strigel, Anton. "Die Fischereipolitik der Bodenseeorte in älterer Zeit mit besonderer Rücksicht auf Überlingen." PhD diss., Universität Freiburg, 1910.
Warde, Paul. *Ecology, Economy and State Formation in Early Modern Germany*. New York: Cambridge University Press, 2006.
Wieland, Georg. "Die Integration der Städte in die neuen Staaten." In *Das Ende reichsstädtischer Freiheit 1802: Zum Übergang schwäbischer Reichsstädte vom Kaiser zum Landesherrn, Begleitband zur Ausstellung "Kronenwechsel: Das Ende reichsstädtischer Freiheit,"* edited by Daniel Hohrath, Gebhard Weig, and Michael Wettengel, 56–110. Ulm: Kohlhammer, 2002.
Zückert, Harmut. *Allmende und Allmendaufhebung: Vergleichende Studien zum Spätmittelalter bis zu den Agrarreformen des 18./19. Jahrhunderts*. Stuttgart: Lucius and Lucius, 2003.

CHAPTER 7

Keep the Water Flowing!

Premodern Swedish Water Management

Eva Jakobsson

In the far north of Sweden, in March 1686, the District Court of Luleå met to negotiate a conflict on the use of Laka Creek.[1] The district judge Oloff Helman presided, while the jury was comprised of the male residents of the Luleå parish. The dispute concerned a complaint made by the inhabitants of three villages, Prästholm, Böhle, and Råneby, against those of three others, Pädersöö, Burilsland, and Sundom, who, they said, had obstructed and, with their mills, built over Laka Creek, which drains into the Gulf of Bothnia. Therefore, fish could not pass during spring, and they did not reach the small lakes further upriver. Furthermore, these obstructions caused flooding and harmed the riparian meadows. The court reminded everyone present what the law said about mills: they should not hinder flowing water. Therefore, the court stated that all milldams had to be arranged "so that upstream waters should have their flow without hindrance and give fish their free passage from the Salt Sea up into the lakes." To strengthen its argument, the court referred to several royal resolutions, which explicitly proclaimed that, "in all streams and rivers, where small water-mills are to be found, the 'King-Vein' [*Kungsådra*] or the deepest section should always be kept open for the use of those who live upstream."

While the water dispute described above is part of a larger human history of managing flowing water, how water is utilized is shaped both by societal premises and by nature itself.[2] This northern water controversy illustrates that, as is still true today, water politics at this time had to balance diverse and interdependent methods of utilizing rivers. Even in this very distant and sparsely inhabited location north of the Arctic Circle in the late 1600s, we encounter a classic river conflict. As this incident suggests, in premodern Sweden, rivers provided both power sources and fishing resources.[3]

In this chapter, I show that political institutions and legal concepts and frameworks served to balance the interests of all those who had the right to utilize flowing water. I also argue that keeping rivers and watercourses open was a vital principle for premodern water conservation in Sweden. I further demonstrate that the apparently natural, free-flowing character of many rivers was the result of concerted, markedly political effort. By no means did such premodern rivers represent a nature untouched by man.[4]

My arguments contribute to a broader scholarly approach that interprets rivers, watercourses, and waterscape as entangled natures, which constitute a blend of nature and society. Metaphoric concepts as an "organic machine," a "socio-natural site," and "enviro-technical landscape" have been introduced to describe the fact that environment constitutes a mix of human actions and nonhuman nature.[5] I argue that this also applies to the premodern waterscape, even though premodern people could not fully control rivers. They belonged to a society that worked with nature. Several historians of premodern environmental history have discussed this relationship with the environment, which Richard Hoffmann refers to as "interactive co-adaptation."[6]

Premodern normative expressions of water management are frequent themes in this chapter. They were part of a large group of by-laws that this agrarian society used for regulating resources (for example, concerning pig feeding, cattle grazing, managing trees and hedges, fencing gates, manuring fields, regular cleaning of watercourses, ditching, and so on).[7] As Paul Warde usefully points out in considering the sustainability of premodern agricultural systems, "collective rule-making . . . sought to prevent *over-use* and excess consumption."[8] Warde characterizes such a system as a place for contesting power relations, "or a set of rules intended to govern social action, rather than a system of ecological management."[9]

The King-Vein Concept in Premodern Swedish Water Politics and Law

The above-mentioned legal institution of the "King-Vein" played a decisive role in the water politics of premodern Swedish rivers. The role of this distinctive legal custom remained remarkably stable from the Late

Middle Ages through the late nineteenth century. The King-Vein was the name of the section of a watercourse—generally measured as a third of the quantity of flowing water—that was to be held open for water to flow freely. This customary idea, whose precise origins remain obscure, evidently served to defend open watercourses and protect the free flow of water. In spite of this expression's use of the Swedish word for "king" (*Kung*) to describe this third part of the river's flow, there is no evidence that any Swedish monarch ever possessed any King-Vein.[10] In fact, somewhat counter-intuitively given its name, the King-Vein should be understood as a type of commons. But it is also true that it was a type of commons that, just as we saw in the Luleå court case, kings and queens almost always upheld. Thus in 1595, the king promised "a free passage up in all rivers and streams in the Kingdom."[11] Royal support for this custom was undoubtedly based partly on the revenues monarchs derived from fisheries' taxation, as well as from the fishing works they owned on Sweden's open rivers. But the monarchs' prominent role in upholding this custom may have involved more than just their fiscal interests. It may also suggest a political or ideological assumption that those claiming sovereign authority were charged with defending the kind of large-scale commons that the King-Vein represented. Indeed, the earliest documentary evidence for the "King-Vein" idea dates from the fifteenth century, when Swedish sovereigns gained greater control over the kingdom, and were thereby in a position to defend open watercourses throughout the realm.

The very first documentary evidence of the King-Vein concept is found in a letter dated 2 August 1442 from Bishop Nils to the residents of the Vadsbo district in the province of Västergötland. The residents complained that a monastery's mill had caused flooding in upstream fields and meadows, and they asked that a King-Vein be established in this watercourse, as no one "recalled or did truly know" if there ever had been one in it.[12] Around the same time, in 1454, the earliest known attempt to define a King-Vein was made by a court that met at Munkeboda rapids, in the Motala River in southeastern Sweden, to settle a long-simmering conflict between the church and the king over eel fisheries there. The court explicitly defined this stream's King-Vein as being open, available, free, and not blocked.[13] Similar wording constantly recurs in later documents, which state that the King-Vein must be held, "open, clean and clear";[14] "the King-Vein is always to be free, clear and transparent as it should be";[15] "the King-Vein has to be free and open";[16] it "should be standing open";[17] and "the King-Vein should not be violated in the middle of the stream."[18] In 1681, a group of lay assessors, considering the Svart Creek, defined the King-Vein this way: in every stream, from "outlet to inlet," there should be an opening where the water was deepest, and the current was strongest. This opening was for fish migration, for the riparian owners, and "for the cause that no fishing or water should be spilled for those who will build downstream." The King-Vein could not even be closed during the night; it had "ever to be kept open."[19]

The King-Vein institution constituted a part of a comprehensive system of Swedish water law whose outlines can be traced back to the thirteenth century. This body of law and custom has been understood as "riparianism," because it closely links water rights to the ownership of land.[20] The main principle of riparianism is that he who owns land adjacent to a watercourse can dispose of and use the water at his shore.[21] Swedish principles of riparianism are manifested already in the straightforward phrasing of the early fourteenth-century *Regional Law of Hälsingland*: "he [who] owns [the] land, owns [the] water."[22] In the early eighteenth century, the same idea was expressed a bit more fully: "Where someone owns land, the water and the waterfall comes with it . . . and so should it remain and last."[23]

In the earliest version of riparianism, the riparian owner had the right to use the water in its natural flow and as it always had flowed, a principle named the "natural flow rule."[24] The oldest preserved book in Swedish, dating from the middle of the thirteenth century, *The Older Regional Law of Västergötland*, formulated this principle as such: "Water shall not be diverted from its old course, so that it harms another man, and not in a different way, than it was running before."[25] In this respect, the King Vein can be interpreted as another way of expressing the "natural flow rule" in the Swedish riparian water law system.

Sources for Premodern Swedish Water History

Sources on Swedish premodern water history are relatively rare, especially for the medieval period. Those that do exist are neither poetic nor literary,[26] but rather consist of minutes, verdicts, and other judicial records, like those mentioned above. Obviously, these normative sources do not necessarily tell us about water management in practice. Still, they can tell us a lot about how those in power wanted the waterscape to be managed.

Although these sources are often hidden deep in regional and local archives, fortunately, scholars have published a collection of documents concerning the King-Vein. This collection, made at the dawn of Swedish hydropower development in the early 1900s, and motivated by a political debate about the possible state ownership of the King-Vein, consists of 180 previously unpublished documents dating from 1399 to 1807.[27] While this collection was made for somewhat different purposes than mine, it provides a wealth of information about the King-Vein's uses and meanings in the preindustrial era. These documents supplement my other sources, which include medieval provincial laws, royal resolutions and decrees, minutes from the national assembly, the Riksdag of the Estates, and the *Civil Code of 1734*.

Open Rivers as a Method of Resource Conservation

The King-Vein was used as a tool for conservation by protecting rivers far out into unpopulated land, where distant spawning grounds were vital for the most valuable fisheries species: Atlantic salmon (*Salmo salar*) and sea trout (*Salmo trutta*). Many texts show that knowledge about how to protect—or conserve—fish as a resource was widespread. People understood that even small mills and individual fishing traps or nets could make changes in the current or otherwise hinder migrating fish. The thirteenth-century *Older Regional Law of Västergötland* states that mills cannot be built so that another man's land, mill, or fishing works can be harmed.[28] The right to erect household mills came with the condition that no other riparian owners were negatively affected.[29] Similarly, the medieval *Regional Law of Dalarna* states: "no one has the right to fish to the detriment of someone else. One can place traps or nettings in the water, where the fish goes up, one night, a second and a third, and then one has to take up one's nettings or traps" to allow the fish to spawn "where it finds it suitable."[30] More briefly and categorically, in 1594 the Riksdag of the Estates mandated that all unlawful crossover obstructions in rapids and creeks should be torn down.[31]

This awareness of migrating fish's need for open access up rivers did not, however, prevent conflicts among riparian fishing stakeholders. In 1690, the Swedish Legal, Financial and Administrative Services Agency recognized this problem, writing, "It is indisputable that when a salmon fishery is built over and blocked, the fish is thereby held back, so that a rather slight part, or rather none, can reach fresh waters and the rapids where they spawn."[32] The Riksdag of the Estates framed such conflicts in the broadest terms in the 1640s, when it discussed how to conserve natural resources in the country. The problem brought to the agenda was "that the animals in the forests, the birds in the air, the fish in the waters are lessening and disappearing." The Clerical Estate explained that this alleged development was caused by God, who was punishing the Swedes for their sinfulness. Further, the clergy cast blame on man for forest clearing, incessant hunting and shooting, fishing at the wrong times, and shooting birds and animals at times when they should be mating. When it came to watercourses, the Clerical Estate emphasized that fishers did not respect the time when fish were swimming up the rivers. In the clergy's view, it was a necessity to protect the King-Veins from all constructions and dams that could block watercourses.[33] It was decisive that the King-Vein was held open all along the watercourse to function as a tool for resource conservation.

Some sources go beyond such general statements to provide very detailed descriptions of how waterscapes changed, which could occur through either human agency or natural processes. Inquiries about whether or not a King-Vein was open or closed, or should be opened, are especially illuminating concerning these waterscape dynamics. These descriptions reveal substantial premodern water system knowledge. Some efforts focused on "opening up" the King-Vein in altered waterscapes. In 1692, Felman, a fisherman, admitted that he had blocked the Gide River: "when the water level falls he makes himself a sheaf of spruce that he binds together with willow branches, [he] undresses, binds a rope around his midriff, [and with] the end of the rope . . . held by another man, sets out towards the river floor and closes the hole with the sheaf."[34] On the other hand, natural obstructions were discussed by a six-man commission in 1603 that inspected a King-Vein near a bridge in Åker Creek, just north of Stockholm. Although the commission determined that the King-Vein had the correct width, they found that some large rocks that had fallen into it. The commission demanded that the parish inhabitants "clear up the bottom and clean the thoroughfare."[35]

At the end of the 1600s, the king ordered an inspection of all watercourses and King-Veins. In a report of June 1698, which resulted from this order, we can follow—meter by meter—the inspectors of the King-Vein in the Svart Creek as they moved downstream.[36] The report tells us that the Börshult fisheries inspectors found "a complete King-Vein, that at present had its full width and a little more." Accordingly, the situation seemed to be satisfactory. Somewhat south of the fisheries, however, they discovered a stone, which had been thrown out into the King Vein. It was assumed that this had been done by farmer Joen to divert water to his fishing works, because he had received earlier orders to clear the King Vein at this location. As the investigators reached the Rövaretorps, Timmerös, and Lönsshullts eel fisheries, they found that the King-Vein in these locations was not hindered and had its "free opening and correct water course." Further on, at Gropa, there was a flourmill, a sawmill, and "a magnificent waterfall." Near the sawmill dam, the commission found big crossties and timber that were placed on the bottom of the river and across it, obstructing the King-Vein to all but a small amount of water, and thus preventing fish from passing. Here, one could also observe parts of old fish traps and poles left behind, causing the water upstream to stand more than a half meter higher than normal. Near the flourmill, they found a big pile of debris, mostly small timber and old wood, some of which had been thrown there to divert water toward an eel-trap. Near the site of another eel-trap and a dam, the commission found it necessary to stop and read aloud from a Court of Appeals verdict of 1685 that had already ordered the clean up of "the King-Vein in this great stream." Now, the commissioners gave all riparian owners from the Börshult fisheries and downstream six weeks to remove all obstructions so that fish could again migrate in the river. As this case shows, the royal commission's role as defender of the commons against individual interests was part of an organized, ongoing, sociopolitical effort to maintain the aquatic infrastructure.

We can find further evidence of this kind of water system knowledge, or what may be considered as a prescientific drainage basin competence, in the discussions that accompanied a new set of river inspections in 1764.[37] As state commissioners toured the southeastern Swedish rivers draining into the Baltic Sea, they "had with grief learned that in none could they find a King-Vein." In some places, the rivers were blocked, while in others, obstructions of timber and stone left open just small streams. Although complaints had led to many water works being removed, they were often rebuilt after a few days. This commission emphasized a systematic approach to river conservation, explaining that to ensure fisheries' easy communication with the Salt Sea, it was vital to clean the rivers and keep water thoroughfares open. The commissioners called for royal action to open them.

Riparian owners around lakes had a similar level of water system knowledge. Lakes, too, were parts of the water system and played significant roles both for migrating fish and for the nobility and farmers who, because of agricultural land on the relatively level lakeshores,[38] depended on the lakes' open outlets to avoid flooding. Disputes over the management of such outlets could go on for centuries, and, at times, provoked dramatic events. In 1422 at an episcopal mill in the outlet of Lake Hjälmaren, Sweden's fourth-largest lake, riparian owners, suffering from spring flooding, gathered to remove a milldam and clear the King-Vein in the lake's outlet. Similarly, mills in the same outlet were again demolished in 1655, 1734, and 1735. The more detailed sources for the last incident explain that the farmers went down to this river at night and damaged the dam in the lake outlet. After being identified and admitting that they had "cleared" the outlet, the farmers claimed that they had only moved debris brought there by the spring flood. When in 1755 Lake Hjälmaren once again flooded and the dam's owners did nothing, some of the farmers went down to the dam and made "some cuts." The dam owner, colonel and Lord Siegroth, arrived "shouting and swearing" and threatened to shoot and kill the farmers. The lord soon assembled blacksmiths armed with iron bars and others armed with rifles, forcing the riparian landholders to take flight. Not until the mid-eighteenth century, when the capacity to control nature began to increase, was a reference mark fixed in the outlet of the lake. When the lake reached this level, the dams should be opened.[39] I will return below to the changes of this period.

The Monarch as Protector of the King-Vein

Although the monarch never owned the King-Vein, he or she was its defender. Therefore, the King-Vein institution was significant whenever the conservation of flowing water was discussed in the Swedish capital, Stockholm—whether at the royal castle or in the Riksdag of the Estates. The monarch's permanent representatives on the regional level (as opposed to ad-hoc commissions) were the county governors, who executed royal directives. Among their duties, they had to help ensure that all King-Veins remained open and prevent any unlawful constructions in them.[40] They were, as state representatives throughout this premodern period, warrants of the King-Vein by law and in fact.[41]

In the middle of the 1590s, the King-Vein appeared for the first time in the Riksdag of the Estates. The King brought this problem to the representatives: "fish are diminishing, especially in the Northern Lands, because of 'crossing obstructions' (*tvärtäppning*) and illegal building."[42] Similarly, in 1680, the king responded to charges that mining works had closed rivers by commanding, as so many times before, that "according to law and resolutions in all streams the King-Vein had to be opened."[43] The king could also act on his own initiative to open a specific King-Vein, as in 1594, when the king ordered Åke Karlson to tear down his mill because he had blocked the King-Vein and thereby damaged meadows. The king wrote: "we order you seriously to take part" in a coming inspection.[44]

The frequency with which the King-Vein was on the royal agenda during the seventeenth century may indicate intensified use of flowing water. Complaints about the state of the waterscapes were raised among representatives in the Riksdag of the Estates and were also addressed directly to the monarch by groups of citizens. In 1650, for example, the nobility brought to the Riksdag of the Estates the question of the outlet of Sweden's larger lakes, claiming that all over the country, the lakes were "no longer naturally drained." In response, the royal government asked the county governors to inspect and tear down any buildings causing obstructions in these outlets.[45] In 1643, Queen Christina, fulfilling a request from the Riksdag of the Estates, proclaimed—and here the water system argument is explicit—that all King-Veins had to be kept open and that wherever they were obstructed, they should be reopened again, especially for the sake of the fisheries.[46] Just a few years later, in 1650, Queen Christina again proclaimed that, where it was necessary, King-Veins could be reopened.[47]

Royal policy became more ambitious in 1680, when the king proclaimed that the King-Vein was to be kept open in *all* rivers.[48] This may have been too far-reaching, as it would interfere with certain water works, for example in the regions where mining was increasing. Thus in 1683, the king ordered investigations to determine in which locations King-Veins had existed "from time immemorial" (*av ålder varit*).[49] This suggests that King-Veins were not found in every river. It also marks a new departure: henceforth, the legal system was charged with finding historical evidence for King-Veins, river-by-river, throughout the country.[50]

Despite some signs of change in the seventeenth century, the discourse and practice of the King-Vein remained essentially stable up to the mid-eighteenth century. Then the central government for the first time approved a unified legal code for Sweden: *The Civil Code of 1734* summarized a body of legal norms that had been developing since medieval times, which included the King-Vein. However, the eighteenth century would also bring new approaches to flowing water conservation.

The Civil Code of 1734 and New Conservation Strategies

The Civil Code of 1734, for the first time, systematically summed up the law of the King-Vein: "where a King-Vein existed since time immemorial, a third of the water should be left open. . . . Where the water is deepest, the County Governor shall watch it, so that it will not be closed."[51] Evidence had to be presented that a King-Vein had been in a river since "time immemorial." Also for the first time, the King-Vein was explicitly defined as a third of the river.[52] This proportion itself can be traced back, however, to the early thirteenth-century *Older Regional Law of Västergötland*, which states that a third of the water should "flow freely" and that the other two thirds could be dammed by mills.[53] The third was an elastic measurement, as it varied according to the quantity of water flowing at a given time. Simultaneously, it secured the other two thirds of the flow for riparian uses.

The *Civil Code of 1734* also synthesized many related but formerly fragmented norms of water management. Thus, it clarified the natural flow principle: "no one can build so that anyone up- or downstream will be harmed."[54] The *Code* further specified the proceedings for building a new water mill, which required that the full range of riparian interests be taken into consideration. First, one had to announce the plans at the district court, thereby informing everyone who might be affected. Next, a commission should conduct an inspection and listen to neighbors and other riparian owners who had mills working up- or downstream. The final task was to find out if a new water mill could be constructed without harming anyone else.[55] If anyone had harmed another by "drawing" the water toward himself, then that person had to "place everything back, as it used to be."[56] Although this entire procedure was centered on principles of riparian usage-rights, it was also impregnated with an awareness that all parties depended on the larger water system. When a dam was finally built, no one had a right to change the water flow: "those upstream cannot restrict water for those downstream, and those downstream cannot [allow] the water to back up on those upstream, [they must] lift the floodgate."[57] All floodgates had to be opened on the First of May, a seasonal adjustment to avoid flooding fields and meadows.[58] All these regulations represent a compromise between the need to protect human uses of aquatic resources and a conservation principle of protecting the free flow of rivers, as, so it was understood, "they had always flowed."

In the 1760s, a new attitude toward the environment becomes visible in the sources: nature itself had to be improved. Although efforts to harness natural processes persisted, the new strategy consisted of a much more systematic campaign to improve rivers by making the water flow more freely. A new term, "stream cleaning" (*strömrensning*), neatly sums up the new approach to Sweden's aquatic infrastructure. The new strategy appears, for example, in the *Fishing By-Laws of 1766.* This text targeted shallows and sandbanks in watercourse outlets that hindered migrating fish. A responsibility was laid on "all" who owned fisheries in a stream to get together, at times of low water, to "dig, dredge or in other ways" make streams flow freely to wash sandbanks away, to benefit both fishing and navigation.[59] A royal decree of 1764 shows that these new efforts to control nature depended on an even greater mobilization of local society to carry them out. The decree, "That Lakes, Streams and Rivers All Over the Kingdom Shall Be Maintained Annually in Order to Prevent Harmful Water Levels," proclaimed that it was local residents' task to keep the watercourses open.[60] It introduces some new features of premodern Swedish water management. First, it integrated river management more fully into the agricultural sector, by linking it to the recently expanded system of ditching fields, which by increasing drainage had increased the need to transport water from the fields, and thus the need for open and free-flowing rivers.[61] Second, it highlighted the will to improve nature by promoting a much more centralized and proactive approach to cleaning rivers. The decree described rivers as obstructed by natural processes, which caused blockages and flooding, especially in those years with large amounts of rain or snow, and this affected a "considerable" part of the kingdom's inhabitants. It was no longer man's actions that were in need of control, but rather nature's. The king therefore ordered: "all obedient subjects of the kingdom must, every year in summer when the water is at its lowest . . . clean and keep to their natural flow all larger and lesser rivers, brooks and creeks." Within this newly centralized and systematic initiative to make all inhabitants responsible for keeping the rivers open, the King-Vein became less important as an instrument of water system conservation.

Conclusion

Early twentieth-century participants in debates about the development of hydropower in Sweden joined discussions of water management that had been going on for centuries. Hydropower developers were the new actors in this arena, not their agricultural opponents. Farmers possessed a deep, nonprofessional knowledge of the water system and consciously used a group of established legal concepts as part of a long-familiar water politics.[62] Similarly, the rivers that were to be harnessed for hydropower had already been shaped by centuries of human manipulation. These observations emphasize, like the argument of this book, the deep roots of nature conservation.

The King-Vein was one of the concepts of Swedish water politics that was still in use in the early 1900s, as it had been for nearly five hundred years. As a type of commons within privately owned watercourses, the King-Vein protected the natural flow of rivers. It balanced up- and downstream interests to make watercourses sustainable resources. The King-Vein also protected *all* riparian owners' rights to use flowing water. As with all commons, riparian landholders have to deal with interdependent resources. In premodern Sweden, the concept of the King-Vein was used effectively to help avoid a river "tragedy."[63] It was part of a normative system governing the Swedish waterscape, where recurrent conflicts arose among different uses of flowing water, including fisheries, energy production, and navigation.[64] In practice, it was complicated to protect the King-Vein, as it had to be protected in every location, at all times, along the river's full stretch. Repeated royal resolutions reflect incessant dispute, reminding us that early conservation politics was inherently a power struggle. Stakeholders in these disputes included mill owners, fishing works owners, farmers using the water system as part of their drainage systems, riparian owners who wanted to be protected from flooding, the state as a tax collector, and navigation interests.

The waterscape is a part of the natural environment that historians often have ignored. Among traditional agrarian historians, this neglect undoubtedly owes something to the fact that the waterscape was similar to most "uncultivated" lands, which, as Paul Warde has observed, have been "seen as the 'other' of agriculture."[65] The materiality of fluid water makes its conservation quite distinctive. Preindustrial societies had a high degree of awareness of interconnections between water in the landscape and agricultural productivity. This formed the backdrop of many arguments against water-related projects, as for example the risk of flooding.[66] At the same time, however, this chapter shows why premodern rivers, like industrial ones, may be seen as "organic machines."[67] Preindustrial Swedes deployed a resource strategy designed to make watercourses function efficiently: each water system had to be open and free flowing. The entire set of water systems had to be protected. This chapter reveals that in Sweden, there was a coherent, albeit prescientific, understanding of rivers and watercourses as systems. This water system knowledge consisted of conceptions about from where the water came, to where it was flowing, and who was using it.[68] In combination with the knowledge of life conditions for the migrating fish and their dependence on open passageways, the concept of the King-Vein played a key role as a tool for resource conservation.

During the eighteenth century, the political will to keep water flowing throughout the landscape became stronger. Together with the emergence of a more instrumental view of nature, this made possible the state-led, systematic cleaning of rivers. Previously, the King-Vein may be understood as part of a strategy of adaptation to natural processes. But from the late eighteenth century onward, new strategies for river conservation gradually increased the potential to create a new waterscape.[69] Nonetheless, with the help of strong agrarian interests in the Riksdag of the Estates and its successor during the nineteenth century, the King-Vein concept continued to protect the natural drainage of rivers and watercourses. Its heyday was reached in the 1880s when the traditional, flexible definition as a third of the river's water was replaced by a new one, as a fixed one-third of the linear span of a watercourse. In practice, this new definition allowed more than a third of the water to pass through it. In its new form, the King-Vein supported extensive nineteenth-century reclamation projects, man-made drainage systems, and the lowering of many lakes.

With the dawn of hydropower and the big dam era in the 1900s, Swedish water management no longer had any interest in free-flowing water and open rivers. Large reservoirs and regulated and tamed "industrialized rivers" were given priority.[70] This was the end of the King-Vein and the natural flow rule. It had been a significant instrument for the sustainable management of flowing water. The King-Vein had served the material and ideological functions of an extensively utilized waterscape in a premodern society.

Eva Jakobsson is an Associate Professor in History at the University of Stavanger in southwestern Norway. Jakobsson's research focuses on different aspects of Swedish water history in a long-term perspective. Her thesis, *Industrialization of Rivers* (*Industrialisering av älvar*), is a study of the political and legal history of efforts to control the Swedish water system for hydro-electricity production. She is at present writing a lake-biography on Lake Vänern—the largest lake in the European Union. Jakobsson participated in the establishment of the European Society for Environmental History.

Notes

1. Karl G. Westman, ed., *Aktsamling till kungsådreinstitutets historia* (Uppsala: Regalrättsutredningen, 1920), 188–89: Bil 109, Aktsamling till kungsådreinstitutets historia / utgiven av K. G. Westman. Utdrag ur Luleå häradsrätts protokoll rörande kungsådra i Lakabäcken, den 12–13 mars 1686. All translations from Swedish were made by the author.
2. Terje Tvedt and Richard Coopey, "A Water Systems Perspective on History," in *A History of Water, Series II*, vol. 2: *Rivers and Society: From Early Civilizations to Modern Times*, ed. T. Tvedt and R. Coopey (London: I. B. Tauris, 2010), 3–28; Richard C. Hoffmann, "Elemental Resources and Aquatic Ecosystems: Medieval Europeans and their Rivers," in *A History of Water, Series II*, vol. 2, *Rivers and Society: From Early Civilizations to Modern Times*, ed. Terje Tvedt and Richard Coopey (London: I. B. Tauris, 2010), 173–85; Richard C. Hoffmann, *An Environmental History of Medieval Europe* (Cambridge, UK: Cambridge University Press, 2014); Ellen Arnold, *Negotiating the Landscape: Environment and Monastic Identity in the Medieval Ardennes* (Philadelphia: University of Pennsylvania Press, 2013).
3. Hoffmann, "Elemental Resources."
4. Martin Schmid, "The Environmental History of Rivers in the Early Modern Period," in *An Environmental History of the Early Modern Period*, ed. Martin Knoll and Reinhold Reith (Vienna: LIT Verlag, 2014), 19–26.
5. Richard White, *The Organic Machine: The Remaking of Colombia River* (New York: Hill and Wang, 1995); Erik Swyngedouw, *Social Power and the Urbanization of Water—Flows of Power* (New York: Oxford University Press, 2004); Eva Jakobsson, "Narratives About the River and the Dam: Some Reflections on How Historians Perceive the Harnessed River," in *Technological Society—Multidisciplinary and Long-time Perspectives*, ed. Åsa Dahlin Hauken (Stavanger: Haugaland Akademi, 2008), 53–61; Tvedt and Coopey, "Water Systems Perspective"; Hoffmann, "Elemental Resources"; Martin Schmid, Verena Winiwarter, and Gertrud Haidvogl, "Legacies from the Past: The Danube's Riverine Landscapes as Socio-natural Sites," *Danube*

News 21: Bulletin of the International Association for Danube Research IAD (June 2010): 2–5; Sara B. Pritchard, *Confluence: The Nature of Technology and the Remaking of the Rhône* (Cambridge, MA: Harvard University Press, 2011); Terje Tvedt, *Water and Society: Changing Perceptions of Societal and Historical Development* (London: I. B. Tauris, 2016); Paula M. Schönach, "River Histories: A Thematic Review," *Water History* 9 (2017): 233–57; Verena Winiwarter, "The Many Roles of the Dynamic Danube in Early Modern Europe: Representations and Contemporary Sources," in *Water in Social Imagination: From Technological Optimism to Contemporary Environmentalism*, ed. Jane Costlow, Yrjö Haila, and Arja Rosenholm, (Boston: Brill Rodopi, 2017), 49–76.

6. Hoffmann, "Elemental Resources," 165. See also Leona Skelton, *Tyne after Tyne: An Environmental History of a River's Battle for Protection, 1529–2015* (Winwick, Cambridgeshire: White Horse Press, 2017), 42, 63; Ellen Wohl, "Wilderness is Dead: Whither Critical Zone Studies and Geomorphology in the Anthropocene?" *Anthropocene* 2 (2013): 5. A good example of wrong conclusions drawn about pristine rivers, when they are perceived as untouched by man, is demonstrated in Robert C. Walter and Dorothy J. Merritts, "Natural Streams and the Legacy of Water-Powered Mills," *Science* 319 (2008): 299–304.
7. Jules N. Pretty, "Sustainable Agriculture in the Middle Ages: the English Manor," *The Agricultural History Review* 38, no. 1 (1990): 1–19; Eva Jakobsson, "Ditching from a Water System Perspective: Draining the Swedish Water Landscape 1200–1900," *Water History* 3 (2013): 349–67.
8. Paul Warde, "The Environmental History of Pre-industrial Agriculture in Europe," in *Nature's End: History and the Environment*, ed. Sverker Sörlin and Paul Warde (Houndmills, Basingstoke: Palgrave Macmillan, 2009), 72, 75–76, and 88.
9. Warde, "Environmental History," 89–90.
10. Karl Westman, *Minnesskrift utgiven av Juridiska fakulteten i Uppsala 1929 till hågkomst av dess första doktorspromotion 1629: Kungsådran inför konung och ständer under nyare tiden intill 1734 års lag* (Uppsala: Lundequistska bokh., 1929), 3; Josef Sandström, "Några konturer kring kungsådreinstitutets historia före 1734," in *Kungl. Humanistiska Vetenskaps-samfundets i Uppsala årsbok 1944* (Uppsala: Kungl. Vetenskaps-samfundet i Uppsala, 1944), 27; Gerhard Hafström, "Kungsådra," in *Kulturhistoriskt lexikon för nordisk medeltid från vikingatid till reformationstid: Bd 9, Konge-Kyrkorummet*, ed. Ingvar Andersson and John Granlund (Malmö: Allhem, 1964); Göran Inger, *Svensk rättshistoria* (Lund: Liber Läromedel, 1980), 45–46, 108.
11. Westman, *Minnesskrift*, 3.
12. Westman, *Aktsamling till*, 12–13: Bil 12, Biskop Nils′ i Linköping och fem världsliga riksråds brev till allmogen i Vadsbo härad rörande södra strömmen i Forsvik, den 2 augusti 1442; Karl G. Westman, *Kungsådran i den svenska rätten under medeltiden* (Uppsala: Lundequist, 1927), 54–57.

13. Westman, *Kungsådran i den svenska*, 57–58; Westman, *Aktsamling till*, 16–17: Bil. 15, Östgöta lagmansrätts dom rörande kungsådran i Munkeboda ström, den 20 september 1454.
14. Westman, *Aktsamling till*, 33–34: Bil. 30. Räftstetingsdom rörande kungsådra i strömmen Helgavärmen (Helgasjöns utlopp), den 27 februari 1530.
15. Westman, *Aktsamling till*, 25–26: Bil 25, Lagmansdombrev rörande rågång mellan Lösings häradsägor samt Ledingshammar m. fl. den 13 maj 1493.
16. Westman, *Aktsamling till*, 89: Bil. 64, Kungl. brev rörande kungsådra i Olands-ån, den 2 april 1642.
17. Westman, *Aktsamling till*, 87: Bil. 60, Grundsunda häradsrätt rörande kungsådra i Husumsån den 23 november 1635.
18. Westman, *Aktsamling till*, 80: Bil. 54, Utdrag ur Tveta härads dombok rörande kungsådra m.m. i Hoslättsån (Tabergsån), den 12 oktober 1620.
19. Westman, *Aktsamling till*, 132–41: Bil 94, Livgedingets lagmansrätts protokollsutdrag och dom rörande kungsådra i Svartån, den 23 juni 1681.
20. Ludwik A. Teclaff, "What You Have Always Wanted to Know about Riparian Rights, But Were Afraid to Ask," *Natural Resources Journal* 12, no. 1 (1972): 30–55; Ludwik A. Teclaff, ed., *Abstraction and Use of Water: A Comparison of Legal Regimes* (New York: United Nations, Department of Economic and Social Affairs, 1972), 8–9. On Swedish water law history, see Eva Jakobsson, "'Wårt till så betydlig del af watten genomskurna Land:' Svensk vattenrättshistoria i ett internationellt perspektiv," in *Miljöhistoria på väg: Artiklar presenterade vid Miljöhistoriskt Möte 1995*, ed. Björn O. Linnér and John Svidén (Linköping: Linköpings Universitet, 1996), 82–93.
21. Ludwik A Teclaff, *The River Basin in History and Law* (The Hague: Martinus Nijhoff, 1967), 89–90.
22. Åke Holmbäck and Elias Wessén, eds., *Svenska landskapslagar. Ser. 3, Södermannalagen och Hälsingelagen* (Stockholm: Geber, 1940), 376.
23. *Förarbetena till Sveriges rikes lag 1686–1736: 1, Lagkommissionens protokoll, 1686–1693* (Uppsala: Almqvist & Wiksells Boktryckeri AB, 1900), 457.
24. Theodore E. Lauer, "Reflections on Riparianism," *Missouri Law Review* 35, no. 1 (1970): 1–25; Allen S. Wisdom, *The Law of Rivers and Watercourses* (London: Shaw and Sons, 1975); Norris Hundley Jr., *The Great Thirst: Californians and Water 1770s–1990s* (Berkeley: University of California Press, 1992), 83–85; Teclaff, *River Basin in History and Law*, 77, 89–90.
25. Åke Holmbäck and Elias Wessén, eds., *Svenska landskapslagar. Ser. 5, Äldre Västgötalagen, Yngre Västgötalagen, Smålandslagens Kyrkobalk och Bjärköarätten* (Stockholm: Geber, 1946), 155.
26. For other types of water related sources, see Arnold, *Negotiating the Landscape*.
27. Westman, *Aktsamling till*.
28. Holmbäck and Wessén, *Svenska landskapslagar. Ser. 5*, 155. For a similar formulation in another regional medieval law, on unlawful use of fishing waters, see Södermannanlagen, Byggningabalken XX, §5, ed. Holmbäck and Wessén, *Svenska landskapslagar, Ser. 3*, 124: "now a village has water on both shores;

then a third of the water should flow freely. . . . If one of the villagers uses downstream and another upstream, no one has the right to stop up for the other."

29. Larger mills were often owned by the kingdom, the church, or by nobility; see Sven B. Ek, "Kvarnlagstiftning och rättspraxis för uppförandeet av nya kvarnar under 1700-talet," *Folkliv: Acta Ethnologica et Folkloristica Europaea* 19–20 (1955/56): 69–84.
30. Åke Holmbäck and Elias Wessén, ed., *Svenska landskapslagar. Ser. 2, Dalalagen och Västmannalagen* (Stockholm: Geber, 1936), 54–55.
31. Anders A. von Stiernman, ed., *Alla riksdagars och mötens beslut 1521–1731: Utgiven av Anders Anton von Stiernman. Stockholm. 1-3. 1728-1733. [Del 1], Alla riksdagars och mötens besluth, samt arfföreningar, regements-former, försäkringar och bewillningar, som, på allmenna riksdagar och möten, ifrån år 1521. intil år 1727. giorde, stadgade och bewiljade äro; med the för hwart och ett stånd utfärdade allmenna resolutioner. Nu til fäderneslandets tienst och nytta tilhopa samlade, och med en fullkomlig förtekning och ordaregister utgifne af And. Anton Stiernman. (Förra delen.)* (Stockholm: med egen bekostnad af Joh. H. Werner, directeur öfwer alle tryckerien i riket, år 1728, 1728), 398.
32. Westman, *Aktsamling till*, 200–202: Bil. 116, Kammarkollegii skrivelse till Kungl. Maj:t rörande kungsådra i Viskan, den 15 september 1690.
33. Forsell, Arne, and Erik Granstedt, eds., *Prästeståndets riksdagsprotokoll: 1, 1642–1660.* (Uppsala: Almqvist & Wiksells Boktryckeri AB, 1949), 56–59.
34. Westman, *Aktsamling till*, 212–16: Bil. 120, Utdrag ur Grundsunda häradsrätts protokoll rörande kungsådra i Gide älv, den 27 augusti 1692.
35. Westman, *Aktsamling till*, 77: Bil. 49, Utdrag ur Åkers skeppslags dombok rörande kungsådra i Åkersån, den 4 maj 1603.
36. Westman, *Aktsamling till*, 229–33: Bil. 128, Utdrag ur Göstrings härads kvarnkommissions protokoll rörande kungsådra i Svartån, den 30 juni 1698.
37. Westman, *Aktsamling till*, 304–306: Bil. 148, Utdrag ur fiskerideputationens i Östergötland betänkande rörande förslaget till fiskeristadga, den 26 november 1764.
38. Eva Jakobsson, "Understanding Lake Vänern: Science History Perspectives on Sweden's Largest Lake, 1600- 1900," in *Transference. Interdisciplinary Communications 2008/2009*, ed. Willy Østreng (Oslo: Centre for Advanced Study, 2010), Internet publication, 7 pp., https://cas.oslo.no/about_cas/cas_reports/seminar_booklets/.
39. *Den stora sjösänkningen: minnesskrift med anledning av Hjälmarens och Kvismarens sjösänkningsbolags 75-årsjubileum, D 1, på dess styrelses uppdrag utarbetad av Bertil Waldén* (Örebro: Hjälmarens och Kvismarens sjösänkningsbolag, 1940): 117–26.
40. Westman, *Aktsamling till*, 98: Bil. 74, Utdrag ur kungl. memorial för landshöfdingarna, den 4 maj 1661.
41. For example, a resolution in 1664 on The Nobility complaints; see Anders A. von Stiernman, ed., *Alla riksdagars och mötens beslut 1521–1731. Utgiven av*

Anders Anton von Stiernman. Stockholm. 1-3. 1728–1733. [Del 2], Alla riksdagars och mötens besluth, samt arfföreningar, regements-former, försäkringar och bewillningar, som, på allmenna riksdagar och möter [!], ifrån år 1633. intil år 1680. giorde, stadgade och bewiljade äro; med the för hwart och ett stånd utfärdade allmenna resolutioner. Andra delen. Nu til fäderneslandets tienst och nytta tilhopa samlade, och med en fullkomlig förtekning och ordaregister utgifne af And. Anton Stiernman. (Stockholm: med egen bekostnad af Joh. Henrich Werner, directeur öfwer alle tryckerien i riket, år 1729, 1729), 1487.

42. Westman, *Minnesskrift*, 3.
43. von Stiernman, *Alla riksdagars och mötens beslut 1521–1731. [Del 2]*, 1834, 1856.
44. Westman, *Aktsamling till*, 68: Bil. 45, Kungl. brev rörande den skada, som genom igenstängande av kungsådra tillfogas Arnold Grothusens landbönder, den 30 april 1594.
45. *Sveriges ridderskaps och adels riksdags-protokoll*. D. 4. H. 2, 1650–1651 (Stockholm: Ivar Hæggströms boktryckeri, 1872), 512.
46. Westman, *Minnesskrift*, 7.
47. Westman, *Minnesskrift*, 9–10. Repeated in 1682 and 1697; Anders A. von Stiernman, ed., *Alla riksdagars och mötens besluth, samt arfföreningar, regementsformer, försäkringar och bewillningar, som på allmenna riksdagar och möten, ifrån år 1521 intill år 1731 giorde, stadgade och bewiljade äro. [D. 3], Ifrån åhr 1682 intil åhr 1731* (Stockholm, 1733), 1176, 2142.
48. Westman, *Minnesskrift*, 19–20.
49. Westman, *Minnesskrift*, 24–25, 34, 42–43.
50. Eva Jakobsson, *Industrialisering av älvar: Studier kring svensk vattenkraftutbyggnad 1900–1918* (Göteborg: Historiska Institutionen, 1996), 119–39.
51. *Sveriges rikes lag: gillad och antagen på riksdagen år 1734. Faks.-utg. till 250-årsdagen av lagens tillkomst efter den första i antikva tryckta upplagan av år 1780* (Stockholm: Inst. för rättshistorisk forskning, 1984), Byggningabalken, XX kap, on mills 3§.
52. Westman, *Minnesskrift*, 24–25, 34; Westman, *Kungsådran i den svenska*, 69.
53. Westman, *Kungsådran i den svenska*, 7–8.
54. *Sveriges rikes lag*, Byggningabalken, XX kap, 2§.
55. *Sveriges rikes lag*, Byggningabalken, XX kap, 4§.
56. *Sveriges rikes lag*, Byggningabalken, XX kap, 7§.
57. *Sveriges rikes lag*, Byggningabalken, XX kap, 6§.
58. *Sveriges rikes lag*, Byggningabalken, XX kap, 6§.
59. *Kongl. Maj:ts Nådige Allmänne Stadga Och Ordning, För Rikets Hafs- Skär- Ström- och Insjö Fiske. Gifwen Stockholm i Råd-Cammaren then 14. Novembr. 1766* (Stockholm: Kungl. tryckeriet, 1766), 9§.
60. *Kongl. maj:ts Nådige Förordning, Huru Sjöar, Strömmar och Åar öfwer alt i Riket årligen böra wårdas, til skadelige wattudämningars afwärjande. Gifwen Stockholm i Råd-Cammaren then 20 februarii 1764* (Stockholm: Kongl. tryckeriet, 1764).

61. Jakobsson, "Ditching from a Water System Perspective," 354–59.
62. Eva Jakobsson, "The History of Flowing Water Policy in Sweden: From Natural Flow to Industrialized Rivers," in *A History of Water, Series II*, vol. 2: *Rivers and Society: From Early Civilizations to Modern Times*, ed. T. Tvedt and R. Coopey (London: I. B. Tauris, 2010): 429–36.
63. Garrett Hardin, "The Tragedy of the Commons," *Science* 162 (1968): 1243–48; Warde, "Environmental History," 74.
64. Stuart Oliver, "Liquid Materialities in the Landscape of the Thames: Mills and Weirs from the Eight Century to the Nineteenth Century," *Area* 45, no. 2 (2013): 223–29, on premodern navigation–mill conflicts on the Thames. Jakobsson, "'Wårt till så betydlig"; Jakobsson, "Ditching from a Water System Perspective"; Eva Jakobsson, "Stockholm's Changing Waterscape: A Long Time Perspective on a City and its Flowing Water," in *Urbanizing Nature: Actors and Agency (Dis)Connecting Cities and Nature Since 1500*, ed. Tim Soens, Dieter Schott, Michael Toyka-Seid, and Bert De Munck (London: Routledge, 2019), 197–216.
65. Warde, "Environmental History," 71.
66. Pierre Claude Reynard, "Charting Environmental Concerns: Reactions to Hydraulic Public Works in Eighteenth-Century France," *Environment and History* 9 (2003): 260. See also Jakobsson, "Ditching From a Water System Perspective"; Brian Donahue, "Dammed at Both Ends and Cursed in the Middle: The 'Flowage' of the Concord River Meadows, 1798–1862," in *Out of the Woods: Essays in Environmental History*, ed. Char Miller and Hal Rothman (Pittsburgh: University of Pittsburg Press, 1997), 227–42.
67. White, *Organic Machine.*
68. Skelton, *Tyne after Tyne*, 56, has similarly observed "a clear understanding of how river systems worked and of how to protect a river´s natural functions from human damage."
69. Winiwarter, "Many Roles of the Dynamic Danube," 72.
70. Jakobsson, *Industrialisering av älvar*, 20–28; Eva Jakobsson, "Industrialization of Rivers: A Water System Approach to Hydropower Development," *Technology, Knowledge and Policy* 14, no. 4 (2002): 41–56; Jakobsson, "History of Flowing Water Policy," 433–36.

Bibliography

Arnold, Ellen. *Negotiating the Landscape: Environment and Monastic Identity in the Medieval Ardennes*. Philadelphia: University of Pennsylvania Press, 2013.

Den stora sjösänkningen: minnesskrift med anledning av Hjälmarens och Kvismarens sjösänkningsbolags 75-årsjubileum, D 1, på dess styrelses uppdrag utarbetad av Bertil Waldén. Örebro: Hjälmarens och Kvismarens sjösänkningsbolags, 1940.

Donahue, Brian. "Dammed at Both Ends and Cursed in the Middle: The 'Flowage' of the Concord River Meadows, 1798–1862." In *Out of the Woods: Essays in Environmental History*, edited by Char Miller and Hal Rothman, 227–42. Pittsburgh: University of Pittsburgh Press, 1997.

Ek, Sven B. "Kvarnlagstiftning och rättspraxis för uppförandeet av nya kvarnar under 1700–talet." *Folkliv: Acta Ethnologica et Folkloristica Europaea* 19–20 (1955/56): 69–84.

Förarbetena till Sveriges rikes lag 1686–1736: 1, Lagkommissionens protokoll, 1686–1693. Uppsala: Almqvist & Wiksells Boktryckeri AB, 1900.

Forsell, Arne, and Erik Granstedt, eds. *Prästeståndets riksdagsprotokoll: 1, 1642–1660*. Uppsala: Almqvist & Wiksells Boktryckeri AB, 1949.

Hafström, Gerhard. "Kungsådra." In *Kulturhistoriskt lexikon för nordisk medeltid från vikingatid till reformationstid: Bd 9, Konge-Kyrkorummet*, edited by Ingvar Andersson and John Granlund. Malmö: Allhem, 1964.

Hardin, Garrett. "The Tragedy of the Commons." *Science* 162 (1968): 1243–48.

Hoffmann, Richard C. "Elemental Resources and Aquatic Ecosystems: Medieval Europeans and their Rivers." In *A History of Water, Series II*. Vol. 2: *Rivers and Society: From Early Civilizations to Modern Times*, edited by T. Tvedt and R. Coopey, 165–202. London: I. B. Tauris, 2010.

_____. *An Environmental History of Medieval Europe*. Cambridge, UK: Cambridge University Press, 2014.

Holmbäck, Åke, and Elias Wessén, eds. *Svenska landskapslagar. Ser. 2, Dalalagen och Västmannalagen*. Stockholm: Geber, 1936.

_____, eds. *Svenska landskapslagar. Ser. 3, Södermannalagen och Hälsingelagen*. Stockholm: Geber, 1940.

_____, eds. *Svenska landskapslagar. Ser. 5, Äldre Västgötalagen, Yngre Västgötalagen, Smålandslagens Kyrkobalk och Bjärköarätten*. Stockholm: Geber, 1946.

Hundley, Norris, Jr. *The Great Thirst: Californians and Water 1770s–1990s*. Berkeley: University of California Press, 1992.

Inger, Göran. *Svensk rättshistoria*. Lund: Liber Läromedel, 1980.

Jakobsson, Eva. "Ditching from a Water System Perspective: Draining the Swedish Water Landscape 1200–1900." *Water History* 3 (2013): 349–67.

_____. "The History of Flowing Water Policy in Sweden: From Natural Flow to Industrialized Rivers." In *A History of Water, Series II*. Vol. 2: *Rivers and Society: From Early Civilizations to Modern Times*, edited by T. Tvedt and R. Coopey, 424–39. London: I. B. Tauris, 2010.

_____. *Industrialisering av älvar: Studier kring svensk vattenkraftutbyggnad 1900–1918*. Göteborg: Historiska Institutionen, 1996.

_____. "Industrialization of Rivers: A Water System Approach to Hydropower Development." *Technology, Knowledge and Policy* 14, no. 4 (2002): 41–56.

_____. "Narratives About the River and the Dam: Some Reflections on How Historians Perceive the Harnessed River." In *Technological Society—Multidisciplinary and Long-time Perspectives*, edited by Åsa Dahlin Hauken, 53–61. Stavanger: Haugaland Akademi, 2008.

_____. "Stockholm's Changing Waterscape: A Long Time Perspective on a City and its Flowing Water." In *Urbanizing Nature: Actors and Agency (Dis)*

Connecting Cities and Nature Since 1500, edited by Tim Soens, Dieter Schott, Michael Toyka-Seid, and Bert De Munck, 197–216. London: Routledge, 2019.

_____. "Understanding Lake Vänern: Science History Perspectives on Sweden's Largest Lake, 1600–1900." In *Transference: Interdisciplinary Communications 2008/2009*, edited by Willy Østreng. Oslo: Centre for Advanced Study, 2010. Internet publication, 7 pp., retrieved 21 January 2020. https://cas.oslo.no/about_cas/cas_reports/seminar_booklets/.

_____. "'Wårt till så betydlig del af Watten genomskurna Land:' Svensk vattenrättshistoria i ett internationellt perspektiv." In *Miljöhistoria på väg: Artiklar presenterade vid Miljöhistoriskt Möte 1995*, edited by Björn O. Linnér and John Svidén, 82–93. Linköping: Linköpings Universitet, 1996.

Kongl. Maj:ts Nådige Allmänne Stadga Och Ordning, För Rikets Hafs- Skär- Ström- och Insjö Fiske. Gifwen Stockholm i Råd-Cammaren then 14. Novembr. 1766. Stockholm: Kungl. tryckeriet, 1766.

Kongl. maj:ts Nådige Förordning, Huru Sjöar, Strömmar och Åar öfwer alt i Riket årligen böra wårdas, til skadelige wattudämningars afwärjande. Gifwen Stockholm i Råd-Cammaren then 20 februarii 1764. Stockholm: Kongl. tryckeriet, 1764.

Lauer, Theodore E. "Reflections on Riparianism." *Missouri Law Review* 35, no. 1 (1970): 1–25.

Oliver, Stuart. "Liquid Materialities in the Landscape of the Thames: Mills and Weirs from the Eight Century to the Nineteenth Century." *Area* 45, no. 2 (2013): 223–29.

Pretty, Jules N. "Sustainable Agriculture in the Middle Ages: the English Manor." *The Agricultural History Review* 38, no. 1 (1990): 1–19.

Pritchard, Sara B. *Confluence: The Nature of Technology and the Remaking of the Rhône.* Cambridge, MA: Harvard University Press, 2011.

Reynard, Pierre Claude. "Charting Environmental Concerns: Reactions to Hydraulic Public Works in Eighteenth-Century France." *Environment and History* 9 (2003): 251–73.

Sandström, Josef. "Några konturer kring kungsådreinstitutets historia före 1734." In *Kungl. Humanistiska Vetenskaps-samfundets i Uppsala årsbok 1944.* Uppsala: Kungl. Vetenskaps-samfundet i Uppsala, 1944.

Schmid, Martin. "The Environmental History of Rivers in the Early Modern Period." In *An Environmental History of the Early Modern Period*, edited by Martin Knoll and Reinhold Reith, 19–26. Vienna: LIT Verlag, 2014.

Schmid, Martin, Verena Winiwarter, and Gertrud Haidvogl. "Legacies from the Past: The Danube's Riverine Landscapes as Socio-natural Sites." *Danube News 21: Bulletin of the International Association for Danube Research IAD* (June 2010): 2–5.

Schönach, Paula M. "River Histories: A Thematic Review." *Water History* 9 (2017): 233–57.

Skelton, Leona. *Tyne after Tyne: An Environmental History of a River's Battle for Protection, 1529–2015.* Winwick, Cambridgeshire: White Horse Press, 2017.

Sveriges ridderskaps och adels riksdags-protokoll. D. 4. H. 2, 1650–1651. Stockholm: Ivar Häggströms boktryckeri, 1872.

Sveriges rikes lag: gillad och antagen på riksdagen år 1734. Faks.-utg. till 250-årsdagen av lagens tillkomst efter den första i antikva tryckta upplagan av år 1780. Stockholm: Inst. för rättshistorisk forskning, 1984.

Swyngedouw, Erik. *Social Power and the Urbanization of Water—Flows of Power.* New York: Oxford University Press, 2004.

Teclaff, Ludwik A., ed. *Abstraction and Use of Water: A Comparison of Legal Regimes.* New York: United Nations, Department of Economic and Social Affairs, 1972.

_____. *The River Basin in History and Law.* The Hague: Martinus Nijhoff, 1967.

_____. "What You Have Always Wanted to Know about Riparian Rights But Were Afraid to Ask." *Natural Resources Journal* 12, no. 1 (1972): 30–55.

Tvedt, Terje. *Water and Society: Changing Perceptions of Societal and Historical Development.* London: I. B. Tauris, 2016.

Tvedt, Terje, and Richard Coopey. "A Water Systems Perspective on History." In *A History of Water, Series II.* Vol. 2: *Rivers and Society: From Early Civilizations to Modern Times*, edited by T. Tvedt and R. Coopey, 3–28. London: I. B. Tauris, 2010.

von Stiernman, Anders A., ed. *Alla riksdagars och mötens beslut 1521–1731: Utgiven av Anders Anton von Stiernman. Stockholm. 1–3. 1728–1733. [Del 1], Alla riksdagars och mötens besluth, samt arfföreningar, regements-former, försäkringar och bewillningar, som, på allmenna riksdagar och möten, ifrån år 1521. intil år 1727. giorde, stadgade och bewiljade äro; med the för hwart och ett stånd utfärdade allmenna resolutioner. Nu til fäderneslandets tienst och nytta tilhopa samlade, och med en fullkomlig förtekning och ordaregister utgifne af And. Anton Stiernman. (Förra delen.)* Stockholm: 1728, 1728.

_____, ed. *Alla riksdagars och mötens beslut 1521–1731: Utgiven av Anders Anton von Stiernman. Stockholm. 1–3. 1728–1733. [Del 2], Alla riksdagars och mötens besluth, samt arfföreningar, regements-former, försäkringar och bewillningar, som, på allmenna riksdagar och möten, ifrån år 1633. intil år 1680. giorde, stadgade och bewiljade äro; med the för hwart och ett stånd utfärdade allmenna resolutioner. Andra delen. Nu til fäderneslandets tienst och nytta tilhopa samlade, och med en fullkomlig förtekning och ordaregister utgifne af And. Anton Stiernman.* Stockholm: med egen bekostnad af Joh. Henrich Werner, directeur öfwer alle tryckerien i riket, år 1729, 1729.

_____, ed. *Alla riksdagars och mötens besluth, samt arfföreningar, regementsformer, försäkringar och bewillningar, som på allmenna riksdagar och möten, ifrån år 1521 intill år 1731 giorde, stadgade och bewiljade äro. [D. 3], Ifrån åhr 1682 intil åhr 1731.* Stockholm: med egen bekostnad af Joh. Henrich Werner, directeur öfwer alle tryckerien i riket, 1733.

Walter, Robert C., and Dorothy J. Merritts. "Natural Streams and the Legacy of Water-Powered Mills." *Science* 319 (2008): 299–304.

Warde, Paul. "The Environmental History of Pre-industrial Agriculture in Europe." In *Nature's End: History and the Environment*, edited by Sverker Sörlin and Paul Warde, 70–92. Houndmills, Basingstoke: Palgrave Macmillan, 2009.

Westman, Karl G., ed. *Aktsamling till kungsådreinstitutets historia*. Uppsala: Regalrättsutredningen, 1920.

_____. *Kungsådran i den svenska rätten under medeltiden*. Uppsala: Lundequist, 1927.

_____. *Minnesskrift utgiven av Juridiska fakulteten i Uppsala 1929 till hågkomst av dess första doktorspromotion 1629: Kungsådran inför konung och ständer under nyare tiden intill 1734 års lag*. Uppsala: Lundequistska bokh., 1929.

White, Richard. *The Organic Machine: The Remaking of Colombia River*. New York: Hill and Wang, 1995.

Winiwarter, Verena. "The Many Roles of the Dynamic Danube in Early Modern Europe: Representations and Contemporary Sources." In *Water in Social Imagination: From Technological Optimism to Contemporary Environmentalism*, edited by Jane Costlow, Yrjö Haila, and Arja Rosenholm, 49–76. Boston: Brill Rodopi, 2017.

Wisdom, Allen S. *The Law of Rivers and Watercourses*. London: Shaw and Sons, 1975.

Wohl, Ellen. "Wilderness is Dead: Whither Critical Zone Studies and Geomorphology in the Anthropocene?" *Anthropocene* 2 (2013): 4–15.

PART III

The Deep Roots of Woodland Conservation

CHAPTER 8

The Medieval Roots of Woodland Conservation

Northern France and Northwestern Europe, ca. 1100–1500

Richard Keyser

> The men of Fays have all usage and deadwood in the woods of Perchois and Bayse, and may take wood to sell in Troyes. . . . The men of Fays cannot cut any oak or beech larger than they can enclose within one hand, unless it is for making lumber to build or repair their houses. But this is allowed only if they first inform the lord of St.-Phal's bailiff or forester.
>
> —Agreement of 1197 between Pierre of St.-Phal and Notre-Dame-aux-Nonnains of Troyes, concerning woods 20 km SW of Troyes

> It is the custom of Champagne that one must preserve the regrowth in forests and woods continually for five years [after wood-harvesting] against all those who have [rights of] pasturage there. And if the woods are on poor soil, then one may give respite for another year or two to ensure the woods are sufficiently protected.
>
> —Judgment of 1271 by Champagne's high court, against the community of Chaource (25 km S of Troyes)

Woods and forests have long provided fertile ground for modern attempts to identify early forms of conservation.[1] In this chapter, I argue, first, that already in the Central Middle Ages (ca. 1000–1300 CE), efforts to conserve resources animated the management of woodlands. For example, as the above agreement of 1197 shows, while local residents usually had usage rights, this was limited in terms of wood collection to small and often dead wood, while more valuable timber, and even medium-sized calibers of oak and beech, were carefully protected under the watchful eyes of the lord's forester. The court judgment of

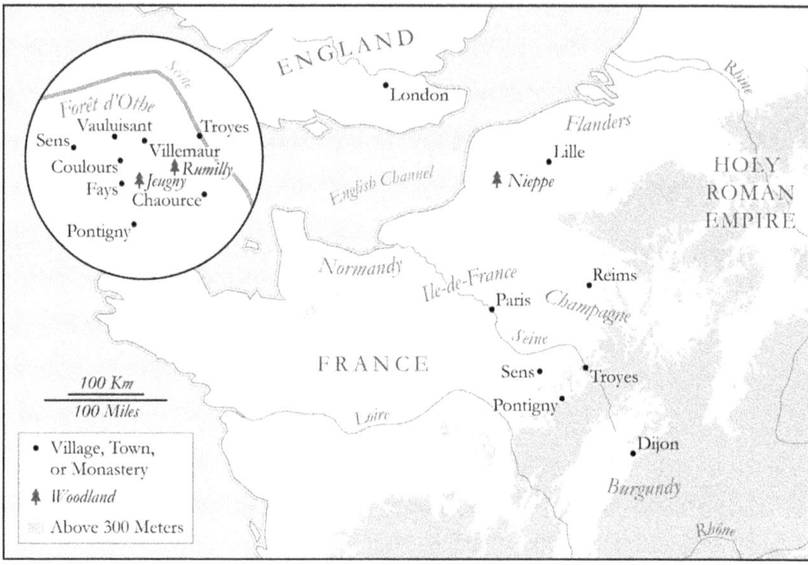

Map 8.1. Northern France. University of Wisconsin–Madison Cartography Laboratory, published with permission.

1271, which proclaims a region-wide prohibition on grazing to allow for woods to regrow after cutting, reflects the pervasiveness of frequent wood-harvesting, which typically took place on short cycles ranging between five and thirty years. This was an intensive, usually commercial form of exploitation that brought in significant revenues for woods-owners and overlords, and which by this time took precedence over local inhabitants' rights of pasturage—the other major use of preindustrial woodlands. All these forms of intensive usage, which often had to coexist within any given woodlot, required robust protective measures and careful management.

Although most of my primary sources, like these two examples, concern Champagne and adjacent regions of northern France, I also examine the royal forest law that developed precociously in late twelfth- and early thirteenth-century England. Further, the secondary literature provides reason to think that similar patterns of managing deciduous woodlands prevailed across much of lowland, northwestern Europe. Sources like those quoted above have in fact survived in large numbers, becoming increasingly abundant after about 1200 CE, and they have long been familiar to medieval historians.[2] Yet until recently, few medievalists emphasized the environmental dimensions of this tradition of resource planning and regulation, and even fewer have considered it as a form of woodland "conservation."[3] This may help explain why this kind of evidence has so far drawn relatively little attention from the broader community of environmental scholars.[4]

Second, I suggest that many forest historians, particularly in France, have over-emphasized the role of later medieval and especially early modern forestry administrations in developing and propagating sophisticated methods of sustainably managing woodlands.[5] These methods mostly concerned harvesting techniques, including, for example, the determination of the number of taller trees to reserve in each cutting cycle within woodland managed as "coppice under standards." I argue that these techniques, which crucially shape a woods' character, were already by the thirteenth century both widespread and subject to painstaking planning and regulation.

It is true, however, that written records that document harvesting practices in detail are sometimes not available until the sixteenth or seventeenth century, and their study has traditionally been the preserve of scholars who focus on the history of specialized forestry administrations. This concentration on early modern forestry administrations, coupled with the record of many early modern European states' increasingly proactive intervention into woodland protection, may leave the impression that previously, woodlands had been poorly administered. Indeed, at least since the 1960s, historians have identified the late seventeenth century, which saw both the publication of John Evelyn's *Sylva* (1664) and Jean-Baptiste Colbert's authorship of the French Forest Ordinance of 1669, as a major historical "divide," one "that separated centuries of abuse from a recognition of the need for positive forest conservation."[6] While scholars sometimes acknowledge that Evelyn and Colbert mark a watershed more in terms of an intellectual confrontation with shortages of woodland resources, in particular at the level of an entire nation, than in terms of changes in practice, it has been easy to assume that before Colbert's ordinance, for example, earlier French forest legislation consisted of a "confused and miscellaneous body of custom, ordinance, and rights of use of the past," or that it "had become confused and unworkable."[7]

Yet accumulating evidence shows not only that sophisticated harvesting methods had developed centuries earlier from widely shared customary practices, but also that medieval authorities, from villages to principalities, had successfully applied them within carefully managed woodlands.[8] Adopting a more local and less intellectual perspective allows us to recognize that, as the two quotations opening this chapter illustrate, practices of resource conservation were typically rooted in local communities before being absorbed into more geographically extensive bodies of customary practice and law.

Finally, my third claim is a corollary of the first two: all of the managerial approaches mentioned above should be understood not only as the roots of what today we call conservation or sustainable management, but also as evidence of enduring practices of "traditional woodland management." This term refers to the ways that woodlands were managed by local communities based on custom and orally transmitted "traditional ecological knowledge."[9] This traditional body of management practices helps to explain the ecological and conservation logic behind the strict limits that were placed on woodland clearance and the careful control of a wide range of woods usages, including grazing livestock, gathering brush and wood for fuel, and commercial harvesting of both small wood and timber.

This chapter begins by briefly expanding on this last point through a discussion of the "new woodland history." The rest of the chapter then argues for the early, widespread existence of both the kinds of woodland protection with which medievalists have long been familiar and those more specialized practices that forest historians have studied so well in later periods.

The New Woodland History

What may be called the "new woodland history" offers alternatives to many familiar narratives of forest history that have been grounded in, and often limited by, the sources generated by forestry administrations. Although the available sources make an administrative focus virtually unavoidable, this focus has often led forest historians to adopt the forest managers' priorities and perspectives. For convenience, we may refer to this scholarly tradition as that of "state forestry," although its assumptions about the goals of forestry were similar to those of large-scale commercial enterprises. These commercial and statist perspectives shaped the learned study of forestry—that is, the management of naturally regenerating woodlands (as opposed to orchards, for example)—from its inception in the later sixteenth and seventeenth centuries. The earliest forestry histories, penned by high-ranking forest officials, consisted almost exclusively of commentary on legal texts from a given region or nation.[10] Within the tradition they established, the history of European woodlands long remained that of state forestry.

The well-known triumph of scientific forestry in the later eighteenth and nineteenth centuries further reinforced the commercial and nationalist goals of large-scale private and state forestry, while providing them with much more effective methods of analysis and accounting.[11] Together, commercial, statist, and scientific priorities tended to eclipse the value of informal or community-based practices. It was only from the 1980s that influential scholars, such as Oliver Rackham in England, Joachim Radkau in Germany, and Andreé Corvol in France, began to question the pertinence of ideas drawn from state and scientific forestry for understanding the history of preindustrial European woodlands.[12] Over the subsequent twenty years or so, these and other scholars, in works too numerous to survey here, not only in history but also in such fields as landscape archaeology and historical ecology, further developed the view that woodland history is not simply that of state forestry.[13]

By making traditional practices central to woodland history, scholars since about the turn of the millennium have thus consolidated an alternative historical paradigm, one that may be considered "new" in comparison to the venerable tradition of state forest history. In this approach, scholars begin from the observation that preindustrial, peasant societies prioritized those products they needed most, especially pasture and small sizes of wood. A focus on tall timber is shared by preindustrial elites and modern forestry, but it is not a feature of early woodlands that can be assumed. Although "high forests" were one type of preindustrial woodland, most were managed with other priorities in mind. Among the more common methods was coppicing—the harvesting of shoots and branches of deciduous trees cut back to the stump—on short rotations of from five to thirty years.[14] Also very common were "wood-pastures"—areas that were managed both for trees and pasture.[15] In practice, many woodlands served multiple purposes at the same time. Thus, when preindustrial sources refer to "woods," "forests," or any similarly generic label for the presence of trees, they present us with a question for research, not a reliable description of any uniform type of vegetative cover. For these reasons, the term "woodland" is more generally applicable to the arboreal landscapes of preindustrial times than that of "forest." The latter word was in fact often used in a technical, legal sense to refer to areas that were placed under special lordly or royal jurisdiction, regardless of the type of vegetation.[16]

Works appearing over the last fifteen years on preindustrial European woodlands, even when they remain focused on state forest administrations, confirm the prevalence of traditional methods of woodland management across western and central Europe. These works also reveal that early forest administrations were conservative in their approach to the practical, material aspects of managing woodlands.[17] As Karl Appuhn remarked of a Venetian forest law of 1476, it "merely formalized what was already a universal practice."[18] The priorities of these nascent bureaucracies instead centered on the flow of revenues, the assertion of authority, and the procurement of certain resources—usually timber—for the lord, ruler, or city-state they served. Moreover, this administrative passivity has broader implications: most methods of woodland management known from the early modern period date back well into the Middle Ages (ca. 500–1500 CE), and they owe their origin less to forest administrations than to customary practices.

Toward a History of Woodland Conservation in the Middle Ages

The problems noted above concerning the bias of administrative sources in fact apply to most early written documents concerning woodland practices. Charters, estate inventories, wood-cutting contracts, and account books were written primarily to record the interests of powerful parties, whether these were lords, rulers, town governments, or individual proprietors. Nonetheless, when their biases are taken into account, these sources have much to teach us. My focus here will be on regulatory texts, which, as normative or prescriptive sources, often express goals and ideals more than actual behavior. Yet, as noted above, early woodland regulations rarely sought to innovate in terms of technique; instead, they promoted customary norms as standards of good practice. Moreover, laws and regulations provide broader perspectives than most other written sources.

My discussion begins by examining medieval protected areas, and then turns in a second section to specific methods of woodland management. A final section addresses a few late medieval French forest administrations, whose sophisticated practices are often overlooked in the English-language scholarship. For each topic, a chronological approach highlights early evidence that often reveals patterns that persisted for centuries thereafter.

Protected Areas in the Middle Ages

Although medieval Christians, like their ancient predecessors, granted certain sites sacred status,[19] these rarely had significant protective implications on any scale larger than a church precinct, cemetery, or small grove.[20] Exceptionally, some areas protected for religious reasons were larger, or might, at least indirectly, entail more proactive landscape management.[21] From the tenth and eleventh centuries, this potential was realized more fully when lords began to invoke similar powers in nonsacred contexts, often describing certain rivers or woodlands, for example, as being "off-limits" (Latin *defensio*) to the local population, or, what was often the same thing, under a lord's exclusive authority (*bannus*).[22]

The most famous type of protected area known from the Middle Ages has already been introduced: the "forest." The Latin term *forestis* first appears from about 650 CE in Merovingian and Carolingian royal documents from northwestern continental Europe (modern Belgium, northern France, and western Germany). From the beginning, medieval forests served for more than just hunting. Although these areas comprised much uncultivated land, they also usually contained farms, villages, and pastures. Their woodlands provided a full array of woodland products, including firewood, timber, honey, and so on, while hunting was often among their least valuable uses in economic terms.[23] Although in much of Europe the forest as a legal institution gradually weakened, the chapters in this volume by Sébastien Poublanc, Sara Morrison, and Péter Szabó show that in many places, it endured, still associated with special lordly or royal jurisdiction, into the early modern period and beyond.

The medieval institution of the forest embodied both a traditional approach to multiuse management and a complex layering of legal claims to property and resources.[24] Among the best early sources, which allow a synoptic view of how medieval forests were organized, are two well-known texts from England: the Edict of the Forest of 1184, issued by Henry II (1154–89), and the Charter of the Forest of 1217, issued by Henry III (1216–72). Although these texts applied only to royal forests, they reveal the typically complex balance between an overlord's supervisory role and the welter of private holdings and local use rights retained by others.

The Edict of the Forest's famous opening passage asserts that penalties for offenses against "the king's forest or venison" should not consist of fines that might be pardoned, but should instead correspond to the "full justice"—presumably capital punishment—that had been applied in the reign of Henry II's grandfather, Henry I (1100–35). Yet almost right away, the text moves on to strike a more flexible note:

> Those who have woods in the king's forest may not give away or sell anything from them that might waste or destroy them [*ad destructionem vel vastum bosci sui*], but they may take what they need for their own use without causing damage [*sine vasto*], under the supervision of the royal forester. . . . All those who have woods within the bounds of the royal forest shall appoint their own foresters, who must give security for any offenses done against the king in these woods. . . . The king's foresters shall take care concerning the forests of knights and others who have woods within the bounds of the royal forest, so that these woods are not destroyed.[25]

Although there is no hint here of anything like an ecocentric perspective, the overall goal of such provisions was to preserve woodlands. The text repeatedly inveighs against any damage or destruction of woods, while trying to limit each party's usage, and insisting on forest boundaries and close supervision by foresters.[26]

The above passage also suggests that interest in preserving woods was widely shared, a conclusion that is reinforced by the Charter of the Forest of 1217. First written to accompany a reissue of the Magna Carta of 1215, the two charters were issued together several more times over the thirteenth century, when they became known collectively as the "charters of liberty."[27] Like its better-known sibling text, the Charter of the Forest was meant to appease political discontent by limiting royal power, in this case by promising to renounce penalties of "life and limb" for forest offenses, by "disafforesting," or removing from royal control, areas recently designated as forests by Henry's predecessors, and by softening and clarifying other features of forest law.[28]

In its first clause, the Charter of the Forest shows that even in the areas of the most concentrated royal power—royally owned woods within the royal forests—a third level of rights merited official recognition. After promising to "disafforest" any woods that Henry III's grandfather Henry II had made into royal forest but that were not part of the royal domain, the text adds: "if he made his own wood forest, it shall remain forest, saving common of pasture and other things in that forest to those who were accustomed to have them previously."[29] This was in fact the typical case: most forests were managed in part as commons.

Finally, these two texts show that early English forest law, despite the ways in which it could enable the king's autocratic power, relied for much of its ordinary proceedings on the cooperation of local people. Both texts spell out the duties of local people who served as "verderers" to guard the king's "vert," or trees, and "agisters," who oversaw the pasturing of royal swine. The charter of 1217 adds a schedule of forest courts that were to be held three times a year, where the verderers and agisters would preside along with the royal foresters in hearing cases involving forest offenses. A "regard," or general inspection, was to be conducted every three years by a separate group of royal "regarders."

In practice, this system of forest law functioned more to limit and tax than to prevent the exploitation of royal forests by other parties.[30] Although the general inspections noted above were apparently conducted only occasionally, the forest courts seem to have met regularly, helping to police the forest by collecting fines and settling disputes. This specialized judicial system began to fade from view toward the end of the Middle Ages as the crown sold off many forests, but it held on in some places, as Sara Morrison's chapter shows for Sherwood Forest, into the seventeenth and eighteenth centuries.

Besides forests, new types of areas organized around the management of natural resources begin to crop up in the documents from about the eleventh century. More focused on hunting and other forms of aristocratic display than forests, these consisted of relatively small, often enclosed preserves, called variously "warrens," "parks," or "chases."[31] Typically wooded to some degree and stocked with deer or other prey animals, as well as exotics like peacocks, these carefully crafted spaces set aside, as Abigail P. Dowling's chapter in this volume discusses, a kind of (superficially) "uncultivated" landscape for elite recreation.

Finally, timber reserves, especially when mandated at the state level, created innumerable, usually small protected areas. In France, this became royal policy in the sixteenth century, when royal edicts began demanding that one-third, later reduced to one-fourth, of all woodlands be set aside as a "reserve" (*la réserve*) for the growth of timber.[32] This mandate proved difficult to enforce, not least because it was interpreted in diverse ways in different regions, as Sébastien Poublanc's chapter in this volume shows for Languedoc.

Woodland Management and Regulation in the Central Middle Ages

While protected areas are important, in preindustrial times, methods for managing woodlands depended only partly on variations in legal status or administrative policy. Similar multiuse ways of managing wooded spaces and harvesting their products may be found in village commons, seigneurial woods, royal forests, and aristocratic parks. Many scattered details about these practices already surface in the early Middle Ages (ca. 500–1000 CE). For example, estate inventories of the ninth and tenth centuries often require tenants to pay fees for pasturing their livestock in woodlands and to deliver to their lords various woodland products, including firewood, vineyard stakes, roofing shingles, construction lumber, pigs, wax, and honey.[33]

But from sometime in the eleventh or twelfth century, depending on the region, many more documents have survived, and they contain a steadily expanding range and quality of evidence. In this period of rapid population growth, two of the most common types of references to woods include grants by lords of usage rights in woodlands, often not further specified, or permission to clear them for cultivation. In England, in such northern French regions as the Ile-de-France, Normandy, and Champagne, in many German-speaking regions, and in other densely populated parts of western Europe's agriculturally rich lowlands, texts containing these authorizations proliferate across the twelfth century and through the first half of the thirteenth.[34]

Not surprisingly, the same lords who granted these authorizations also claimed the power to prohibit clearance and to limit usage. Already by the mid twelfth century, these restrictions begin to multiply. These may be considered some of the first well-documented forms of woodland protection to be applied at the scale of the village or even of specific woodlots. For example, when in 1156 Guillaume, the Count of Nevers, recorded an exchange of lands that he had made with the Cistercian monks of Pontigny, he added the stipulation that the monks could not clear four small woods that were all located very near their monastery and that were called, respectively, the woods (*nemus*) of Revisy, of Vaupayen, of St.-Etienne, and the "forest" (*foresta*) of Contais.[35]

The expanding documentation concerning limits on woods usage allows us to group such rights into two categories: wood extraction and pasturage. In the first place, grants of wood-collecting rights began to include precise limits, usually based on daily or weekly amounts, for example the "one cartload of wood" that in 1153 a donor retained for his lifetime from a gift of land to a religious community, or the "five donkey-loads of wood" that in 1154 were allotted to the priory of St.-Denis of Reims from a specific woodland.[36]

By the thirteenth century, extensive silvo-pastoral ways of using woodlands, typical of the early Middle Ages, were becoming less predominant. One of the most well-known medieval uses of the woods, as Dolly Jørgensen discusses in her chapter in this book, was for grazing pigs, especially during the fall "pannage" season, when highly nutritional acorn and beechnut mast became available. Although this practice continued into early modern times, in both England and Northern France, it became increasingly restricted in terms of the timing, numbers of animals allowed, fees charged per animal, any supplementary collecting of nuts by hand, and so on.[37] In regions subject to the most intense anthropogenic pressures, it became marginal as a form of woods usage. This is clear from the small share of revenues, typically less than 10 percent, that pannage represented in princely and royal forest accounts in Champagne and the Ile-de-France, and more generally from the pattern of regional specialization in pig raising that characterized northern France by the late Middle Ages.[38]

By the late twelfth century, limitations in terms of the quantity of wood that was allowed to be harvested had become standard, as had the distinction between firewood or other small wood (*lignum*) and lumber (*marrenum*), often further specified as a contrast between "dead wood for burning" and "live wood for building." Typically, as seen in the first quotation at the beginning of this chapter, villagers had regular access only to dead wood or to small branches of the less valuable species, while lords and sometimes the village communities themselves carefully controlled the use of timber, especially when this consisted of the two most prized species: oak and beech.[39] Virtually the same kinds of limitations and from roughly the same period are found widely across northwestern Europe, for example in both England and western Germany.[40]

In Champagne and many other densely populated regions, the outlines of a new system gradually become clear. In the evocative phrase used by French historians, the thirteenth century saw a shift from the "nourishing forest" (*la forêt nourricière*), from which people derived food and much else for themselves and their livestock directly, to the "commercial forest" (*la forêt marchande*), when markets for woodland resources grew.[41] While virtually all woodland resources became commercialized and more tightly controlled, arguably the linchpin of this development was wood production. This system produced timber, but the bulk of the demand was for the small wood that supplied fuel, fencing, vineyard stakes, and the raw material for tools, furniture, and most nonelite construction. The tools used to collect this harvest were billhooks and axes, not saws, which were in fact often prohibited.[42]

We can follow this shift in the Forêt d'Othe, in the uplands between Sens and Troyes (about 120 km southeast of Paris), as rights of pasturage were restricted and measures were taken to allow for frequent wood harvesting. Between 1193 and 1215, two nearby religious houses, the Cistericans of Vauluisant and the Knights Templars of Coulours, made a series of agreements that set precise limits on the number of pigs that could be sent into the woods, with each animal charged a fee based on its age, that excluded all grazing in meadows during the hay-growing season between March and July, and that banned all livestock from the "coppices" (*cospeicii*), "as dictated by the custom of Villemaur," which was the local "castellany" or administrative district.[43] That this protection for coppices was claimed as a custom of the entire district is indicative of the new prevalence of wood production. Indeed, as the second quotation at the beginning of this chapter shows, by 1271, Champagne's high court claimed that a similar custom applied throughout the principality, specifying an obligation to allow at least five years before reopening a cut-over woods to grazing. Similar customary norms emerged in this period across northern France.[44]

These examples also suggest that by the mid-thirteenth century, most woodland users, even privileged lords and religious communities, were subject to local custom—in a sense this era's "public" law.[45] Thus, an accord of 1245 allowed a religious community to graze pigs and other livestock in the woods of Jeugny, 20 km south of Troyes, but only when and where other local users had these rights, and all according to "the customs of the forest" (*consuetudines foresta*).[46] Of course, claims that something was customary must be taken with a grain of salt. Because rulers, lords, and owners of woods stood to gain from it, they were no doubt quick to reinforce any norms that facilitated the lucrative commerce in wood and timber. If one considers only the techniques involved, however, it seems clear that these derived not from any bureaucratic, top-down planning, but rather from community-based knowledge and practices.

One purportedly customary rule directly benefited the powerful: overlords and rulers across northern France claimed a distinctive form of "woods lordship" (*gruerie*). This included the power to authorize the "sale" of woods (i.e., contracts for commercial wood-harvesting) both of coppice and timber. This woods lordship, whose origins are obscure, but which resembles the natural-resource-oriented overlordship of forest law, is occasionally mentioned in the twelfth century and then becomes commonplace over the thirteenth.[47] Whatever it may have owed to early medieval forest customs, it also reflects the thirteenth century's growing recourse to law to express the subdivision of rights. Even as local inhabitants fought to maintain their usage rights against mounting restrictions, owners of woodlands benefitted from increasingly well-defined rights as proprietors of the soil, which in the new legalese was now called the "ground" or "bottom" (Latin *fundus*, French *treffonds*). In contrast, the trees growing on the land were distinguished as its "surface" or "tonsure" (*superficia*, *tonsure*), and it was over them that overlords and rulers claimed their right of woods lordship. The very high rates charged for the tax on wood-harvesting, which commonly ranged between 10 and 50 percent of the total revenue, shows that in effect, trees were subject to shared ownership.[48] Although motivated by material gain rather than ecology, this legal custom may be considered an original, "vertical" form of woodland conservation.

Intensive Harvesting under Late Medieval French Forest Administrations

All of the different ways of managing and limiting woodland usage that I have reviewed so far have been well known for years, at least to many medieval historians. Other aspects of woodland management that also emerged by the thirteenth or fourteenth centuries are less familiar to historians. These consist especially of harvesting methods, which determine much about how woods will subsequently regrow, thereby enabling the careful management that is necessary in order to couple high productivity with sustainability over the long term.

Here, I will note just two, interrelated methods. The first is a form of planned harvesting that divides a woodland into a number of parcels, each of which represents one year's harvest. The number of parcels is determined by the total number of years in that woodland's harvesting cycle. An early example of this approach dates to 1219, when the abbey of Molesme sold to two men the right to cut 1000 arpents of its woods of Rumilly, about 20 km south of Troyes, over a period of ten years. The contract also requires that the two woodmongers were to cut any one part of the woods only once over these ten years, so presumably, they would cut one parcel of about 100 arpents each year. A coppicing cycle of ten years was in fact quite typical in this area, so it is quite possible that these woods were already being harvested on a continuous rotation. Other thirteenth-century cutting contracts add that woodcutters should move from one contiguous parcel to another, a procedure that facilitated supervision and allowed for easy access to the next year's parcel. It also gave this form of intensive harvesting its standard early-modern French name, which emphasized the idea of "contiguous cutting" (*coupe à tire et aire*).[49]

A second technique was often coupled with the first, because it likewise required harvesting within carefully delimited parcels. It consisted of the retention, at each cycle of harvesting of smaller trees, of a limited number of taller trees—these retained trees were called "standards" or "staddles" (*étalons* or *baliveaux*). Both of these techniques, contiguous cutting and the retention of standards, could be applied in woods exclusively devoted to timber. In the Middle Ages, however, these methods are best documented in mixed woodlands that combined coppice and timber, a system that was called "coppice under standards" (*taillis sous futaie*). The retention of standard trees may be considered as a type of selective thinning, even if the trees left standing were usually scattered rather widely. The precise number of trees retained varied according to many parameters, both ecological and economic. Moreover, after a first set of standards were retained and grew through another harvest cycle, some of them might be cut while others were again reserved, thus creating two or more age-classes of taller trees within a given woodlot. Nor was the density of taller trees merely a technical matter.[50] As Sébastien Poublanc shows in his chapter in this book, peasant communities typically wanted more open woodlands that favored both the production of small wood and grazing, while lords often preferred densities that would close the forest canopy and shade out coppices and pasture.

Although this kind of complex micromanagement of woodlots is, as far as I know, only documented thanks to administrative institutions of some importance, it long predates the early modern bureaus of state forestry with which it is often identified. My research on Champagne shows that systems of this kind were already in operation by the mid-thirteenth century. Over extensive areas of this region, both comital and lordly woodlands were harvested on rotations varying between four and twenty years for small wood, and between forty and one hundred years for timber. Given that subsequently very similar rotations continued to be applied in these woodlands, which retained roughly the same boundaries for more than five hundred years, these practices may be considered highly sustainable.[51]

Although in other regions such detailed evidence is rare for the thirteenth century, by the fourteenth, more examples become available. One well-studied case is the Count of Flanders' forest of Nieppe, about 30 km west of Lille, where penalties for damage to both coppices and standards (*estalons*) of oak are recorded from about 1300. From the late fourteenth century, account books show that these woodlands were exploited through a tightly organized system of coppice-under-standards, with the coppice cut every twenty to twenty-five years and with three or four age-classes of oak timber.[52] Just as in the case of Champagne, these practices owed nothing to any known learned study of forestry.

By the late thirteenth and early fourteenth centuries, the French monarchy had also established its bureau of "Waters and Forests" (*Eaux et Forêts*) through a series of edicts that focused mostly on questions of jurisdiction and personnel. These decrees inaugurated an extremely long-lived tradition of royal proclamations whose core of woodland management techniques scarcely evolved, even as its administrative content was repeatedly revised and expanded throughout the preindustrial era. As the above examples from Champagne and Nieppe show, it is clear that the royal forestry administration was not innovating when it mandated the retention of standards in an ordinance of 1376:

> Because in times past the masters of the forests, when making their sales of the forest, forgot, by inadvertence or otherwise, to retain standards to repopulate the forest, . . . it is ordered that from now on that eight or ten standards should be retained in each arpent that is sold, and this the masters must include in their letter [of sale], so that the merchants cannot make any excuses.[53]

Rather than seeing this primarily as a first step in a long process of technical and regulatory progress, as state-centered traditions of forest history would have it,[54] it would make better historical sense to emphasize that in this provision, the bureau of Waters and Forests was making explicit a customary norm.

The best-documented of all late medieval forest administrations may be the Duke of Burgundy's bureau of Waters and Forests, headquartered in Dijon and modeled on the similar royal office. This forestry service has left behind rich archives for the period between about 1350 and

1500, superbly studied in a recent monograph by Corrine Beck. Wood sales usually made up about half of the total revenues from ducal forests, and in those woods subject to regular harvesting, we find the same patterns noted above for Champagne, Nieppe, and the royal domain, including coppice under standards, the reservation of specified numbers of standards, and often a few different age-classes of taller trees.[55]

The Burgundian records also provide more explicit evidence that, at least in some woodlots, harvests were organized by proceeding systematically from one contiguous plot to another. And from the mid-fifteenth century, localized efforts to replant oak woodlands were begun, both by sowing acorns and by planting saplings. It is true that medieval forest administrations lacked the kind of unifying vision that would emerge under Jean-Baptiste Colbert, Louis XIV's financial minister and author of the French forest ordinance of 1669. But as Beck argues, all the basic forest know-how on which he drew—and then some, because his ordinance did not include the replanting of trees—was already available by the late Middle Ages.[56]

Conclusion

Forest and environmental historians have overlooked the importance of much of the early evidence for woodland conservation reviewed above. This is understandable when it comes to such techniques as the retaining of "standard" trees in mixed-age woodlands, which is apparently not well documented in many regions before the fifteenth or sixteenth centuries. But this neglect is more surprising in the case of the abundant, long-familiar evidence of twelfth- and thirteenth-century woodland protections.

Among the possible reasons for the scholarly neglect of central-medieval woodland conservation, I can suggest two. The first is that before the full significance and prevalence of preindustrial commons, or commons-like ways of using resources, became better appreciated in the wake of work by Elinor Ostrom and other social scientists,[57] it was easy to dismiss many early protections because of their lack of an explicitly "public" dimension. While this kind of ideological packaging is by no means an essential criterion for sustainable management, it is true that it was only toward the end of the Middle Ages that many governments began invoking their role as public authorities in order to legitimate their natural resources policies.

However, recent research has found that this tradition began quite early. Thus, Richard Hoffmann has pointed to the impressive body of what he calls "late medieval state environmental regulation."[58] From the thirteenth century onward, urban councils, principalities, and kingdoms all began to justify their expanded regulatory efforts with claims that they were acting in "the public interest," or for "the common good" (for example, in order to safeguard fish stocks).[59] By the early sixteenth century, similar state oversight and its legitimation in terms of the "public good" were extended to woodlands, when rulers in both France and Spain began to speak of their edicts concerning forests in this way.[60]

A second possible reason for the relative neglect of twelfth- and thirteenth-century woodland conservation may be the centrality of Germany in traditional forest history, coupled with the timing of the central medieval cycle of economic and demographic growth. This surge in human numbers, which led to widespread land clearance and woodland regulation, began first as early as the eleventh century in southern and western Europe, including many parts of Italy, Spain, France, and England. This phase of growth gradually extended eastward into central Europe, affecting many German-speaking lands somewhat later than those further west.[61] Although in many parts of western and southern Germany, localized woodland regulations have survived from as early as the late eleventh century, the early evidence is sparse. Woodland regulations survive in more significant numbers for many Germanophone lands only from the late thirteenth century and, especially, from the fourteenth and fifteenth centuries.[62]

Moreover, by the late Middle Ages, the story of woodland regulation in many German-speaking regions becomes inseparable from both the growth of the mining and metallurgical sectors and increasing state control over forests. As Siegfried Epperlein noted, the earlier, more locally based phase of woodland regulation is usually mentioned, if at all, only as a prelude to the more significant story of the fifteenth and later centuries.[63]

The problem with minimizing the central-medieval phase of woodland management and protection is less about dating these phenomena than it is about understanding their character and the kinds of contexts in which they developed. The needs of villages and small towns for small wood and, to a lesser degree, lumber were ultimately more pervasive and important for the origins of conservative woodland management than those of princes, industrial users, emerging nation–states, or colonial enterprises. Instead, the most typical contexts for preindustrial environmental regulation were various kinds of lordships and commons, and over the later Middle Ages and into early modern times, most of these overlapped with or were contained within village territories. Preindustrial woodland conservation emerged first and was most securely anchored at the local level.

It remains true, however, that administrative sources often provide the best written evidence for early conservation. To conclude, I turn to one of them, an ordinance of May 1346, in which King Philip VI of France ordered the officers of the bureau of Waters and Forests "to make inquiry and visit all the forests and woods where there are or will be sales, ensuring that the said forests and woods can be perpetually sustained in good condition."[64] This statement sums up key principles that animated most preindustrial conservation: resources were regulated to ensure they could continue to be exploited—"sustainably"—for human needs.

Richard Keyser is Senior Lecturer in Legal Studies in the Center for Law, Society, and Justice at the University of Wisconsin-Madison, where he teaches classes on American legal history, law and environment, and related topics. Trained in European history (PhD Johns Hopkins University, 2001), he has published articles on the legal, religious, and environmental history of medieval France and England. His 2009 essay on "The Transformation of Traditional Woodland Management" won the American Society for Environmental History's Alice Hamilton prize.

Notes

1. On these efforts, and on the use of the term "conservation," see this volume's Introduction. For the above epigraph, quote one: Henri d'Arbois de Jubainville, *Histoire des ducs et comtes de Champagne depuis le VIe siècle jusqu'à la fin du XIe*. 8 vols. (Paris: A. Durand, 1859–69), 3: 480–2 (pièce justificatif no. 166): "Homines de Fay in nemoribus de Perchoi et de Bayse habent omne usuarium et nemus mortuum; si voluerint, in lignis Trecis ad vendendum ducere poterunt. Abbatissa similiter in eisdem nemoribus usum suum habet ad ardendum. Sciendum tamen, quod alicui hominem de Fay scindere non licebit neque fagum neque quercum majoris grossitudinis, quem manu claudere possit, nisi pro marremio faciendo aut ad edificationem domum

suarum aut ad reficiendum. Nec hoc etiam ipsis licebit, nisi hoc prius majori vel forestario domini Sancti Fidoli dixerint." Quote two: Paulette Portejoie, ed., *L'ancien coutumier de Champagne (XIIIe siècle)* (Poitiers: Oudin, 1956), 174 (article XXI): "Encor us'on en Champagne que on doit garder les revenues des forez et des bois .v. anz continuez contre toutes gens qui ont vainne pasture. Et si li bois seoit en mauvais treffons, on y pourroit mettre respite .i. an ou .ii., par quoy li bois se puest deffendre suffisenment."

2. Georges Duby, *Rural Economy and Country Life in the Medieval West* (Columbia: University of South Carolina Press, 1968), 141–45, 159–61; Siegfried Epperlein, *Waldnutzung, Waldstreitigkeiten und Waldschutz in Deutschland im hohen Mittelalter, 2. Hälfte 11. Jahrhundert bis ausgehendes 14. Jahrhundert* (Stuttgart: Franz Steiner Verlag, 1993).
3. Charles R. Young, "Conservation Policies in the Royal Forests of Medieval England," *Albion* 10, no. 2 (1978): 95–103; Jules N. Pretty, "Sustainable Agriculture in the Middle Ages: The Medieval Manor," *Agricultural History Review* 38 (1990): 14–16; and James Latham, "The Development of Forest Conservation in Europe," in *Europe's Changing Woods and Forests: From Wildwood to Managed Landscapes*, ed. Keith J. Kirby and Charles Watkins (Boston: CAB International, 2015), 309–25.
4. Prominent recent books on sustainability in preindustrial Europe focus on later periods, without discussing this early evidence: Ulrich Grober, *Sustainability: A Cultural History* (Totnes, UK: Green Books, 2012); Jeremy Caradonna, *Sustainability: A History* (New York: Oxford University Press, 2014); and Paul Warde, *The Invention of Sustainability: Nature and Destiny, c. 1500–1870* (New York: Cambridge University Press, 2018).
5. *Les Eaux et Forêts du 12e au 20e siècle* (Paris: Centre National de la Recherche Scientifique, 1987), 77–82, 113–16; Xavier Rochel, "Forest Management and Species Composition: A Historical Approach in Lorraine, France," in *Europe's Changing Woods and Forests*, ed. Kirby and Watkins, 279–89.
6. Michael Williams, *Deforesting the Earth: From Prehistory to Global Crisis* (Chicago: University of Chicago Press, 2003), 203. See also this volume's Introduction.
7. Clarence Glacken, *Traces on the Rhodian Shore: Nature and Culture in Western Thought from Ancient Times to the End of the Eighteenth Century* (Berkeley: University of California Press, 1967), 491; Williams, *Deforesting the Earth*, 204; cf. Peter Savill, "High Forest Management and the Rise of Even-Aged Stands," in *Europe's Changing Woods and Forests*, ed. Kirby and Watkins, 94.
8. Jérôme Buridant, "Du 'modèle' à la pratique: La gestion des peuplements caducifoliés dans la France moderne, XVIe–XVIIIe siècle," *Flaran* 24 (2002): 203–20; Richard Keyser, "The Transformation of Traditional Woodland Management: Commercial Sylviculture in Medieval Champagne," *French Historical Studies* 32 (2009): 353–84; Peter Buckley and Jenny Mills, "Coppice Silviculture: From the Mesolithic to the 21st Century," in *Europe's Changing Woods and Forests*, ed. Kirby and Watkins, 77–92.

9. Charles R. Menzies, ed., *Traditional Ecological Knowledge and Natural Resource Management* (Lincoln: University of Nebraska Press, 2006).
10. For England and France, see John Manwood, *A Treatise and Discourse of the Lawes of the Forrest* (London: Thomas Wight and Bonham Norton, 1598); and Louis de Saint-Yon, *Les edicts et ordonnances des roys, coustumes des provinces, reglemens, arrests et jugemens notables, des eaues et forets* (Paris: Veuve Abel Langelier, 1610).
11. S. Ravi Rajan, *Modernizing Nature: Forestry and Imperial Eco-Development, 1800–1950* (New York: Oxford University Press, 2006); and K. Jan Oosthoek and Richard Hölzl, eds., *Managing Northern Europe's Forests: Histories from the Age of Improvement to the Age of Ecology* (New York: Berghahn Books, 2018).
12. Oliver Rackham, *Ancient Woodland: Its History, Vegetation, and Uses in England* [1980], 2nd ed. (Colvend: Castlepoint Press, 2003); Joachim Radkau, "Holzverknappung und Krisenbewußtsein im 18. Jahrhundert," *Geschichte und Gesellschaft* 9, no. 4 (1983): 513–43; and Andrée Corvol, *L'homme et l'arbre sous l'Ancien Régime* (Paris: Economica, 1984).
13. Keith J. Kirby and Charles Watkins, eds., *Europe's Changing Woods and Forests: From Wildwood to Managed Landscapes* (Boston: CAB International, 2015).
14. Buckley and Mills, "Coppice Silviculture," 77–92; Savill, "High Forest Management," 93–115.
15. Tibor Hartel and Tobias Plieninger, eds., *European Wood-Pastures in Transition* (New York: Routledge, 2014).
16. Dolly Jørgensen, "The Roots of the English Royal Forest," *Anglo-Norman Studies* 32 (2009): 114–28; and Ellen F. Arnold, *Negotiating the Landscape: Environment and Monastic Identity in the Medieval Ardennes* (Philadelphia: University of Pennsylvania Press, 2013), 22–30, 46–56.
17. Recent books in English include: Bo Fritzbøger, *"A Windfall for the Magnates": The Development of Woodland Ownership in Denmark c. 1150–1830* (Odense: University Press of Southern Denmark, 2004); Péter Szabó, *Woodland and Forests in Medieval Hungary*, British Archaeological Reports, International Series 1348 (Oxford: Archaeopress, 2005); Paul Warde, *Ecology, Economy, and State Formation in Early Modern Germany* (New York: Cambridge University Press, 2006); Karl Appuhn, *Forest on the Sea: Environmental Expertise in Renaissance Venice* (Baltimore: Johns Hopkins University Press, 2009); Arnold, *Negotiating the Landscape*; Kieko Matteson, *Forests in Revolutionary France: Conservation, Community, and Conflict, 1669–1848* (New York: Cambridge University Press, 2015); and John T. Wing, *Roots of Empire: Forests and State Power in Early Modern Spain, c. 1500–1750* (Boston: Brill, 2015).
18. Appuhn, *Forest on the Sea*, 112.
19. J. Donald Hughes, *Environmental Problems of the Greeks and Romans: Ecology in the Ancient Mediterranean*, 2nd ed. (Baltimore: Johns Hopkins University Press, 2014), 183–95.

20. John Howe, "Creating Symbolic Landscapes: Medieval Development of Sacred Space," in *Inventing Medieval Landscapes: Senses of Place in Western Europe*, ed. John Howe and Michael Wolfe (Gainesville: University Press of Florida, 2002), 208–23; Latham, "Development of Forest Conservation," 309–11.
21. Barbara Rosenwein, *Negotiating Space: Power, Restraint, and Privileges of Immunity in Early Medieval Europe* (Ithaca: Cornell University Press, 1999); Arnold, *Negotiating the Landscape*; Martine Chalvet, "La forêt domaniale de la Sainte-Baume: un espace exceptionnel et protégé en Provence," *Les Cahiers de Framespa* 13 (2013), accessed 9 August 2019, http://journals.openedition.org/framespa/2250; DOI: 10.4000/framespa.2250.
22. Jan F. Niermeyer and C. Van de Kieft, eds., *Mediae Latinitatis Lexicon Minus*, ed. J. W. J. Burgers, 2 vols. (Boston: Brill, 2002).
23. Rackham, *Ancient Woodland*, 175–88; Jørgensen, "Roots of the English Royal Forest"; Arnold, *Negotiating the Landscape*, 22–30, 46–56; and Szabó, *Woodland and Forests*, 20–24, 87–92.
24. Cf. Louis Warren, "Owning Nature: Toward an Environmental History of Private Property," in *The Oxford Handbook of Environmental History*, ed. Andrew C. Isenberg (New York: Oxford University Press, 2014), 398–424.
25. William Stubbs, ed., *Select Charters and Other Illustrations of English Constitutional History* (Oxford: Clarendon Press, 1870), 150–51; David C. Douglas and George W. Greenaway, eds., *English Historical Documents*, vol. 2: *c. 1042–1189*, 2nd ed. (New York: Oxford University Press, 1981), 450–52. See Charles R. Young, *The Royal Forests of Medieval England* (Philadelphia: University of Pennsylvania Press, 1979), 27–32.
26. Young, "Conservation Policies in the Royal Forests."
27. Peter Linebaugh, *The Magna Carta Manifesto: Liberties and Commons for All* (Berkeley: University of California Press, 2009), 21–45.
28. Young, *Royal Forests*, 60–73.
29. Stubbs, *Select Charters*, 339: "Et si boscum suum proprium afforestaverit, remaneat foresta, salva communa de herbagio et aliis in eadem foresta illis qui eam prius habere consueverunt"; Harry Rothwell, ed., *English Historical Documents*, vol. 3: *1189–1327* (London: Eyre & Spottiswoode, 1975), 337–40.
30. Young, "Conservation Policies."
31. John Fletcher, "The Impact of Hunting on European Woodland from Medieval to Modern Times," in *Europe's Changing Woods and Forests*, ed. Kirby and Watkins, 116–26.
32. For Venetian parallels, see Appuhn, *Forest on the Sea*, 101, 111, 263.
33. Arnold, *Negotiating the Landscape*, 63–68.
34. Duby, *Rural Economy*, 61–87; Epperlein, *Waldnutzung*; Danny Lake-Giguère, "La livrée et le contrôle des droits d'usage dans les forêts royales de Normandie du XIIe au XIVe siècle," *Annales de Normandie* 67 (2017): 27–51.
35. Martine Garrigues, ed., *Le premier cartulaire de l'abbaye cistercienne de Pontigny (XIIe-XIIIe siècles)* (Paris: Bibliothèque Nationale, 1981), 155.

36. John Benton and Michel Bur, eds., *Recueil des actes d'Henri le Libéral, Comte de Champagne (1152–1181)* (Paris: De Boccard, 2009), 54–55: *solam bigam*, in land at Arembercourt (Département de l'Aube); and 87–88: *ad ligna . . . V asinos*, in the *nemus* at Nanteuil-la-Forêt (Département de la Marne).
37. For England, see Jean Birrell, "Common Rights in the Medieval Forest: Disputes and Conflicts in the Thirteenth Century," *Past and Present* 117 (1987): 37–40.
38. Keyser, "Transformation of Traditional Woodland Management," 378; Jean-Marc Moriceau, *Histoire et géographie de l'élevage français, du Moyen Âge à la Révolution* (Paris: Fayard, 2005), 345–52.
39. Beech = *Fagus sylvatica*; the most common species of oak are sessile (*Quercus petraea*) and pedunculate (*Q. robur*).
40. Birrell, "Common Rights," 27–36; and Bernd Fuhrmann, "Holzversorgung, Waldentwicklung, Umweltveränderungen und wirtschaftliche Tendenzen in Spätmittelalter und beginnender Neuzeit," *Vierteljahrschrift für Sozial- und Wirtschaftsgeschichte* 100 (2013): 311–27.
41. Corrine Beck, *Les eaux et forêts en Bourgogne ducale (vers 1350–vers 1480): Société et biodiversité* (Paris: L'Harmattan, 2008), 125–225, quoted at 159.
42. Christiane Raynaud, *A la hache! Histoire et symbolique de la hache dans la France médiévale (XIIIe–XVe siècles)* (Paris: Le Léopard d'Or, 2002), 102–26.
43. William O. Duba, ed., "The Cartulary of Vauluisant: A Critical Edition," (MA Thesis, University of Iowa, 1994), 364–66 (Aug. 1193), 322–24 (1215): *custodient pecora sua ab ingressu cospeiciorum secundum consuetudinem Villemauri.*
44. Keyser, "Transformation of Traditional Woodland Management," 375–76.
45. John Hudson, "Customs, Laws and the Interpretation of Medieval Law," in *Custom: The Development and Use of a Legal Concept in the Middle Ages*, ed. Per Andersen and Mia Münster-Swendsen (Copenhagen: DJOF Publishing, 2009), 1–16.
46. Charles Lalore, ed., *Cartulaire de Montier-la-Celle* (Paris: Thorin, 1882), 137–38.
47. Michel Devèze, *La vie de la forêt française au XVIe siècle*, 2 vols. (Paris: Ecole Pratique des Hautes Etudes, 1961), 1: 59–61, 165–67, 173–81; Heinrich Rubner, *Untersuchungen zur Forstverfassung des mittelalterlichen Frankreichs* (Wiesbaden: Steiner, 1965), 19–42; Michel Bur, *La formation du comté de Champagne, v. 950–v. 1150* (Nancy: Université de Nancy, 1977), 292–305, 322–38.
48. Keyser, "Transformation of Traditional Woodland Management," 371.
49. Keyser, "Transformation of Traditional Woodland Management," 372.
50. Rochel, "Forest Management."
51. Richard Keyser, "Wood for Burning: The Continuity of Woodland Management in Medieval and Early Modern France," in *Environmental History in the Making*, vol. 1: *Explaining*, ed. Estelita Vaz, Cristina Joanaz de Melo, and Lígia M. Costa Pinto (Basel, Switzerland: Springer Publishing, 2017), 307–20.
52. Monique Sommé, "Règlements, délits et organization des ventes dans la forêt de Nieppe (début XIVe-début XVIe siècle)," *Revue du Nord* 72 (1990): 511–28.

53. François-André Isambert et al., eds., *Recueil général des anciennes lois françaises depuis l'an 420 jusqu'a la Révolution de 1789*, 29 vols. (Paris: Belin-Leprieur, Plon, 1826–1833), 5: 462 (article 21): "Et pour ce que ou temps passé, les maistres en faisant et vendant ventes de bois, ont par inadvertence ou autrement, oublié à faire retenue de baiviaulz ou estallons, pour la repueple des forez . . . est ordené que d'oresenavant en toutes ventes qui seront faites sera entendue la retenue des bayveaulx ou estallons de dix ou huit en l'arpent; et ce seront tenus les maistres de mettre en leurs letres, par quoy les marchans ni puissent trouver excusation."
54. Devèze, *La vie de la forêt française*, 1: 64–67, 128, 175–80; 2:77, 91–93; *Les Eaux et Forêts du 12e au 20e siècle*, 8, 29–80, 114–16, 121–22; I have been unable to consult Anne-Marie Bocquillon, *Le roi dans ses forêts de Cuise, Laigue, Retz du XIIIe au XVe siècle* (PhD thesis, Université de Paris 1, 2000).
55. Beck, *Les eaux et forêts en Bourgogne*, 161–62.
56. Beck, *Les eaux et forêts en Bourgogne*, 143–74.
57. Elinor Ostrom, *Governing the Commons: The Evolution of Institutions for Collective Action* (New York: Cambridge University Press, 1990).
58. Richard Hoffmann, *An Environmental History of Medieval Europe* (New York: Cambridge University Press, 2014), 265.
59. See Dolly Jørgensen, "'All Good Rule of the Citee': Sanitation and Civic Government in England, 1400–1600," *Journal of Urban History* 36 (2010): 300–15; and the chapter in this volume by Richard Hoffmann.
60. For a French example, see the Introduction to this volume. For Spain, see Wing, *Roots of Empire*, 2–3, 19–29, 38, 44, 54–56, 68, etc.
61. Robert Bartlett, *The Making of Europe: Conquest, Colonization, and Cultural Change, 950 - 1350* (Princeton: Princeton University Press, 1993), 106–66.
62. Kurt Mantel, *Forstgeschichte des 16. Jahrhunderts unter dem Einfluß der Forstordnungen und Noe Meurers* (Berlin: Universität Freiburg im Breisgau, 1980), 873–923; Epperlein, *Waldnutzung*.
63. Epperlein, *Waldnutzung*, 9–11. For recent examples, see Joachim Radkau, *Wood: A History*, trans. Patrick Camiller (Malden, MA: Polity Press, 2012), 56–134; and Fuhrmann, "Holzversorgung."
64. Isambert, *Recueil général*, 4: 523, no. 133 (Brunay, 29 May 1346, paragraph 4): "Les mestres des forez dessudiz, selon ce qu'il sont ordenez, enquerront et visiteront toutes les forez et bois qui y sont, et seront les ventes, qui y sont à faire, eû regart à ce que lesdittes forez et bois se puissent perpetuellement soustenir en bon estat."

Bibliography

Appuhn, Karl. *Forest on the Sea: Environmental Expertise in Renaissance Venice*. Baltimore: Johns Hopkins University Press, 2009.

Arnold, Ellen F. *Negotiating the Landscape: Environment and Monastic Identity in the Medieval Ardennes*. Philadelphia: University of Pennsylvania Press, 2013.

Bartlett, Robert. *The Making of Europe: Conquest, Colonization, and Cultural Change, 950–1350*. Princeton: Princeton University Press, 1993.

Beck, Corrine. *Les eaux et forêts en Bourgogne ducale (vers 1350–vers 1480): Société et biodiversité*. Paris: L'Harmattan, 2008.

Benton, John, and Michel Bur, eds. *Recueil des actes d'Henri le Libéral, Comte de Champagne (1152–1181)*. Paris: De Boccard, 2009.

Birrell, Jean. "Common Rights in the Medieval Forest: Disputes and Conflicts in the Thirteenth Century." *Past and Present* 117 (1987): 22–49.

Bocquillon, Anne-Marie. *Le roi dans ses forêts de Cuise, Laigue, Retz du XIIIe au XVe siècle*. PhD thesis, Université de Paris 1, 2000. (Not consulted.)

Buckley, Peter, and Jenny Mills. "Coppice Silviculture: From the Mesolithic to the 21st Century." In *Europe's Changing Woods and Forests: From Wildwood to Managed Landscapes*, edited by Keith J. Kirby and Charles Watkins, 77–92. Boston: CAB International, 2015.

Bur, Michel. *La formation du comté de Champagne, v. 950–v. 1150*. Nancy: Université de Nancy, 1977.

Buridant, Jérôme. "Du 'modèle' à la pratique: La gestion des peuplements caducifoliés dans la France moderne, XVIe–XVIIIe siècle." *Flaran* 24 (2002): 203–20.

Caradonna, Jeremy. *Sustainability: A History*. New York: Oxford University Press, 2014.

Chalvet, Martine. "La forêt domaniale de la Sainte-Baume: un espace exceptionnel et protégé en Provence." *Les Cahiers de Framespa* 13 (2013). Retrieved 9 August 2019. http://journals.openedition.org/framespa/2250; DOI: 10.4000/framespa.2250.

Corvol, Andrée. *L'homme et l'arbre sous l'Ancien Régime*. Paris: Economica, 1984.

D'Arbois de Jubainville, Henri. *Histoire des ducs et comtes de Champagne depuis le VIe siècle jusqu'à la fin du XIe*. 8 vols. Paris: A. Durand, 1859–69.

De Saint-Yon, Louis. *Les edicts et ordonnances des roys, coustumes des provinces, reglemens, arrests et jugemens notables, des eaues et forets*. Paris: Veuve Abel Langelier, 1610.

Devèze, Michel. *La vie de la forêt française au XVIe siècle*. 2 vols. Paris: Ecole Pratique des Hautes Etudes, 1961.

Douglas, David C., and George W. Greenaway, eds. *English Historical Documents*. Vol. 2: *c. 1042–1189*, 2nd ed. New York: Oxford University Press, 1981.

Duba, William O., ed. "The Cartulary of Vauluisant: A Critical Edition." MA Thesis, University of Iowa, 1994.

Duby, Georges. *Rural Economy and Country Life in the Medieval West*. Columbia: University of South Carolina Press, 1968.

Epperlein, Siegfried. *Waldnutzung, Waldstreitigkeiten und Waldschutz in Deutschland im hohen Mittelalter, 2. Hälfte 11. Jahrhundert bis ausgehendes 14. Jahrhundert*. Stuttgart: Franz Steiner Verlag, 1993.

Fletcher, John. "The Impact of Hunting on European Woodland from Medieval to Modern Times." In *Europe's Changing Woods and Forests: From Wildwood to Managed Landscapes*, ed. Keith J. Kirby and Charles Watkins, 116–26. Boston: CAB International, 2015.

Fritzbøger, Bo. *"A Windfall for the Magnates": The Development of Woodland Ownership in Denmark c. 1150–1830*. Odense: University Press of Southern Denmark, 2004.

Fuhrmann, Bernd. "Holzversorgung, Waldentwicklung, Umweltveränderungen und wirtschaftliche Tendenzen in Spätmittelalter und beginnender Neuzeit." *Vierteljahrschrift für Sozial- und Wirtschaftsgeschichte* 100 (2013): 311–27.

Garrigues, Martine, ed. *Le premier cartulaire de l'abbaye cistercienne de Pontigny (XIIe-XIIIe siècles)*. Paris: Bibliothèque Nationale, 1981.

Glacken, Clarence. *Traces on the Rhodian Shore: Nature and Culture in Western Thought from Ancient Times to the End of the Eighteenth Century*. Berkeley: University of California Press, 1967.

Grober, Ulrich. *Sustainability: A Cultural History*. Totnes, UK: Green Books, 2012.

Hartel, Tibor, and Tobias Plieninger, eds. *European Wood-Pastures in Transition*. New York: Routledge, 2014.

Hoffmann, Richard. *An Environmental History of Medieval Europe*. New York: Cambridge University Press, 2014.

Howe, John. "Creating Symbolic Landscapes: Medieval Development of Sacred Space." In *Inventing Medieval Landscapes: Senses of Place in Western Europe*, edited by John Howe and Michael Wolfe, 208–23. Gainesville: University Press of Florida, 2002.

Hudson, John. "Customs, Laws and the Interpretation of Medieval Law." In *Custom: The Development and Use of a Legal Concept in the Middle Ages*, edited by Per Andersen and Mia Münster-Swendsen, 1–16. Copenhagen: DJOF Publishing, 2009.

Hughes, J. Donald. *Environmental Problems of the Greeks and Romans: Ecology in the Ancient Mediterranean*, 2nd ed. Baltimore: Johns Hopkins University Press, 2014.

Isambert, François-André, et al., eds. *Recueil général des anciennes lois françaises depuis l'an 420 jusqu'à la Révolution de 1789*. 29 vols. Paris: Belin-Leprieur, Plon, 1826–1833.

Jørgensen, Dolly. "'All Good Rule of the Citee': Sanitation and Civic Government in England, 1400–1600." *Journal of Urban History* 36 (2010): 300–15.

_____. "The Roots of the English Royal Forest." *Anglo-Norman Studies* 32 (2009): 114–28.

Keyser, Richard. "The Transformation of Traditional Woodland Management: Commercial Sylviculture in Medieval Champagne." *French Historical Studies* 32 (2009): 353–84.

_____. "Wood for Burning: The Continuity of Woodland Management in Medieval and Early Modern France." In *Environmental History in the Making*. Vol. 1:

Explaining, edited by Estelita Vaz, Cristina Joanaz de Melo, and Lígia M. Costa Pinto, 307–20. Basel, Switzerland: Springer Publishing, 2017.

Kirby, Keith J., and Charles Watkins, eds. *Europe's Changing Woods and Forests: From Wildwood to Managed Landscapes*. Boston: CAB International, 2015.

Lake-Giguère, Danny. "La livrée et le contrôle des droits d'usage dans les forêts royales de Normandie du XIIe au XIVe siècle." *Annales de Normandie* 67 (2017): 27–51.

Lalore, Charles, ed. *Cartulaire de Montier-la-Celle*. Paris: Thorin, 1882.

Latham, James. "The Development of Forest Conservation in Europe." In *Europe's Changing Woods and Forests: From Wildwood to Managed Landscapes*, edited by Keith J. Kirby and Charles Watkins, 309–25. Boston: CAB International, 2015.

Les Eaux et Forêts du 12e au 20e siècle. Paris: Centre National de la Recherche Scientifique, 1987.

Linebaugh, Peter. *The Magna Carta Manifesto: Liberties and Commons for All*. Berkeley: University of California Press, 2009.

Longnon, Auguste, ed. *Documents relatifs au Comté de Champagne et de Brie, 1172–1361*. 3 vols. Paris: Imprimerie Nationale, 1901–14.

Mantel, Kurt. *Forstgeschichte des 16. Jahrhunderts unter dem Einfluß der Forstordnungen und Noe Meurers*. Berlin: Universität Freiburg im Breisgau, 1980.

Manwood, John. *A Treatise and Discourse of the Lawes of the Forrest*. London: Thomas Wight and Bonham Norton, 1598.

Matteson, Kieko. *Forests in Revolutionary France: Conservation, Community, and Conflict, 1669–1848*. New York: Cambridge University Press, 2015.

Menzies, Charles R., ed. *Traditional Ecological Knowledge and Natural Resource Management*. Lincoln: University of Nebraska Press, 2006.

Moriceau, Jean-Marc. *Histoire et géographie de l'élevage français, du Moyen Âge à la Révolution*. Paris: Fayard, 2005.

Niermeyer, Jan F., and C. Van de Kieft, eds. *Mediae Latinitatis Lexicon Minus*, edited by J. W. J. Burgers. 2 vols. Boston: Brill, 2002.

Oosthoek, K. Jan, and Richard Hölzl, eds. *Managing Northern Europe's Forests: Histories from the Age of Improvement to the Age of Ecology*. New York: Berghahn Books, 2018.

Ostrom, Elinor. *Governing the Commons: The Evolution of Institutions for Collective Action*. New York: Cambridge University Press, 1990.

Portejoie, Paulette, ed. *L'ancien coutumier de Champagne (XIIIe siècle)*. Poitiers: Oudin, 1956.

Pretty, Jules N. "Sustainable Agriculture in the Middle Ages: The Medieval Manor." *Agricultural History Review* 38 (1990): 1–19.

Rackham, Oliver. *Ancient Woodland: Its History, Vegetation, and Uses in England* [1980], 2nd ed. Colvend: Castlepoint Press, 2003.

Radkau, Joachim. "Holzverknappung und Krisenbewußtsein im 18. Jahrhundert." *Geschichte und Gesellschaft* 9, no. (1983): 513–43.

_____. *Wood: A History*. Translated by Patrick Camiller. Malden, MA: Polity Press, 2012.

Rajan, S. Ravi. *Modernizing Nature: Forestry and Imperial Eco-Development, 1800–1950*. New York: Oxford University Press, 2006.

Raynaud, Christiane. *A la hache! Histoire et symbolique de la hache dans la France médiévale (XIIIe–XVe siècles)*. Paris: Le Léopard d'Or, 2002.

Rochel, Xavier. "Forest Management and Species Composition: A Historical Approach in Lorraine, France." In *Europe's Changing Woods and Forests: From Wildwood to Managed Landscapes*, edited by Keith J. Kirby and Charles Watkins, 279–89. Boston: CAB International, 2015.

Rosenwein, Barbara. *Negotiating Space: Power, Restraint, and Privileges of Immunity in Early Medieval Europe*. Ithaca: Cornell University Press, 1999.

Rothwell, Harry, ed. *English Historical Documents*. Vol. 3: *1189–1327*. London: Eyre & Spottiswoode, 1975.

Rubner, Heinrich. *Untersuchungen zur Forstverfassung des mittelalterlichen Frankreichs*. Wiesbaden: Steiner, 1965.

Savill, Peter. "High Forest Management and the Rise of Even-Aged Stands." In *Europe's Changing Woods and Forests: From Wildwood to Managed Landscapes*, edited by Keith J. Kirby and Charles Watkins, 93–106. Boston: CAB International, 2015

Sommé, Monique. "Règlements, délits et organization des ventes dans la forêt de Nieppe (début XIVe-début XVIe siècle)." *Revue du Nord* 72 (1990): 511–28.

Stubbs, William, ed. *Select Charters and Other Illustrations of English Constitutional History*. Oxford: Clarendon Press, 1870.

Szabó, Péter. *Woodland and Forests in Medieval Hungary*. British Archaeological Reports, International Series 1348. Oxford: Archaeopress, 2005.

Warde, Paul. *Ecology, Economy, and State Formation in Early Modern Germany*. New York: Cambridge University Press, 2006.

_____. "Fear of Wood Shortage and the Reality of Woodland in Europe, c. 1450–1850." *History Workshop Journal* 62 (2006): 29–57.

_____. *The Invention of Sustainability: Nature and Destiny, c. 1500–1870*. New York: Cambridge University Press, 2018.

Warren, Louis. "Owning Nature: Toward an Environmental History of Private Property." In *The Oxford Handbook of Environmental History*, edited by Andrew C. Isenberg, 398–424. New York: Oxford University Press, 2014.

Williams, Michael. *Deforesting the Earth: From Prehistory to Global Crisis*. Chicago: University of Chicago Press, 2003.

Wing, John T. *Roots of Empire: Forests and State Power in Early Modern Spain, c. 1500–1750*. Boston: Brill, 2015.

Young, Charles R. "Conservation Policies in the Royal Forests of Medieval England." *Albion* 10, no. 2 (1978): 95–103.

_____. *The Royal Forests of Medieval England*. Philadelphia: University of Pennsylvania Press, 1979.

CHAPTER 9

Managing Southern French Forests under—and before—Colbert

Between Law and Custom, ca. 1500–1700

Sébastien Poublanc

The early modern era was characterized by an unprecedented shortage of wood.[1] It was also defined by the accompanying, frantic search for sources of wood, particularly within the French kingdom. All circumstances contributed to make the shortage of forests and their products felt, be it for domestic needs (including construction and heating); for industrial demand (forges and furnaces); or for the state's developing needs regarding transatlantic commerce. Perceptions of this shortage differed according to whether one was in the provinces or a large urban center; although city dwellers felt the lack of wood keenly, rural communities located near the forests benefited from privileged access to silvicultural products and actively participated in their management.

The forests were the collateral in a battle between multiple, competing interests—lay and ecclesiastical communities, individuals, wood merchants and barons, and, above all, the king of France. Beginning in the sixteenth century, the king effectively asserted himself as a player of the first order; his many orders and laws were aimed at controlling the forests via the intervention of the increasingly powerful royal bureau of "Waters and Forests" (*Eaux et Forêts*). The king's laws sought to establish a harmony among various modes of exploitation and management, with varying success in different provinces. Those provinces farthest away from the centers of power effectively maintained autonomy over woodland management. Because of these two facts—the desire to coordinate forest-related practices and the sustained autonomy of the provinces—the king decided to enact a large-scale homogenization of forest management in 1661. Within the span of a few years, all of the Waters and Forests departments were reformed and given a single legal code, the "Great Ordinance" (*Grande Ordonnance*) of 1669. The overall reform, which became known as the Great Reformation (1661–1685), has often been perceived by scholars as the culmination of public forest management: rationalized, efficient, and accomplished under the leadership of a gifted administrator. Yet, a study of the archives demonstrates, to the contrary, the importance of agreements—dating back centuries—between the royal administration and southern French populations. The polar opposite of an effective and radical reformation, the reform engaged in a game of compromise and negotiation characteristic of older traditions.

A close study into the concrete impacts of the reform is illuminating. Rather than the radical new approach touted by the reformers, the Midi's reformation archives demonstrate the longevity of the practical accords between the king, the local populations, and their forests. The archive from the *Grande Maîtrise* of Toulouse is a uniquely complete source, with twenty-seven linear meters of archived material, comprising 2,000 court cases, 1,000 forest maps, and information about 1,623 forests. It was also the only archive from the reform to have been directed by Louis de Froidour, who, because he began to take into account biological and geographical factors, is considered to be one of

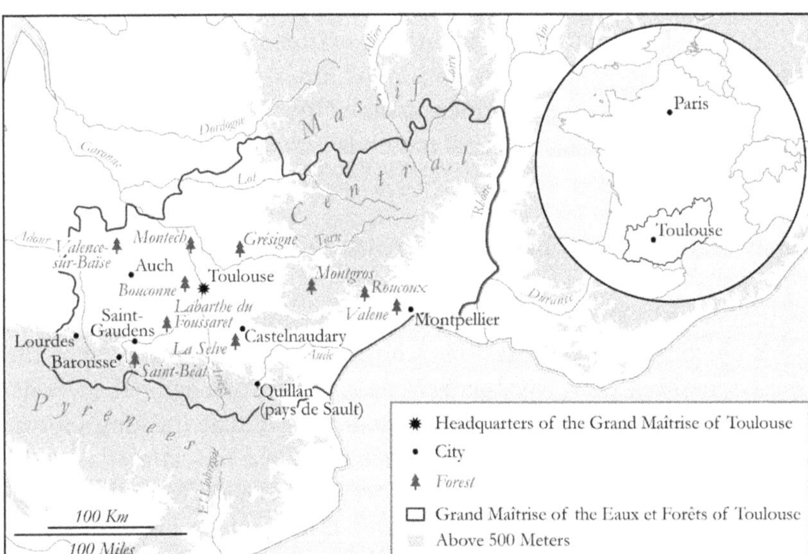

Map 9.1. Grand Maîtrise of Toulouse. University of Wisconsin-Madison Cartography Laboratory, published with permission.

the pioneers of the modern, scientific silviculture that began to develop in the eighteenth century.

De Froidour, Lord of Cerizy, who came from a family of Picard foresters, had a forestry as well as a legal background. Originally assigned to the Ile-de-France in 1662, he owed his appointment as reform commissioner of the Midi as much to his relationship with Jean-Baptiste Colbert as to his skills. From 1665 to 1673, he surveyed the forests of the *Grand Maîtrise* of Toulouse, prosecuted offenders, and drafted forest regulations for each forest throughout the large region this office administered. At the end of the reformation, he was appointed Grand Master of the Waters and Forests of Toulouse, a position he held until his death in 1685.

The Midi, here understood as the province of Languedoc and part of the province of Guyenne, was also one of those out-of-the-way regions that maintained autonomy over woodland practices. Although it had been attached to the royal domain in 1271, the royal bureaucracy had difficulty establishing itself outside large cities, such as Toulouse and Montpellier. Yet locals here had adopted royal forest practices as early as the fourteenth century, long before the Great Reformation. Thus, the reformation in the Midi between 1665 and 1673 was more about enforcing centralizing power structures and systematic organization than overhauling resource management practices. Above all, the historical documents allow us to demonstrate the existence of a complex system of forest management, run by both royal foresters and village inhabitants. They show that the reformers adapted the Great Ordinance, incorporating local proto-silvicultural methods into individual regulations.

Forest Management before de Froidour's Reformation

Beginning in 1250, the Count of Toulouse appointed his own foresters in his personal estates. After the reunification of the county with the king's domain in 1271, royal seneschals took over managing the woods. They spearheaded an ambitious project to consolidate the king's forest holdings via land exchange, inheritance, purchase of scattered parcels, and even arbitrary appropriations. Not limited to the Midi, their goal was to limit the fragmentation of the woods to create viable, exploitable units rather than innumerable small plots.[2] Consolidation was in part a response to the shrinking of woodlands at the turn of the fourteenth century, when an administration dedicated to controlling the forests became necessary.[3] Part of this process was the creation of a specialized forest official: the forest "guard" (*custos*). Distributed according to the size and number of the royal forests, fifty-two *custos* charged with guarding the royal forests thus appeared under the authority of the first masters of the Waters and Forests. Invested with the power to police the woods, these officers protected royal interests in the harvesting of wood and timber. Their task remained the same from the first half of the fourteenth to the end of the fifteenth century. This deliberate creation of extensive areas of forest and their officers redrew the map of the region's woodlands in favor of the monarchy's territorial and economic interests.

From the late fifteenth century, "forester-captains" took over the role of the forest guards; similar officers are mentioned in the charters of the Ile-de-France, Orleans, and the bailiwick of Sens, as well as Normandy and Anjou.[4] Local noble families used the office to augment their social status and their incomes and progressively monopolized the captaincies. Thus, local elite interest allowed for the maintenance of an official presence in the royal woods and guaranteed the survival of the forester office; it also allowed the captains to write the first forest regulations specific to the Midi.

According to the reformers' reports, the High and Low Pyrenees regions had long held the attention of the *Grande Maîtrise* of Toulouse. These strategically important areas contained important timber forests from which logs were floated to several cities on the plains—the Garonne for Toulouse, the Aude to resupply the Carcassonnais (forest district of Quillan), and the Ardèche and the Rhone to supply the arsenals of the Levant (Villeneuve-de-Berg). The *Grande Maîtrise* officers had two primary objectives in these regions: to protect those species deemed "useful" and to limit harvesting, whether commercially or by local users, especially when the traditional method of "foraging" (*furetage*), cutting one tree at a time wherever the best specimens might be found, was used.

Legislators were aware of the extent of the harvesting done in this way, and their regulations demonstrate a desire, if not to eliminate it entirely, at least to limit it. Their first task was to identify the most endangered types of trees and understand their use. Oak woods, which were rare, verged on disappearing entirely from continuous harvesting by local inhabitants, while sawmills and charcoal burners rapidly devoured the pines and firs. By the beginning of the seventeenth century, conifers had all but disappeared from the lower margins of mountain forests.

Both young and old trees were felled for boards and planks. As beech trees quickly filled in the resulting openings in the canopy, the younger firs and pines below went unpollinated for want of mature fruiting trees above them.[5] The export of conifers enabled the construction of buildings throughout lower Languedoc, particularly in the Razès, the Carcassonnais, and the Lauragais (Castelnaudary and Quillan), while acorns from the oaks were essential to fatten up hogs.[6] The disappearance of these trees would have significant economic repercussions, both at the local and regional levels, and so they had to be preserved.

To reduce the volume of wood used, the amounts that could be harvested were regulated: in Sault (Quillan), strict limitations meant that many traditional uses of wood were virtually prohibited. For example, the felling of young firs, whose supple wood was prized for making oars and wooden barrels, was thus reduced to the bare minimum: "Employing a young fir tree for each curved stave used in the construction of these vessels . . . it is impossible that we can do without it, without causing the public significant inconvenience, [so] we had to allow it going forward, as he considers that it should be done under certain conditions and to prevent the abuse that is committed there."[7]

Based on a policy of limiting waste that was meant to go into effect right away in order to assure the forests' growth, regulations were distributed throughout the districts of the Pyrenees. For example, one directive stated, "because those who aim to work on great or small projects in the said forests cut large quantities of trees all at once, a great portion of which they then leave and abandon . . ., new growth is impeded, which afterwards causes great clearings and empty areas in the said forests."[8]

Waste of resources is a persistent theme in the archives. In the royal foresters' view, meridional populations thought of the forest as inexhaustible and did not attempt to optimize their use of its products. Acting on this belief, the regulations preceding the reformation attempted to curb harvesting by local inhabitants. Typically standing almost alone against the force of custom, the foresters did their best to enforce royal orders in their assigned forests; however, in order to succeed, they had to compromise with the political powers already in place. In this way, local officials, like the village and town executives known as "consuls" (*consuls*), became assistants to the forest administration in its efforts to combat local destructive practices. As incentive to respect the regulations, they received payment in kind, just like the foresters. When they too were confronted with wood shortages, the consuls adopted their own measures to guarantee wood to their communities.

Communal Proto-Silviculture

Although the reformers dwelled at length on the destruction caused by the inhabitants, their letters and testimonies also frequently mentioned protective measures and forest management plans. By the end of the Middle Ages, certain species—such as fir trees—were protected due to their economic value. From Antiquity to the Middle Ages, intensive silvopastoral exploitation favored the growth of beech trees, while the increase in the harvesting of beech wood resulting from fifteenth-century developments in metallurgy—coupled with diminishing forest resources—forced nearby villagers and metallurgists to favor fir for building and framing out mining tunnels. Locally, these developments prompted protections for stands of fir, which were placed off-limits. The French toponyms, "Bedats" or "Devèzes," which were derived from contemporaneous Occitan, designate those as restricted areas. These pressures also led to the presence of fir woods in locations that were bioclimatically aberrant, but that allowed for both exploitation and surveillance: near villages and further down valleys than they would otherwise be.[9]

The fir tree was not the only one to benefit from such protection. In the Pyrenees, local communities actively protected oak trees to safeguard the acorn harvest for pigs. An economic resource of the first order—salable and consumable—pigs occupied a special place at the heart of meridional societies. In fact, these societies did not hesitate to protect oak woodland when needed. The reformers noted, "There are some regions planted with oak, but poorly growing, stunted, and of no value . . . which are nevertheless preserved rather carefully by the inhabitants of the areas, without their selling it or even cutting from it except in cases of extreme necessity for heating, because of the acorn harvest which serves for the grazing of their hogs."[10]

The same was true in Sault (Quillan) where the increasing scarcity of oaks since at least the middle of the sixteenth century led the forest administration to take protective measures.[11] Further to the west, de Froidour and his companions discovered "replanting," the transplanting of trees into long deforested mountainous areas.[12] In the plantations (*plantades*) of Bigorre and Béarn (Tarbes), plantations of widely spaced oaks were established "after the ruin of ancient woods which the communities who wanted to have them had taken the pains to replant; and they had had this custom for several centuries."[13] This socioeconomic system allowed the inhabitants to find wood for their needs, grazing for their animals, forest litter, which was collected after it had been enriched by the pasturage of the sheep, as well as acorns for the hogs. The same principle was found in the silviculture of coppice-under-standards (*taillis-sous-futaie*) in Béarn and Soule (Tarbes), "which the people of this region had found that, thus cut or pruned, they normally put out a good quantity of branches and were good for the acorn harvest."[14]

Protecting the woods was not always due to economic concerns: protecting habitable spaces was also taken into account. Such was the case in Saint-Béat, where de Froidour related that the inhabitants "did not dare chance going to cut any of the trees which had grown from the rocks, fearing that by weakening their roots, which seemed to embrace and retain the rocks, they would make the rocks fall."[15] All of the above measures together helped to shape the wooded landscape that the reformers found at the beginning of their enterprise in 1666.

The Reformation: Inscribing Modernity in the Long Term

Before getting into the details of the reformation's regulations, and how they brought together ancient texts and proto-silviculture, it is necessary to provide an overview of the mid-seventeenth-century forests. This will allow us to understand why the reformers largely broke with the Grande Ordonnance of 1669 and took local conditions into account. But even before introducing the forests, we should consider several contextual elements in order to understand the differences between the writings of the seventeenth century and the traditional "proto-silviculture" that I have presented above. The foresters' alarmist testimony aimed to show a gradual and continuing process of anthropization:

> The said forest [Montech] had been reduced to such a state that the king could take no profit from it, and it was so stripped of trees by the flagrant infractions and overgrazing that are done there every day, and even in the state in which it is in, one should not expect it to reestablish itself unless it is cut back to the roots [to restart its growth]. . . . The said forest having been found in such a poor state that one could practically traverse it entirely in a carriage.[16]

But the reformers' actions must be placed in their sociocultural context: de Froidour, primary author of the operation in the Midi, remained a city-dweller, as was Jean-Baptiste Colbert, the chief architect of the reformation as a whole. Both were indelibly marked by the townsfolk's belief in a forest crisis. It is therefore not surprising to find many "bastardized" (*abatardis*) forests in de Froidour's proposals, although they were only considered degraded because of his expectations and prescriptions for management. The locals had no difficulty finding the resources they needed the most in those same forests. His misrepresentation stands out so clearly because the reformation needed to justify its existence. This is undoubtedly one of the reasons that de Froidour standardized his surveyors' reports: to discover only beautiful, tall forests where one expected to discover damaged woods would not have made any sense.[17]

From Small Woods to Vast Forests

What did one see in the mid-seventeenth-century woodlands of the Midi, then? To start, the view as a whole was not nearly so disastrous as the reformers made it out to be. Forests were found throughout the *Grande Maîtrise*, although they were widely dispersed and unevenly distributed. On the plains, the Midi's principle demographic reservoir, large forests remained the exception among a multitude of small woods, most of which belonged to local communities. The rare large forests still present today, like the Grésigne, Bouconne, or the forests located not far from Castelnaudary, mostly belonged to the king; they corresponded to those areas supervised by the *custos* mentioned above.[18] A few rare communal forests, such as those of Valene at Montpellier or of Montgros, complete the panorama.

The ensemble shared certain characteristics—notably a significant fragmentation of the woodland cover. This was the result of the sustained drive by local communities and elites, which was renewed in the sixteenth century, to clear these areas, particularly in Lower Languedoc. Indeed, the immediate environs of the Mediterranean littoral were quickly cleared. Most of the woods in the *Grande Maîtrise* were rather small: half covered less than 66 hectares; and a quarter, less than 24 hectares. The region around Toulouse had the most thoroughly fragmented woodlands and the greatest percentage of small woods interspersed with cultivated areas, which is hardly surprising given its geographic

Table 9.1. Extent of woodlands (in hectares) by Forest District (*maîtrise*).

Forest District (*maîtrise*)	Number of Forests	Average Area	Median Area
L'Isle-Jourdain	183	80.54	28.50
Tarbes	124	128.87	43.61
Comminges	489	181.88	51.3
Saint-Pons-de-Thomières	199	143.95	50.44
Rodez	152	144.40	61.56
Villemur	80	213.55	83.22
Villeneuve-de-Berg	132	197.73	95.19
Castelnaudary	63	226.89	99.15
Montpellier	31	241.15	117.71
Pamiers	177	470.39	173.85
Quillan	111	415.20	200.64

proximity to the most populated areas of the Midi: here, the *silva* gave way to the plow. Bouconne and, to a lesser extent, the forest of Labarthe du Foussaret remained the only vestiges of the medieval forest that had formerly extended up to the suburbs of Toulouse.[19]

In contrast, the less-populated mountainous spaces still had vast forests, as Georges Bertrand has noted, "where humans had confined them: often degraded, occasionally safeguarded, or even 'gardened,' in a word, remodeled."[20] Their locations had been determined by the population growth that marked the sixteenth and early seventeenth centuries, which increased the demand for arable land.[21] This situation exacerbated the antagonism between the rural communities and the ironworkers who were the primary contributors to deforestation, first with mills, then with Catalan-style forges.[22] Although industrial demand encouraged the creation of protected areas, such as for firs, which were used to frame out mining tunnels, it also led to deforestation around villages. The mountains were in fact a crowded world, and in the competition for woodland resources, the forge-masters had the advantage over the villagers. Local scarcity of wood in the region of Foix (Pamiers) led, for example, to an accord of 1347–1348 between the Count of Foix and the inhabitants of Couserans, "who organize the supply of firewood for [iron-smelting in] Vicdessos from the forests of Comminges."[23] This agreement to bring in wood, which would have been mostly beechwood, from a distance of about 20 km was still in effect at the time of the reformation. Thus, the preservation of the county's highland forests had helped to maintain, and no doubt even to expand, iron-smelting in Vicdessos.[24] But the only ones to benefit from this arrangement were the ironworkers; the local communities were excluded.

Calculating the area covered by woodlands does not suffice, however, to understand how the reformers judged a forest's quality; other considerations played a part, including its management, the age of its trees, and the species present.

Behind the Commissioners' Words: The Maintenance of Southern French Timber

The reformation commissioners' choice of words is important: it painted the image of destroyed forests, of ravaged trees eaten by animals whose "teeth, breath, and wool . . . were pestiferous and inappropriate to the woods."[25] These descriptions strike the reader's imagination, suggesting that the forests consisted of miserable coppices crowned from time to time by some scattered tall trees and populated mostly by small trees with spreading canopies.

The heavy symbolic weight of these words resulted from two factors, one emotional and the other economic. From the economic point of view, the taller "standard" trees (*baliveaux*) represented the principal capital of the forests, which is what the reformers wanted, especially the naval timbers so dear to the state. Timber stands also presented a certain majesty in the trees' rising toward the heavens: their height recalled man's erect posture and the "old witnesses of the ages past," as opposed to coppices, scraggy shrubs in the making, with little monetary value and no majesty.[26] Seventeenth-century foresters' contrasting perceptions of these two possible states of the forest imperceptibly influenced their writing.

For them, the type of management corresponded to the predicted maturation date of a given tree population, which was estimated within a certain age range. Thus, the precise age of individual trees was a fact of less importance than the observed method of management and the nature of the land. Together, these allowed foresters to prepare for the forest's future. Within the *Grande Maîtrise*, the most common management style favored the short harvesting cycles of coppicing (*le taillis*). The duration of these cycles varied: foresters called those woods that were cut every ten years "young coppices" (*jeunes taillis*), while those that reached twenty to thirty years, provided that they were not destined to grow into timber, were described as "tall coppices" (*hautes taillis*). Altogether, coppices covered over 43 percent of the surface area in the *Grande Maîtrise*. In contrast to timber stands (*la futaie*), coppices were mentioned across the whole territory, and the forests that had none were quite rare, since coppicing typically accompanied population expansion. Its rapid harvest cycle provided firewood and space to pasture animals. At the same time, however, as people and animals took their share, they impeded woodland regeneration.

The choice to manage the forests to favor timber (*la futaie*) instead corresponded to woods that had not been "cut in the ordinary ways of coppices, [and which] had been left to grow up to 30, 40, 50, 100, and 200 years, and even more."[27] The terminology that the reformers used to refer to various age classes reflects their overriding focus on the production of what they considered to be full-grown timber. For trees slated to grow into timber, those aged fewer than twenty years were called "regrowth of timber" (*revenu de futaie*); between twenty and thirty years, "quarter-timber" (*quart de futaie*); between thirty and sixty years, "half-timber" (*demi futaie*); between sixty and 120 years, "high timber" (*haute futaie*, or *jeune haut futaie*); and over 120 years, "old high timber" (*vieille haute futaie*). By mapping these descriptions and other information provided by the reformers, we find that timber stands represented at minimum a third of the total wooded surface. In fact, there was even more timber than that, because we also need to take into account the significant percentage (16 percent) of woodlands that were under mixed management, supporting coppices and timber together.[28]

As expected, the timber stands were mostly located in the steepest areas of the *Grande Maîtrise*, such as the Pyrenees (Quillan, Pamiers, Comminges et Tarbes), the Causses of Quercy (Villemur), Montagne Noir (Saint-Pons), and the Haut-Vivarais (Villeneuve-de-Berg). The report notes that in the latter forest, the difficulty in extracting logs was such that "in local terminology, they call this place *cul d'Iher*, which means the 'ass of Hell.' The rest of the valley is very beautiful and quite pleasant . . . [and] the beech trees are very young and beautiful timber, and the firs are so beautiful and in such great number that one could take three to four thousand masts."[29]

A few extensive timber stands, mostly under royal control, were nevertheless successfully maintained on the plains; such was the case of the Grésigne and of the forests of Castelnaudary. However, the lay communities of villages and towns also safeguarded their lumber, particularly on the plains. It is interesting to note that 200 (of 859) still possessed timber stands, thereby disproving the view that ordinary people only concerned themselves with their immediate interests. These timber stands also show that the local inhabitants' protective measures were effective. The entire issue for the reformers therefore consisted of combining this local silviculture with their regulations—without, however, crediting the creators of the earlier methods.

A Quarter, a Third, or No *Réserve*?

The arrival of the reformers led to a new bureaucratic approach toward forest regulation. In contrast to most foresters' limited education and narrow focus on their own practical experience, de Froidour had a well-developed knowledge of forestry that had proved itself in the woodlands of the Cardinal Mazarin, principal adviser of Louis XIV.[30] Moreover, the norms of forest management evolved considerably during the reformation.

While the Great Ordinance was meant to guide management of forests throughout the kingdom, it was above all a text built on the reformation of the forests of the Ile-de-France. It did not consider the vicissitudes of management in each province, whether these concerned the feasibility of cutting on a regular cycle, the regeneration of conifers, or the utility of allowing special privileges for mountain communities. It was left to de Froidour and other commissioners, therefore, to respect the spirit of the Ordinance while managing the forests in a way that accommodated the interests of all involved.

The reform regulations in the *Grande Maîtrise* of Toulouse were a patchwork begun in 1665; the first texts were delivered in 1667 and continued in fits and starts as the reformation progressed. After the Great Ordinance of 1669 was registered in Toulouse's Parlement in 1670, new rules were superimposed onto the commissioners' orders. The commissioners continued, however, to draft provisional regulations that took into consideration the specifics of each local region, breaching the Ordinance's unitary law. There was therefore no singular text of the law, but a plurality of opinions concerning each space under local forest management, "entirely in accordance with the king's order, except in two or three small points which can easily be reconciled."[31] This process of adaptation allowed for many regional specificities in forest management.

The first difference concerns the question of the "reserve" (*la réserve*), an area set aside for the growth of timber. In each forest, the Ordinance called for a reserve equivalent to a quarter of the total surface area. Far from privileging this system, de Froidour and the other reformers preferred to integrate each woodland reserve into its environment. In order to accomplish this innovative choice, they established a typology of different types of woodlands.

For the reformers, there was no reason to transform the scrubland (*garrigue*) of the Lower Languedoc and the Causses into timber stands. Composed essentially of shrubby vegetation, profoundly impacted by the exploitation of those with local use rights, these extensive zones did not have enough trees to make a reserve, particularly since the majority of its species could not produce useful timber. The reformers ordered that these spaces be managed in such a way as to rapidly produce small wood, especially for fuel, for the inhabitants (*bois d'affouage*).

Small and thin, trees of ten to twenty years of age formed an ideal coppice for everyday usage by local inhabitants. Furthermore, the proper age for harvesting was correlated to surface area: the larger the woodland, the greater the possibility of aging the coppices so as to diversify their use. The system was, however, not quite as uniform as it appeared. In the royal forests, reserves were established where timber stands had long existed. The forest of Roucoux was "previously planted with tall timber of the same type and quality, as it was told to us, and according to what we have been able to learn from the sorts of tree that we have found there."[32] From it, they endowed a reserve of eighty-five hectares and employed a slower rhythm of harvesting, about every twenty-five years. Few in number, these forests demonstrate the rationality of the commissioners for whom forest management did not have to be uniform but was understood to embrace a range of particular cases.

Due to their recognition of local conditions, the list of woods that could be converted into timber stands was drastically reduced. Far from insisting that each forest have its own reserve, the reformers based their decisions in part on anthropogenic pressures:

> Moreover, given that the said forest [La Selve] is planted with many "white woods" (*bois blancs*) [broadleaf trees like maple and poplar that were less valuable than oak or beech], which are not at all appropriate for growing timber (*a estre mis en futaye*), and besides which is nothing more than a shrub which, during a century or the six

> score years at least that must be given to the woods in order to allow them to grow as timber, being exposed to the pillaging of four or five villages which surround it, is hardly in a state to defend itself from their offenses, timber stands should only be reserved in the large bodies of forest. And given also that in the course of time, it would be irritating to incessantly and over many years pay wages and other rights to the officers of a forest from which the king takes no profit, we are of the opinion that for the good management of the said forest, harvests should be set at the age of forty years, at which time the beech and maple wood are good not only for the charcoal burners, as is hazel, but for all sorts of construction work and for the work of cartwrights.[33]

The challenges of increasing human pressure and the need to protect the woods are summed up perfectly here: converting coppices into timber stands is a long-term operation, and one that necessitated a fallow time that required enforcement. Uncompressible, this span of time allowed the trees to grow and eventually furnish precious timber standards. To that end, the forest needed to fulfill two conditions: it had to have defensive structures and guards, and it needed to be situated in a space of lower anthropogenic pressure. These characteristics excluded small wooded areas, in which the needs of the locals would push them to attack the reserved trees, condemning the operation to failure.

The commissioners were pragmatic: the flexible principles articulated for the Selve forest were applied to all of the forests in the *Grande Maitrise*. Thus, only those woods were endowed with reserves that had the proper soil and terrain for growing timber, quality species, outlets for its products, a substantial surface area, and which were not subject to excessive anthropogenic pressures. Rather than fall prey to the lure of profit above all, the reformers moderated their regulations, even in cases where a woodland could have produced fine timber stands. In opposition to the dogma of standardization, officials proposed a model of management based on societal needs; however, in order to assure the production of construction lumber and thereby avoid having their reforms undone, they referred to King Charles IX's ordinance of 1561, which had prescribed that a third of the woodlands belonging to commoners and ecclesiastical communities be set aside for timber.[34]

In reality, the institution of a third of the forest as a reserve essentially concerned the ecclesiastical forests. The regulation thus attempted to compensate for the small number of reserves in communal wooded areas, which far outnumbered ecclesiastical woodlands. An adept politician, de Froidour authorized religious communities to fell mature timber without first paying for the royal patent letters that were normally required, a measure that he hoped would assure their submission. In the end, only the royal forests conformed to the order of 1669: just this once, the king set the example for his domain.

Assuredly, the commissioners wandered rather far from the prescriptions of the Great Ordinance; rather than putting a quarter of every forest in reserve, they promoted an assortment of organizational measures destined to satisfy needs for construction lumber as well as those of the neighboring inhabitants. The choice of whether or not to establish a reserve reflected local realities.

On the Proper Economy of Woods: Planting Trees

The application of the above reserve of one-third was not the only exemption to the Great Ordinance proposed by the commissioners; other peculiarities appeared in their regulations. Among them, a policy of replanting, which was intended to remediate the clearing of the woods with new plants and accompanied by specific defensive measures, was remarkable as much for its breadth as for its discretion.

Recognizing the benefit of this communal practice, which benefitted hogs and other grazing animals, the commissioners adopted its precepts while regularizing the procedure. The reformers' sense of superiority again appears here: they claimed that although the locals acknowledged the gravity of their situation, they did not act on it and neither knew nor chose "lands to acquire and use according to their nature."[35] The reformers therefore mandated the planting of four to five saplings for each old tree felled. The number was deliberate: it took into account the inevitable losses that diminished the density of the coppices. Additionally, the communities were required to watch over the saplings for three years and to replace any losses that might occur. Even so, the measure did not suffice to repopulate the forests: it was an expedient solution to limit the impact of current usage and did nothing for the already deforested spaces.

In order to repopulate those, the reformers required some communities to sow acorns or seeds. Although this measure was less costly than transplanting saplings, it still required the governing bodies of villages or towns characteristic of the Midi (*consulats*) to maintain the seedlings so that their density corresponded to the commissioners' prescriptions. The reformers mandated replanting mostly within an area that formed a triangle of which Lourdes, Auch, and Saint-Gaudens formed the vertices; however, the measure was not limited to this area. It was also applied to other forests, notably in Rouergue.

Whereas replanting aimed to replace or increase the number of trees within wooded areas, a separate set of procedures, a kind of reforestation (*enfrichement*), promoted planting trees, whether by transplanting saplings or by sowing seeds, in areas that had been deforested or were for other reasons considered to be "useless and empty lands." The experts indicated, however, that clearings outside of the royal forests should not be replanted—only those within the perimeter of the forest, in order to avoid usurpation of these areas by other parties. To this end, the commissioners indicated "that they should be sown or planted with seeds of the best and most appropriate type of tree for each site in the said forests, such as fir in the Pyrenees, and oak in the other mountains, or beech or chestnut, at least in the soils which will not be strong enough to grow oak or fir."[36] Ecological preoccupations thus warred with economic considerations: it was necessary to optimize the parcels' productivity by excluding all "white wood" in favor of the most marketable resources. Religious communities and commoners had to replant the empty spaces inside their woods in the same way.[37]

Far from limiting themselves only to the question of useless and empty places, the commissioners also directed landholders on how best to "mix" (*panacher*) their timber stands by replanting timber trees within stands that had become too old. Beyond the benefits of regeneration, the principal concern was to wrest revenue from these wooded areas in bad shape, where the wood was sold cheaply.

Regulating the Harvest

For replanting, the reformers added new material on a topic not treated by the Great Ordinance. In other areas, they directly contradicted its directives. They denounced its prescriptions to only harvest via a method of "contiguous cutting" (*tire et aire*), which consisted of thoroughly harvesting from one contiguous plot to another. In fact, this method had a rather shadowy presence in the seventeenth century: though officially favored, it was seldom applied in a standard form, even at the heart of the royal domain. In Burgundy, Picardy, and the Ile-de-France, foresters employed numerous variants.[38] The Ordinance was neither set in stone nor definitive: in practice, its prescriptions formed a framework that was adapted to a given region. Things were no different in the southern part of the realm, where de Froidour at first fervently opposed "foraging" (*furetage*), or the harvest of individual trees based on their size wherever they were found (*la coupe par pied d'arbres*).[39]

His negative view of foraging reflected the standard view of seventeenth-century foresters. De Froidour tried to replace it in the southern forests with the practice of contiguous cutting; however, the first fir forests in which it was applied did not provide the hoped-for results: nothing regrew! The southern regions were not the North, and the administrative considerations that guided the reformer had, in the end, less of an impact than the silvicultural results.[40] De Froidour rapidly came around to what he had previously condemned:

> No felling by *arpent* [a measure of area, roughly equivalent to an acre], given that the said trees [firs] were not regrowing at all from the roots and were coming only from seeds. It would be dangerous to establish felling by acre and by contiguous cutting, and there would be danger that [the fir] would not come back, or at least not until after very many years, as experience has shown in all the places where such harvests have been made.[41]

A certain number of forests thus escaped the prescriptions of the Great Ordinance: to the 249 forests of fir in mountainous areas can be added all those located in the plains, but whose steep local terrain made them difficult to exploit. Locals were stakeholders in the process. It fell to them to choose which trees to cut, while saving those believed to be pollinators or that had economic value. Essentially, they were restricted to the crooked, poorly growing trees that were nevertheless sufficient for domestic needs. Foraging thus remained the solution reserved for the most difficult woodlands. Along with resprouting from stumps cut back to the base (*le recépage*), clearing or expurgation, and contiguous cutting, it rounded out the reformers' options.

The choice of method depended on each forest's state: any woods that were too damaged to be exploited immediately were given over to resprouting by cutting down to the base. Woodlands made up of overly dense stands of young trees were "purged," while the establishment of scheduled harvests (*la coupe réglée*) with contiguous cutting depended on the volume of wood that could be harvested within a given surface area. Cutting back to the stumps was a treatment of last resort applied when woodlands were deemed too degraded for other treatments. It aimed to restore the space to a tabula rasa so that, over the course of several seasons, the tree population might regenerate and be harvested by contiguous cutting. They described it as the same as "a head whose hair has been ravaged and thinned by illness," and for "which it is necessary to shave so that the hair might grow back in great abundance." They urged that it was the same for degraded forests. "It is necessary to 'resprout' (*receper*) them so that the wood might come back" to reestablish the woodland.[42] The process was also employed in stands of excessively old timber. It allowed for the regeneration of the stock of trees to improve exploitability. The marked parcels were entirely harvested by contiguous cutting, but without the usual preservation of scattered "standard" trees (*baliveaux*). Thereafter, the forest could be managed normally.

Purgation was also employed to regenerate a forest. But de Froidour and his deputies' understanding of that term differed. De Froidour believed it to be "a monster" that had to be eradicated, all the more so because it "was very common in this Province [Languedoc], and notably in the king's forests."[43] It consisted of cutting coppice woods every ten to twenty years, but leaving one sapling on each stump. Cut in this fashion, the forest becomes trapped in a state that impedes any development, as only a few strong saplings of the same age are left at each harvest, surrounded by younger branches. The wood harvested in this way had only minimal commercial value, serving mainly for firewood or for making small objects. Even so, purgation was particularly easy for those populations with only the simplest tools, which doubtless explains its widespread application.

Among all these techniques, contiguous cutting remained the preferred method. It underpinned the reign of "coppice under standards," a hybrid method that attempted to straddle the fence by reconciling the production of firewood and other small wood for domestic needs with that of timber for construction. Economic necessity determined the age of future coppicing: the greater the anthropogenic pressure, the shorter the harvesting cycle and the younger the coppice would be, thereby providing precious firewood. Only in larger woodlands was it possible to lengthen the harvest cycle enough to increase the trees' age and thus the quality of wood. Generally, coppices remained young and the wood harvested from them was of a mediocre quality. The same went for standards: with only young saplings to compete with for light, they grew short and stocky, with spreading canopies—the complete opposite of the beautiful tall trunks for which foresters hoped. Of a mediocre quality at best, their trunks were generally disdained by builders. In the best case, they were bought by naval commissioners, who used a great deal of twisted wood in the hulls of their vessels. Moreover, even when timber stands were well protected in a reserve, the results were often no better. If the "standard" trees (*étalons*) grew continuously, the various age-classes that were left to grow after each thinning, including the sixty-to-eighty-year-old "modern standards" (*baliveaux modernes*), the one-hundred-or-more-year-old "ancient standards" (*baliveaux anciens*), the "twice-ancients" (*bisanciens*), which were four times the age of the surrounding coppice woods, and the "old barks" (*vieilles écorces*), which were five to six times the age of the coppice, would eventually close the canopy and smother both coppice and all the younger standards. Despite these inconveniences, coppice-under-standards remained the preferred method for reformers, who used it to raise the age at harvest. In the reformers' thinking, the short harvesting cycles of the coppices in the *Grande Maîtrise* needed to be gradually lengthened to at least ten to twenty-five years in order to provide more ample resources to the public.

Conclusion

All throughout the reformation, the royal commissioners took into account the ways of life of the Midi's inhabitants. From the plains of the Garonne to the peaks of the Pyrenees, they recognized people's profound attachment to and identification with their forests. Their regulations reflected the peculiarities born of the long anthropogenic modification of the southern woodlands. In crafting them, the reformers reprised the principal of inquiry as a fundamental element of the *Waters and Forests*: careful investigation allowed royal agents to understand the management of these immense territories.

While the Great Ordinance of 1669 remained the referent text par excellence in matters of management, the reformers dedicated themselves to understanding local needs in order to protect the forests from any excess. This was no altruistic decision: achieving the objectives of the state required walking a fine line between coercive management and respecting regional practices. It required an older layer of medieval structures, reinforced and multiplied, to watch over them.

Sébastien Poublanc earned a doctorate in modern history from the University of Toulouse, Jean Juarès (2015). His thesis examines the history of state forests between the thirteenth and seventeenth centuries. He currently works as an independent research engineer in Toulouse, specializing in the numerical humanities, and he is an editor of the digital magazine *Mondes Sociaux*. He has published articles on both the history of forest conflicts and digital history.

Notes

Translated from the French by S. C. Kaplan with assistance from Richard Keyser and Abigail P. Dowling.

1. Jérôme Buridant, "Du 'modèle' à la pratique: la gestion des peuplements caducifoliés dans la France Moderne, XVIe-XVIIIe siècle," *Flaran* 24 (2002): 203.
2. Philippe Jéhin, *Les forêts des Vosges du Nord du Moyen Âge à la Révolution: Milieux, usages, exploitations* (Strasbourg: Presses universitaires de Strasbourg, 2005), 31.
3. Michel Devèze, *La vie de la forêt française au XVIe siècle*, 2 vols. (Paris: Ecole Pratique des Hautes Etudes, 1961), 1: 64–70.
4. Edouard Decq, "L'administration des Eaux et Forêts dans le domaine royal en France aux XIVe et XVe siècles: Mémoire posthume d'Edouard Decq," *Bibliothèque de l'Ecole des chartes* 83/84, no. 1 (1922): 78.
5. Christian Fruhauf, *Forêt et société: de la forêt paysanne à la forêt capitaliste en pays de Sault sous l'Ancien régime (vers 1670–1791)* (Paris: Centre national de la recherche scientifique, 1980), 72.
6. Fruhauf, *Forêt et société*, 63.
7. Archives départementales de la Haute-Garonne (ADHG), fonds de la réformation de Froidour, 8 B 025, f°300 v, *Procès-verbal d'avis des commissaires portant règlement général pour les forêts de la maîtrise de Quillan avec ordonnance provisoire pour l'exécution du règlement contenu en l'avis cy dessus jusqu'à ce qu'il ait plu au roi autrement y pourvoir*, 8 May 1670.
8. "Et parce que ceux qui ont dessein de travailler en gros ou menus ouvrages auxdittes forests couppent tout a la fois tres grande quantité d'arbres dont ils en laissent apres et abandonnent une grande partie et ainsy outre la perte desdits arbres a cause qu'ils demeurent et se pourrissent sur les lieux, ils occupent de grandes places auxquelles les rejets et renaissances sont empeschées, qui causent apres de grandes clairieres et vaccans esdittes forests"; ADHG, fonds de la réformation de Froidour, 8 B 009, f°121 v, *Reglement concernant les forests de la vallée de Barousse fait le dernier octobre 1625 par le lieutenant general de la Table de Marbre*, 31 October 1625.
9. Jérôme Bonhôte et al., "Histoire de l'environnement et cartographie du temps dans la moitié est des Pyrénées: Pour une 'chrono-chorologie,'" in *Les*

temps de l'environnement (Toulouse: Presses universitaires du Mirail, 2000), 502.

10. "Il y a quelques contrées plantées de chesne [*Quercus pyrenaica*], mais mal venant, rabougry et de nulle valleur . . . sont conservez neantmoins assez precieusement par les habitans des lieux sans qu'il s'en vende et mesme qu'ils s'en couppe que dans les extremes necessitez pour le chauffage à cause du glandage qui sert pour la paisson de leurs porcs"; ADHG, fonds de la réformation de Froidour, 8 B 009, f°288 r, *Proces verbal d'avis pour le reglement des coupes des forests dependantes de la maistrice de Commenge*, 8 May 1670.
11. Christian Fruhauf, "De la forêt paysanne à la forêt capitaliste en pays de Sault sous l'Ancien Régime," *Revue forestière française* 31, numéro spécial "Sociétés et forêts" (1980): 160–65. The oaks in question were mostly pubescent oaks (*Quercus pubescens*) and some green oaks (*Quercus ilex*).
12. Michel Bartoli and Bernard Gény, *Histoire des forêts du Béarn jusqu'en 1789: Découverte du règlement forestier de Louis de Froidour (1673)* (Orthez: Société des sciences, lettres et arts de Pau et du Béarn, 2015).
13. "Après la ruine des anciens bois que les communautés qui en ont voulu avoir aient pris le soin d'en replanter et il y a plusieurs siècles qu'on y est dans cet usage"; Michel Bartoli, *Louis de Froidour (1626?–1685): notre héritage forestier*, Les dossiers forestiers 23 (Paris: Office national des forêts, 2011), 104.
14. "Dont les gens de cette contrée ont expérimenté que, ainsi tronçonnés ou écimés, poussent ordinairement quantité de branches et sont bons pour le glandage"; Bartoli, *Louis de Froidour*, 106.
15. "N'osent se hasarder d'aller coupper aucuns des quelques arbres qui sont venus sur ces rochers, crainte qu'esbranlant leurs racines qui semblent embrasser et retenir ces rochers, ils ne les fassent tomber"; Paul de Castéran, *Lettres écrites par M. de Froidour, grand maître enquêteur et général informateur des Eaux et Forêts au département de la grande maîtrise de Languedoc à M. de Hericourt, son procureur général à Toulouse et à M. de Medon, conseiller au présidial de Toulouse, publiées avec des notes par M. Paul de Casteran* (Auch: Imprimerie et lithographie G. Foix, 1899), 350.
16. "Ladite forest [*Montech*] estoit reduitte en tel estat que le roy n'en tiroit aucun proffit & qu'elle se despeuploit de bois par les grands delits & abroutissemens que l'on y faisoit tous les jours & mesme qu'en l'estat auquel elle estoit, il ne falloit pas attendre qu'elle se peut restablir a moins qu'elle ne fut receppée . . . ladite forest s'estant trouvée en sy mauvais estat qu'on auroit peu la traversser presques par tout en carosse"; ADHG, fonds de la réformation de Froidour, 8 B 002, f°278 v, *Proces verbal de la visitation de la forest de Montech*, 28 September 1666.
17. Sébastien Poublanc, "Compter les arbres: une histoire des forêts méridionales à l'époque modern," (PhD Thesis, Université Toulouse 2 Jean Jaurès, 2015), 89.
18. Henri de Coincy, "Quelques précisions sur les grandes maîtrises des Eaux et Forêts en Languedoc," *Annales du Midi* (1924): 39–57, 184–203.

19. Louis Vié, "Louis de Froidour commissaire député pour la reformation puis grand maître des eaux et forêts," *Revue de Comminges* (1913): 198.
20. "Là où les sociétés humaines les ont cantonnées: souvent dégradées, parfois sauvées voire 'jardinées,' en un mot remodelées"; Georges Bertrand, "La forêt pyrénéenne réinventée," *Revue de Comminges* 2 (1998): 254.
21. Emmanuel Le Roy Ladurie, *Les paysans de Languedoc* (Paris—La Haye: Mouton, 1974), 190.
22. Jérôme Bonhôte, Jean Cantelaube, and Catherine Verna, "De la mouline à la forge à la catalane: Six siècles de sidérurgie directe dans les Pyrénées ariégeoises (XIIIe-XIXe siècles)," *Revue de Comminges* 4 (2000): 768.
23. Bonhôte et. al, 768.
24. Bonhôte et. al, 764.
25. "La dent, l'haleine et la laine . . . sont pestifères aux bois et malvenants"; Louis de Froidour, *Reglement fait par le sieur de Froidour, grand maistre des Eaux & Forests au département de Languedoc, Haute Guyenne, Bearn, Basse Navarre, Soule & Labourd, concernant les forests du païs de Bigorre: Les observations faites contre ledit reglement et les motifs dudit reglement avec les réponses dudit sieur de Froidour ausdites remarques & observations. Fait à S. Jean de la Castelle, le 29 decembre 1684* (Toulouse: J. Pech, 1685), 63.
26. Alain Corbin, *La douceur de l'ombre: L'arbre, source d'émotions, de l'Antiquité à nos jours* (Paris: Fayard, 2013), 56.
27. "Coupés en coupes ordinaires de taillis, [*et qu'*] on laisse croître jusqu'à 30, 40, 50, 100, & 200 ans, & au delà"; Louis de Froidour, *Instructions pour les ventes des bois du roy* (Toulouse: Raymond Bosc, 1668), 9.
28. E.g., the surface area of forest management types in the *Grande Maîtrise* of Toulouse in hectares: coppice, 122,371 h.; timber, 91,014 h.; hybrid, 54,361 h.
29. "L'on appelle cet endroit en terme du pays cul d'Iher, qui signiffie cul d'Enfer. Du reste la vallée est fort belle et fort agréable, contient presque une lieue de pays. Ce qui est à main gauche qui est le revers des montagnes qui sépare la France et l'Espagne est entièrement planté de bois, . . . ; les hestres [*Fagus sylvatica*] sont très jeunes et belle fustaye, et les sapins [*Abies alba*] sont si beaux et en si grand nombre qu'on peut y prendre trois à quatre mille matz"; de Castéran, *Lettres écrites par M. de Froidour*, 149.
30. Jérôme Buridant, "Froidour avant Froidour: l'action de Louis de Froidour dans les forêts du nord de la France, 1651–1666," *Les Cahiers de Framespa: Nouveaux champs de l'histoire sociale* 13 (15 March 2013).
31. "Entierement conforme à son ordonnance, sauf en deux ou trois petits points qui se peuvent facilement concilier"; ADHG, fonds de la réformation de Froidour, 8 B 001, f°393 r, *Remarques sur l'ordonnance du roy*, 5 May 1670.
32. "Autrefois plantée d'une haute fustaye de mesme essence et qualité, comme on nous l'a dit et selon que nous l'avons peu connoistre par les sortes de bois que nous y avons trouvé"; ADHG, fonds de la réformation de Froidour, 8 B 020, f°107 r, *Procez verbal de visitation des bois de Masux Barrot Fabret et Roucoux du mas de Malet*, 11 October 1669.

33. "Du reste, comme ladite forest [*La Selve*] est plantée de beaucoup de bois blancs qui ne sont point propres a estre mis en fustaye, et que d'aillieurs ce n'est qu'un buisson qui pendant un siecle ou six vingt ans au moins qu'il fault donner au bois pour le laisser croistre en fustaye, estant exposé au pillage de quatre ou cinq villages dont il est environné, n'est point en estat de se deffendre de leurs delicts, les fustayes ne devant estre reservées que dans les grands corps de forests, et veu aussy que dans la suitte des temps, il seroit fascheux de payer incessamment et pendant longues années des gages et autres droicts a des officiers d'une forest dont le roy ne tireroit aucun proffit, nous sommes d'advis que pour le bon mesnagement de ladite forest, les coupes en doivent estre reglées a l'aage de quarante ans, auquel temps les bois de hestre [*Fagus sylvatica*] et erables [*Acer campestre*] non seulement sont bons pour charbonner, ainsy que le bois de coudre, mais pour toutes sortes d'ouvrages, de charonnage"; ADHG, fonds de la réformation de Froidour, 8 B 051 A 1, *Proces verbal d'avis pour le reglement des couppes des forests de Lauragois*, 2 January 1670.
34. ADHG, fonds de la réformation de Froidour, 8 B 001, f°396 r, *Remarques sur l'ordonnance du roy*, 5 mai 1670.
35. "Les terres pour les acquérir et employer selon leur naturel"; ADHG, fonds de la réformation de Froidour, 8 B 011, f°209 v, *Procès contre les habitants de Lourette*, 27 July 1668.
36. "Qu'elles doivent estre semées ou plantées de la semence et de l'essence de bois la meilleure et la plus convenable pour chacun fonds desdites forests, comme de sapin [*Abies alba*] dans les Pirenées & de chesne [*Quercus pubescens*] dans les aultres montagnes ou de hestre [*Fagus sylvatica*] ou de chasteigner [*Castanea sativa*] au moins dans les fonds quy ne seront pas asses forts pour pousser du chesne ou du sapin"; ADHG, fonds de la réformation de Froidour, 8 B 001, f°501 r, *Proces verbal d'avis de Messieurs les commissaires deputez pour la reformation generale des Eaux et Forests au departement de la grande maistrise de Toulouse donné a sa maiesté pour l'infeodation ou repeuplement des terres desfrichées dans les forests des maistrises de Quillan, Pammiers, Saint Gaudens et Saint Pons*, 7 May 1670.
37. ADHG, fonds de la réformation de Froidour, 8 B 001, f°325 r, *Proces verbal d'advis pour les reglements des bois des beneficiers, des communautes ecclesiastiques & seculiers et des particuliers mesmes*, 27 July 1668.
38. Buridant, "Du 'modèle' à la pratique," 213.
39. de Froidour, *Instructions pour les ventes*, 16.
40. Gustave Huffel, *Histoire des forêts françaises, de l'origine jusqu'à la suppression des maîtrises des Eaux et Forêts* (Nancy: Ecole Nationale des Eaux et Forêts, 1925), 75.
41. "Aucune couppe par arpent, joint que lesdits arbres [*Abies alba*] ne repoussant point par les racines et venant seulement de semance, il seroit dangereux d'en establir les couppes par arpent et a tire et aire & qu'il y auroit peril qu'il n'en reviendroit plus, du moins que par un tres long cours d'années ainsy que l'experience l'a fait voir en tous les lieux ou l'on a fait de pareilles coupes";

ADHG, fonds de la réformation de Froidour, 8 B 009, f°310 r, *Proces verbal de la reformation generalle des Eaux et Forests de la maistrise particuliere de Commenge establie a Saint Gaudens*, 19 July 1673.

42. "Pour bien entendre cette sorte de vente, il faut se figurer une teste que quelque maladie a dépoüillée & dégarnie de cheveux, laquelle il faut raser pour les y faire revenir & les faire recroître avec abondance; il en est de mesme des forests en ce que lors qu'elles ont esté pillées & depeuplées de bois ou abrouties par les bestiaux, il faut les receper pour y faire revenir le bois; c'est un ménagement que l'on pratique ordinairement pour le restablissement des bois, & la coupe qui se fait pour cela s'appelle recepage"; de Froidour, *Instructions pour les ventes*, 18.
43. "Est tres-commune en cette Province, & notamment dans les forests du roy"; de Froidour, *Instructions pour les ventes*, 17.

Bibliography

Bartoli, Michel. *Louis de Froidour (1626 ?-1685): notre héritage forestier*. Les dossiers forestiers 23. Paris: Office national des forêts, 2011.

Bartoli, Michel, and Bernard Gény. *Histoire des forêts du Béarn jusqu'en 1789: Découverte du règlement forestier de Louis de Froidour (1673)*. Orthez: Société des sciences, Lettres et arts de Pau et du Béarn, 2015.

Bertrand, Georges. "La forêt pyrénéenne réinventée." *Revue de Comminges* 2 (1998): 253–58.

Bonhôte, Jérôme, Jean Cantelaube, and Catherine Verna. "De la mouline à la forge à la catalane: Six siècles de sidérurgie directe dans les Pyrénées ariégeoises (XIIIe-XIXe siècles)." *Revue de Comminges* 4 (2000): 749–92.

Bonhôte, Jérôme, Bernard Davasse, Claude Dubois, Didier Galop, Véronique Isard, and Jean-Paul Métailié. "Histoire de l'environnement et cartographie du temps dans la moitié est des Pyrénées: Pour une 'chrono-chorologie.'" In *Les temps de l'environnement*, 501–15. Toulouse: Presses universitaires du Mirail, 2000.

Buridant, Jérôme. "Du 'modèle' à la pratique: la gestion des peuplements caducifoliés dans la France Moderne, XVIe-XVIIIe siècle." *Flaran* 24 (2002): 203–20.

_____. "Froidour avant Froidour: l'action de Louis de Froidour dans les forêts du nord de la France, 1651–1666." *Les Cahiers de Framespa: Nouveaux champs de l'histoire sociale* 13 (15 March 2013).

Cassard, Jean-Christophe. *L'âge d'or capétien: 1180–1328*, edited by Jean-Louis Biget. Paris: Belin, 2012.

Corbin, Alain. *La douceur de l'ombre: L'arbre, source d'émotions, de l'Antiquité à nos jours*. Paris: Fayard, 2013.

Davasse, Bernard. "Forêts charbonniers et paysans dans les Pyrénées de l'Est du Moyen âge à nos jours une approche géographique de l'histoire de l'environnement." *GEODE* (2000).

de Castéran, Paul. *Lettres écrites par M. de Froidour, grand maître enquêteur et général informateur des Eaux et Forêts au département de la grande maîtrise de Languedoc à M. de Hericourt, son procureur général à Toulouse et à M. de Medon, conseiller au présidial de Toulouse, publiées avec des notes par M. Paul de Casteran*. Auch: Imprimerie et lithographie G. Foix, 1899.
de Coincy, Henri. "Quelques précisions sur les grandes maîtrises des Eaux et Forêts en Languedoc." *Annales du Midi* (1924): 39–57, 184–203.
Decq, Edouard. "L'administration des Eaux et Forêts dans le domaine royal en France aux XIVeet XVesiècles: Mémoire posthume d'Edouard Decq." *Bibliothèque de l'École des chartes* 83/84, no. 1 (1022): 65–115.
de Froidour, Louis. *Instructions pour les ventes des bois du roy*. Toulouse: Raymond Bosc, 1668.
_____. *Reglement fait par le sieur de Froidour, grand maistre des Eaux & Forests au département de Languedoc, Haute Guyenne, Bearn, Basse Navarre, Soule & Labourd, concernant les forests du païs de Bigorre: Les observations faites contre ledit reglement et les motifs dudit reglement avec les réponses dudit sieur de Froidour ausdites remarques & observations. Fait à S. Jean de la Castelle, le 29 decembre 1684*. Toulouse: J. Pech, 1685.
Devèze, Michel. *La vie de la forêt française au XVIe siècle*. 2 vols. Paris: Ecole Pratique des Hautes Etudes, 1961.
Fruhauf, Christian. "De la forêt paysanne à la forêt capitaliste en pays de Sault sous l'Ancien Régime." *Revue forestière française* 31, numéro spécial "Sociétés et forêts" (1980): 160–65.
_____. *Forêt et société: de la forêt paysanne à la forêt capitaliste en pays de Sault sous l'Ancien régime (vers 1670–1791)*. Paris: Centre national de la recherche scientifique, 1980.
Gilles, Henri. "L'administration royale des Eaux et Forêts en Languedoc au Moyen-Âge." *Bulletin philologique et historique (jusqu'à 1610)* (1966): 331–73.
Huffel, Gustave. *Histoire des forêts françaises, de l'origine jusqu'à la suppression des maîtrises des Eaux et Forêts*. Nancy: Ecole Nationale des Eaux et Forêts, 1925.
Jéhin, Philippe. *Les forêts des Vosges du Nord du Moyen Âge à la Révolution: Milieux, usages, exploitations*. Strasbourg: Presses universitaires de Strasbourg, 2005.
Ladurie, Emmanuel Le Roy. *Les paysans de Languedoc*. Paris—La Haye: Mouton, 1974.
Mayade-Claustre, Julie. "Lydwine Scordia, 'Le roi doit vivre du sien': La théorie de l'impôt en France (xiiie-xve siècles)." *Médiévales: Langues, Textes, Histoire* 52 (1 June 2007): 197–99.
Poublanc, Sébastien. "Compter les arbres: une histoire des forêts méridionales à l'époque modern." PhD Thesis, Université Toulouse 2 Jean Jaurès, 2015.
Vié, Louis. "Louis de Froidour commissaire député pour la reformation puis grand maître des eaux et forêts." *Revue de Comminges* (1913): 191–209.

CHAPTER 10

Conserving the "Vert" in Early Modern Sherwood Forest

Sara Morrison

During the seventeenth century, trees in the royal forest of Sherwood grew by natural regeneration, not in man-made plantations of acorns and saplings. Arriving in the forest in 1662, the royal surveyor of woods Thomas Corbin observed a large expanse of mature trees called Birkland and Bilhagh covering two thousand acres with nearly 42,000 oaks. Corbin described one "old wood" (*great hagge*) where the trees "grow upon ould Rootes" and had become "the biggest sort of Timber."[1] Corbin's valuable observations provide a rare glimpse into traditional woodland practice in two mature, heavily grazed woods within a vast royal forest. Writing to the Lord Treasurer, the Earl of Southampton, Corbin provides an account of the natural regeneration process in Birkland and Bilhagh, the Crown's major oak woods within the forest. Concluding that these woods were useful for the navy, he proposed felling the trees in winter because the roots "are very good and sound" and will put forth "very strong suites" if preserved from sheep and cattle. To limit grazing and prevent local townspeople sending in their pigs, he recommended fencing the woods in 1662. Corbin was confident that if temporarily protected, the new shoots would "not only be trees but keep up the glory of the forest."[2] Here, Corbin is balancing the long-term management for the "glory of the forest" with short-term supply of grazing, explaining how to do both through the very strong shoots growing upon the roots of Sherwood's ancient oaks.

In the late seventeenth century, private and royal woods buttressed England's security, providing timber for building ships, "the wooden walls" of the realm.[3] The royal forest of Sherwood furnished a small yet secure supply of naval timber consisting of oaks.[4] The reliability of

Sherwood's timber production demonstrates a high degree of stability in management practices over the long term, because it took at least eighty to one hundred years to produce serviceable naval timber, and many of this forest's oaks were three hundred years old. As Corbin noted, this timber was produced entirely from the regrowth of cut stumps and old roots, an approach he thought would perpetuate indefinitely.[5]

Woodland practices in seventeenth-century England reflected enduring methods of sustainable management, used by village foresters and royal forest officials alike. The pre-modern wood economy relied on good husbandry of woodlands, which was handed down by customary experience and reinforced by early state protection from at least the late fifteenth century.[6] Different techniques were used depending on the desired outcome: small wood and poles, mature timber trees, or multiple-use under wood pasture. Foresters employed coppicing to allow for periodic felling of small wood. Coppice with standards had the same effect of producing new shoots by natural regeneration, but also ensured that more mature trees were mixed in with young ones. Foresters also used a wood-pasture system, which accommodated grazing animals, including the royal deer, alongside timber management. Within the royal forests, all of these management practices were upheld by local forest officials under the auspices of the royal forest courts.

Historians have instead seized, however, on the widely publicized plantation schemes proposed by Arthur Standish and John Evelyn, as well as the Parliamentary acts they helped inspire for making plantations in the Forest of Dean (1668) and the New Forest (1698). Scholars have no doubt gravitated to these plans and acts because they provide highly visible indications of "improvement" and modernity, which assumes inevitable "progress" in managing the environment. Yet even in Dean and New Forest, the long-growth cycle needed for naval timbers meant that it was not until the later eighteenth century that the navy yards received plantation-grown timber. Moreover, this new source of supply was limited, because these early plantations were not very successful. They encountered both unforeseen difficulties in setting acorns and new saplings and resistance from forest commoners who resented loss of access for their animals.[7]

This chapter argues that during the seventeenth century, many of England's royal forests, such as Sherwood in Nottinghamshire, enjoyed three levels of protection, all tending toward the conservative or sustainable management of resources. The first and most important of these consisted of the customary methods of traditional woodland management just noted above, practices that are now familiar to scholars at least in general terms. A second level of protection, also relatively well known, derived from increasingly proactive intervention by the Tudor-Stuart state. Scholars have often overlooked, however, a third layer of stewardship that served to sustain trees, woods, understory, and ground cover in royal forests: a revived system of forest law and its attendant local courts. Medieval kings developed this system of forest law, as Richard Keyser's chapter in this volume discusses; since the sixteenth century, a new series of decrees placed it under the authority of high-ranking royal officials.[8] Its operation, however, was largely in the hands of local people. Perhaps most important among local officials were verderers, prominent men elected by the county's freeholders who were sworn to protect the woods and report offences to forest courts.

Scholars have underestimated the role of these local forest courts and their associated personnel in part because this legal system declined at the end of the Middle Ages. There was an almost three-hundred–year gap between the last of Sherwood's medieval "regards," or royally supervised forest inquests, which was carried out in 1393, and the next one, conducted in 1662–63.[9] Partially compensating for the intervening period's lack of an inspection on the scale of the regard, which was supposed to list all known offences committed against "vert" (trees) and "venison" (deer), were a series of Tudor-Stuart statutes that extended royal protection over all woods, whether in forests or not. Tudor monarchs also continued to supervise the royal forests through the Exchequer Court; two Exchequer commissions reported on Sherwood in 1539 and 1598.[10] It was not until the 1630s, however, that Charles I ushered in changes that proved more significant at the local level by reviving the traditional system of forest courts. Previously, historians have been interested in the economic and political aspects of Charles I's forest policy rather than any local impact on traditional woodland management.[11]

This chapter explores early conservation measures that operated within the English royal forests, beginning with an introduction to the royal forests' legal system and the ways it enhanced the protection given to the vert. The second section summarizes Tudor and Stuart state intervention through legislation that applied to all English woods. As fears of fuel scarcity mounted in the sixteenth century, the state introduced measures to sustain fuel supply and protect timber reserves over the long term. The final section illustrates traditional woodland practices employed in seventeenth-century Sherwood, which contrasts with the experimental forestry projects that have received so much previous attention.[12] I argue that the revival of forest law in Sherwood enabled forest courts to both prosecute vert offences and support compliance with early modern woodland statutes. Three case studies illustrate a variety of management: (1) Thorneywood Chase, near Nottingham in southern Sherwood Forest, had a number of small coppiced woods in private ownership; (2) Clipstone Park, an example of private wood pasture, belonged to the Cavendish family, later dukes of Newcastle, who were leading forest landowners with woods in the northern and central forest; and (3) finally, in contrast to private woods, the royal woods of Birkland and Bilhagh were open wood-pasture under silvo-pastoral management.

Conserving the Vert in Royal Forests

Dating from the Norman Conquest (1066), the royal forests were set up as hunting preserves. As John Manwood summed up this traditional view in 1598, a royal forest was "a certaine Territorie of wooddy grounds and fruitfull pastures" where wild beasts of the forest rested under "the safe protection" of monarchs.[13] These legally defined forests were man-made ecosystems to promote specific crops of deer, wood, timber, and fuel for the royal household. Forest officers managed the vert (trees and covert) and venison (deer) for the king's delight; those in charge of vert were aptly named verderers.[14] They kept a record of all wood felled within the forest by royal warrants, and they presented any cases of illegal felling or "waste" (destruction of woods) to the attachment court, which met every forty days, and referred more complex cases to the highest court, the forest eyre or Justice Seat.[15] This body of forest law, applied in forest courts by specially appointed officers, provided an extra layer of protection for trees and woods in areas designated as forest.

Medieval historian Charles Young has drawn attention to what he called the "conservation policies" of English royal forests. He detailed the checks and balances of this legal system, arguing that tightly enforced rules protecting vert "conserved" timber and woods and provided cover for the deer. He believed forest law's prohibitions "at least delayed the destruction of the forests."[16] Citing the numerous "disafforestations," or releases from forest law, of 1327, the disappearance of the regard, and infrequent meetings of forest eyres, Young referred to the sixteenth-century destruction of woods as "a negative illustration" of withdrawing medieval forest policies.[17] In assessing the efficacy of medieval forest law for conservation, other scholars too have noted that this system underwent a long eclipse until the early seventeenth century.[18]

Although both forest law and early modern statutes often referred to the "preservation" of woods, the intention was never to prevent use in the modern sense of this term. Rather, the goal was to regulate use, to ensure the wise use of vert resources, or, in a word, to promote their conservation. The monarch needed access to a regular supply of timber for construction and repairs for royal residences, park fences, hunting lodges, mills, and bridges, as well as fuel wood to heat castles, manors, and hunting lodges. Good stewardship of all woods, not just royal forest vert, was imperative in the premodern, wood-based economy. Both time-honored customary practice and state intervention were vital to the survival of the English woods in the seventeenth century as demands escalated. The revival of forest law in seventeenth-century Sherwood added a further layer of protection for the vert by reinforcing and enhancing methods of sustainable management that were already practiced throughout medieval and early modern England.

Early Modern State Intervention

As the strength of forest law weakened in the later fifteenth century, the Tudor-Stuart state stepped in to preserve vert, first in the remaining royal forests and by the sixteenth century, throughout all of England's woods.[19] These laws built on earlier ones, such as an act of 1482 that protected all woods in royal forests, chases, and purlieus; after felling trees, forest officers had to fence out animals from these woods for seven years, which increased from three years previously. By increasing the period for fencing after cutting, the young shoots or "springs" were more likely to survive, reducing the risk from nibbling by deer and cattle.[20] Tudor monarchs continued to extend the enclosure period to aid with natural regeneration of coppices and woods.

Coppice with Standards

In 1543, Henry VIII's Act for the "preservation of woods" recognized coppice with standards as the accepted practice to ensure protection to the growth of small trees, or "springwood." In addition, the Act stipulated that for each wood over two acres, the owners who cut them must retain at least twelve "standard" (or maturing trees) trees per acre. Standard trees were not to be felled until they reached a girth of ten inches circumference; any infringement came with steep fines of three shillings and four pence for every great tree removed.[21] One outcome was to build up supplies of mature trees, known as "standards" or "storers," in manors across England. This represented a long-term commitment to protecting timber for the navy, which was to become invaluable during the seventeenth century. In 1570, Elizabeth I strengthened the original 1543 act, extending the enclosure period of coppices after felling from seven to nine years.[22] These Tudor acts encouraged mixed aged woods, prevented clear cutting, and supplied constant fuel from springwood growth. In the later sixteenth century, Elizabeth introduced a statute to restrict Sussex iron producers and Wealden dye makers, both because their massive fuel demands affected prices near London and because they were a potential threat to mature standard trees.[23] However, as some have argued, including Andrew Yarranton already in 1677, the iron industry protected coppices with standards, since ironmasters' demand for fuel deterred the conversion of woods to pasture and tillage.[24] Ironmasters were not interested in timber trees, which are an inefficient source of fuel; they wanted small poles from coppices for making charcoal. By adopting the practice of coppice-with-standards from the mid-sixteenth century, small trees and timber trees grew simultaneously within one wood.[25]

The intentional spoil or destruction of trees, called "waste," was among the most commonly noted offences within the royal forests. In the late sixteenth century, Elizabeth's officials in Sherwood, lacking specialized forest courts, prosecuted this and other forest offenses in the Exchequer court.[26] As the revived forest law became more active over the seventeenth century, forest courts helped to document—and presumably to limit—the widespread waste offences that were recorded during the twenty years of the English Revolution and Interregnum (1640–60). Partly in response to these disorders, in 1663, Charles II passed an Act for the punishment of unlawful cutting or stealing or spoiling of wood and under-wood and of destroyers of young timber-trees. After 1661, the restored monarchy also took a new interest in protecting timber trees for naval building. Navy purveyors used wood from a variety of sources in the late seventeenth century: a small but secure supply from various royal forests supplemented by timber from private owners nearer the dockyards.[27] The attraction of "free" royal timber offset the travel costs, but it was also a way to leverage local timber prices.

Bark, Ferns, and Ling

From the early seventeenth century, an expanding tanning industry, with its need for tannin-rich tree bark, put additional pressure on woodland resources. In 1604, James I introduced a statute setting down regulations for taking tree bark for tanning.[28] This act applied to the whole country, setting out the traditional felling season in spring when the sap rose, between 1 April and 30 June, and restricted all other felling for this purpose. Once the trees were on the ground, the bark was removed; however, the process was often messy.

In the 1680s, in response to increased demands for both shipbuilding timber and tanning bark, Robert Plot, an Oxfordshire naturalist, developed a new method for debarking, called *pilling,* which stripped the bark from standing timber trees in spring but left them until the following winter before felling. Much easier and less messy than debarking felled trees, pilling also promoted the seasoning of standing timber for the navy. As Plot explained in a treatise of 1687 that he sent to Samuel Pepys and the Navy Board, the quality of the stripped timber was improved by standing throughout the summer months to dry out, and could be harvested unharmed the following winter.[29]

Early Stuart statutes also targeted heath burning and the cutting of "ling," or heather (*Calluna vulgaris*), in order to manage both vert and game birds. In 1610, James I introduced a statute to prohibit setting fires to promote regeneration of moorland grasses and ling in heathlands within northern counties, including Nottinghamshire.[30] Forest courts upheld this statute.[31] Subsequent statutes of 1706 and 1708 include provisions that specifically concern Sherwood Forest. The first sought to protect ling and the heathlands that harbored Queen Anne's red deer herds, while the second banned the burning of ferns for ashes, which damaged ground cover used by deer as covert and disturbed the royal game birds.[32] Protecting vert for deer habitat was a major concern for Queen Anne.[33]

Stuart Forest Revival

All of the above examples reflect the early modern English state's intervention to promote the growth of trees and manage woodland resources across the country. At the same time, Tudor and Stuart monarchs still held many royal forests as part of the crown estate. A greater interest in the financial potential of these forests drew Elizabeth and the early Stuarts into a reassessment of their forest assets, although Elizabeth and James I both rejected projects to "farm" or lease out the royal forests for short-term profit.[34] James I's interest in his woods included countywide appraisals of royal coppices and timber woods, including in selected forests such as Sherwood.[35] There is strong evidence of renewed activity in Sherwood's local forest courts even during the troubled decades of the 1640s and 1650s.[36] Even then, the verderers of the forest court continued to prosecute infringements of both forest law and Tudor statutes.[37]

Historians have usually emphasized, however, the late medieval and early modern decline of local forest law, drawing in particular on some of the statements made by John Manwood in his famous treatise, *A Treatise and Discourse of the Lawes of the Forest* (1598). In documenting the royal forest institution and legal system, he reflected on its decay in the late sixteenth century and presented his work as an epitaph to the lost age of royal forest law. However, neither later editions of Manwood's book nor the seventeenth-century forest record support this demise. Several new editions of his work across the seventeenth and early eighteenth centuries suggest more interest in the royal forests and their management than its author observed in 1598. For example, the 1717 edition incorporated recent forest records, including specific legal cases, statutes, and decisions from the seventeenth century. Apparently, forest law had not died out and was still in operation.[38] This and other later editions of Manwood's book signified an ongoing interest in forest governance and the need to update the text as a working manual for those affected by forest law.

Similarly, while historians have noted the revival of forest government under the early Stuarts, they have focused on the period before 1640 and the monarchs' political and financial motives.[39] Research on the Stuart royal forests shows that Sherwood had a system of forest law that remained in force even during the Interregnum (1649–60) and then continued to function well into the eighteenth century.[40] In Sherwood, the revival of forest law took place at all levels. At the lowest level were the swainmote and attachment courts, which had jurisdiction over forest grazing and the vert respectively. More serious cases, or those that were not resolved in the first instance, were referred to a forest eyre, a specialized higher court, which last sat intermittently between 1663 and 1676.

Furthermore, Sherwood's forest courts were able to draw on the information gathered by the systematic surveys of royal forests that were begun under James I in the early seventeenth century. Verderers working for Sherwood's attachment courts built on this record by documenting all felling activities: royal warrants for felling wood for building or gifts, fees paid to forest officers, licenses to cut coppices, and illegal actions, all of which were presented at the next session. Their records provided the necessary accountability for forest keepers guarding the different "walks" or regions of Sherwood to explain who, why, and when vert was cut in the forest. Without this record, individual keepers were liable. Various forest books compiled specifically for Sherwood have survived from the early seventeenth century.[41]

After the Restoration in 1660, Charles II called a forest regard in Sherwood between 1662 and 1663; the regard roll listed offences committed in Sherwood since the last Justice Seat or forest eyre, in preparation for the next session held between 1663 and 1676.[42] There were very few forest eyres after 1660; the only other courts besides Sherwood were held in New Forest and Waltham Forest.[43] While the Justice Seat was a rare event, verderers' monthly attachment courts recorded the daily infractions on woods. In Sherwood, these records provide a valuable picture of the Justice Seat's important role in policing the vert during the seventeenth and early eighteenth centuries. Verderers presided over attachment courts to oversee the vert: timber, woods, coppices, and surface vegetation. They heard petitions for and granted licenses to the owners of private woods who wished to cut their coppices inside the forest boundary, viewing all woods to ensure enclosures did not harm the deer. Licenses granted by the verderers not only protected the royal right to vert as covert for game in the royal forest, but also reinforced and upheld the Tudor statutes. Practices for sustainable management included fencing coppices, keeping cattle out, and regulating the felling season, length of time for enclosure, and retaining standard trees.[44]

Sherwood Forest

Woodlands and Environment

The Nottinghamshire forest of Sherwood lay at the heart of midland England. It was the largest royal forest north of the River Trent, which formed its southernmost boundary. When many forests were either sold or their bounds reorganized during the seventeenth century, Sherwood remained as a royal forest; however, it was not a contiguous stretch of woodland. The 1609 Crown survey shows that Sherwood Forest contained a large hunting area with pockets of woodland, islands of cultivation and forest villages, some grassy lawns, and large areas of forest waste and heathlands.[45] There were no royal coppices, like in some royal forests, such as the Hampshire forests of Buckholt and New Forest.[46] In the heart of the forest, the Crown owned four ancient "hays," or woods. The most important were Birkland and Bilhagh, which provided shade for red deer, mature timber, and vital forest commons to local villages. There were also many privately owned coppices and woods.

The extensive Crown survey revealed a number of startling truths about Sherwood in 1609. The forest occupied 95,000 acres, which included 89,000 acres of land in wastes, woods, and cultivation, plus 6,000 acres in parks. The woods covered under 10,000 acres (9486 acres), which was only a tenth of the area under forest law. Large parts of the forest lay as wastes (35,080 acres) with permanent cultivation occupying just under 50,000 acres (49,839 acres) in 1609.[47] The system of temporary cultivation, or "breck," used in Sherwood Forest may account for additional woods in waste areas with poor sandy soils in the north and central forest, where villages petitioned the forest courts to enclose waste parcels for seven years.[48] The survey also showed that in 1609, the Crown estate in Sherwood was small; there were many red deer but no longer any royal houses nearby for hunting. The Crown owned roughly 2,000 acres of oak woods, mostly in the center of the forest in two ancient woods called Birkland and Bilhagh, measuring about 1,000 acres apiece. A survey by James I in 1608 returned the total number of trees in these woods as 49,900 oaks; most were very old or aging timber trees that needed immediate felling.[49] Several of these ancient oaks still survive in Birkland, with the most famous Major Oak estimated at over eight hundred years old. Although the area of royal woods in Sherwood was quite small, its management as High Forest allowed it to provide valuable timber, building materials, and wood rights and other commons to local inhabitants.

Since the medieval period, Sherwood Forest was traditionally known as a hunting forest and was not covered with trees like a modern-day plantation, nor was all land and wood owned by the Crown. In 1609, there were a number of privately owned woods within Sherwood Forest, but this was nothing new. The medieval regard in 1393 listed private woods, especially in the southern forest near Nottingham. Only

Figure 10.1. The Major Oak. © Anthony Parkes, reproduced under a Creative Commons Attribution-ShareAlike 2.0 Generic license.

a tenth of the legal forest territory of Sherwood was wooded in 1609; besides the Crown's 2,000 acres in Birkland and Bilhagh, another 8,000 acres were privately owned. Manorial woods belonged to lords and forest townships in the north and center, whereas the southern forest, called Thorneywood Chase near Nottingham, comprised small parcels of privately owned coppices.[50]

Statutes in Sherwood

While I argue that forest law offered an extra layer of protection within royal forests, it is important not to underestimate the role of early modern statutes in the preservation of woods by natural regeneration. By the mid-seventeenth century, some areas of Sherwood were protected by statutes alone. For example, manorial woods belonging to the Cavendish estate were exempt from forest jurisdiction through a special royal grant of March 1640. This permitted William Cavendish, earl and later duke of Newcastle, to cut woods, make charcoal, and fence out the royal deer.[51] Newcastle owned a large part of central Sherwood in the manors of Mansfield, Mansfield Woodhouse, Edwinstowe, Clipstone, Sutton, and Kirkby in Ashfield.[52] Returning from exile in 1660, Newcastle leased his woods to local ironmasters for fuel.

Between the 1660s and the early eighteenth century, manorial agreements between the Newcastle family and local wood merchants and ironmasters showed the duke upheld woodland practice according to the Tudor statutes. Woods were coppiced and fenced after felling and remained enclosed for nine years. Ironmasters were permitted to take dead roots from the woods, but prohibited from digging up live roots, which halted natural regeneration. They were to return the woods in good condition after the term, usually ten years. In any single wood, the felling occurred in thirds, so that from years one to three, a third of a wood was cut, and so on. Newcastle reserved the right to reenter his woods to fence them immediately after cutting, following the Tudor statutes, to preserve springwoods.[53] Newcastle's woods attracted some of the most influential ironmasters of this period,[54] including Humphrey Jennens of Erdington Hall, Warwickshire, and his sons John and Charles, who set up Kirkby furnace in 1671 where they made cast iron in 1693; also the Sitwell family who ran Clipstone Park forge in 1656 and operated forges and furnaces in Derbyshire and South Yorkshire.[55] This case study in Sherwood offers support for Oliver Rackham's idea that industrial usage tended to preserve woods, even though they were mainly coppices, because of the value placed on fuel.[56]

Sherwood Forest Law

Forest governance in Sherwood reinforced woodland practices, particularly in coppicing and protecting standard trees in private woods. Local forest courts of attachment recorded coppice license petitions, verderers viewing woods, fee trees awarded to forest officers as payment for service, and fines for offences such as theft and waste, or illegal cutting.[57] Where court records have not survived, forest books belonging to officers list coppice petitions, royal warrants for trees, and offences between 1614 and 1660.[58] A book kept by William Clay, forest clerk, during the 1660s and 1670s records petitions for licenses to cut coppices in Thorneywood Chase.[59] Since it was possible to enclose private woods, they were easier to manage than the Crown's open wood-pastures, which were subject to a multitude of rights.

The forest courts in Sherwood applied forest law for vert, while reinforcing early modern statutes, in both Crown and private woods. This is illustrated by three examples: (1) private coppices in Thorneywood Chase, where owners regularly petitioned attachment courts for a license to cut and fence woods; (2) private woods belonging to Clipstone

Park, which Sherwood's verderers protected in the 1650s through the forest courts; and (3) the royal woods of Birkland and Bilhagh, which in the 1660s were managed under wood-pasture, supporting multiple demands for forest commons, royal deer, and naval timber.

Private Coppice: Thorneywood Chase

The location of coppices in Thorneywood Chase appear on a rare map for southern Sherwood, which accompanied the 1609 survey.[60] Throughout the early modern period, the same family names in forest villages of Lambley, Lowdham, Gedling, Woodborough, Bulcote, and Burton Joyce reappear in petitions to enclose coppices in swainmote and attachment courts. All coppice enclosures had nine-year licenses, although the total acreage of wood enclosed varied over time.[61] Verderers viewed all woods prior to licensing, which was always conditional on sufficient available deer browse. These woods were located in the south of Sherwood Forest near Nottingham, shown on Richard Blome's 1673 Nottinghamshire county map with the forest boundary of 1609 (see Map 10.1).

John Gosling's forest book contains numerous petitions for coppice licenses and warrants to cut wood, or fee trees awarded between 1614 and 1660.[62] Many but not all coppice petitions were from Thorneywood, and a few specified the number of trees in a wood.[63] Letters between the verderers and the Lord Warden in the 1650s show no break in laws regulating coppice licenses. In February 1650, Lady Foljambe had permission to cut woods in Lowdham, and promised "to leave sufficient standills." The following year, verderers allowed the Marquis of Dorchester to fell Marshall Hills in Thorney Woods, but required him to make a wall and fences to "preserve the spring" for nine years.[64] The swainmote court at Mansfield continued to charge offenders for unlawful felling and theft. In June 1648, the widow Frances Wood of Lambley had cut her wood but failed to preserve the spring, while Richard James of Nottingham took away thirty-two oak standards from Parson's Coppice (Lambley) worth over fifteen pounds. In June 1650, the Mansfield swainmote charged Sir William Dalston for putting his cattle into Horleywood in Lowdham, which had destroyed the springwoods.[65] In these cases, forest courts added extra support to the early modern statutes.

During the post-Restoration forest eyre of 1663–76, William Clay's forest book contained petitions, warrants, and presentments related to vert. In 1666, Charles Lacocke of Woodborough petitioned for a license to fell and coppice his twelve-acre wood of Mardales, which verderers viewed and allowed.[66] In September 1674, the Chief Justice in eyre allowed John Overbury of Lambley to cut three acres of his spring-woods and enclose it for nine years.[67] In January 1674 as the court was ending, there was a flurry of petitions for licenses from coppice owners in Lambley, Lowdham, and Woodborough.[68] Coppice licenses were usually granted as long as the enclosure was not prejudicial to the deer. For this reason, crab apple trees, whose leaves and fruit were much favored by deer, were often excluded, especially during the very cold winters of the early eighteenth century. On 30 September 1715, the attachment court excluded from any cutting all holly trees and crab apple trees, which were reserved for deer browse, to prevent them leaving the forest.[69]

Private Wood-Pasture: Clipstone Park

In the 1650s, Sherwood's foresters and verderers zealously used the attachment court to safeguard private woods from poor stewardship practices. The Commonwealth and Protectorate government had sold Clipstone Park and its extensive woods, which had belonged to a leading royalist, William Cavendish, earl of Newcastle.[70] The new owners contravened almost every safeguard protecting these woods for a quick profit. The Trustees for the Sale of Royalists' estates had sold Clipstone without knowledge of obligations to commoners for *estovers*, or wood rights, or the need to petition forest courts before felling. In 1652, the forester in charge of Clipstone, Mr. Firbourne, brought a charge against Captain Barrett in the swainmote court for failing to notify the court of his plans to fell in Clipstone woods.[71] William Barrett, a former Parliamentary officer, had already felled 160 trees and set up an iron forge in Clipstone Park, unaware of his neglect in forest matters. In 1653, a correspondence ensued among the verderers, the Lord Warden of Sherwood Forest, and the London-based Trustees for land sales to make Captain Barrett aware of his responsibilities under forest law. Barrett promised to attend the next sessions and petition the court for a license to cut his woods.[72] For Clipstone Park, see Map 10.1.

By 1655, Clipstone Park woods had passed to a London draper, Mr. Clarke, who was either oblivious or unconcerned about forest law. Clarke felled without a license contravening forest law. Reporting to the Lord Warden of Sherwood in April 1655, the verderers found Clarke

1. Nottingham, 2. Carleton, 3. Burton Joyce, 4. Lambley, 5. Lowdham, 6. Woodborough, 7. Clipstone, 8. Edwinstowe, 9. Ollerton, 10, Budby, 11. Thoresby, 12. Walesby, 13. Welback Park, 14. Warsop, 15. Mansfield-Woodhouse, 16. Mansfield, 17. Sutton-in-Ashfield, 18. Kirkby-in-Ashfield, 19. Annesley, A. Birkland, B. Bilhagh

Map 10.1. Nottinghamshire County (Richard Blome, 1673), showing Sherwood Forest's 1609 boundary and places mentioned in the text. University of Nottingham Libraries, published with permission.

had cut many trees, including standards; this violated the 1543 statute requiring twelve standards per acre in woods over two acres. He also paid local laborers huge salaries to remove the trees quickly and had not enclosed the woods. Clarke had felled in March, which was too early and before the tanners were able to strip or pill the bark. This is an example of Sherwood's verderers using the forest attachment court to uphold Tudor statutes as well, charging Clarke with forest infractions.[73] Verderers were concerned because Clarke had access to 28,000 trees in Clipstone woods. While verderers intervened to stop the worst infringements against the vert, Clarke's indiscriminate felling also endangered the remaining deer.[74] His methods were totally at odds with vert preservation by natural regeneration, which made Clarke's actions a greater threat to sustainability than Captain Barrett's oversight. However, by November 1655, Clarke no longer held the woods, which were recovered by Cavendish's brother. He granted the timber and other trees to a Derbyshire ironmaster, George Sitwell, in September 1656. Sitwell coppiced Clipstone woods to fuel the forge. In 1660, the park returned to the Cavendish family and Sitwell continued to use the woods for fuel during the 1660s, which boded well for their future sustainability.[75]

Open Wood-Pasture: Birkland and Bilhagh

In contrast to previous examples of private coppices and woods, the royal woods of Birkland and Bilhagh were not enclosed, but managed as open wood pasture for multiple users, including deer. Forest villages had rights to cattle grazing and swine pannage at certain times of year, and by 1660, the timber oaks were used for shipbuilding.[76]

After the Restoration in 1660, the stewardship of royal woods in Sherwood Forest became a priority for Charles II when he realized that his royal forests provided cheap timber supplies for shipbuilding.[77] In 1662, Thomas Corbin, the Surveyor General of woods, was dispatched to Sherwood to survey Birkland and Bilhagh for navy use. Corbin counted nearly 42,000 trees in a very detailed woods survey. In his painstaking account of timber oaks, he priced and listed each tree individually. The density of oaks was twenty-one per acre in 1662, which is only slightly lower than that recorded in two previous surveys of 1608 (24.5 per acre) and 1630 (22 per acre). This indicates that forest conditions were surprisingly stable across fifty-four years and during decades that saw both civil war and financial crises.[78] In December 1662, Corbin was joined by John Russell, the naval purveyor assigned to

Sherwood, so they could sample the quality of a few trees. The original sample swelled to 5,550 timber oaks as they concluded, "there may be many more useful trees."[79] Sherwood Forest continued to send timber oaks to the London dockyards at Woolwich during the 1660s, especially between 1665 and 1667 during the second Anglo-Dutch war.[80]

In 1662, Thomas Corbin urged fencing Birkland and Bilhagh immediately after felling to promote natural regeneration.[81] Under wood-pasture, natural regeneration was difficult even when limiting animal numbers with careful regulation of pastures, or *agistment*. Oliver Rackham described the oaks in Birkland and Bilhagh as "giraffe pollards," because their boughs were cut high above the reach of browsing deer.[82] Besides swine and other commonable beasts, royal deer also had free passage, so at times, the dual mandate of forest law to protect venison and vert favored deer browse over the regeneration of springwood. Luckily for Corbin, in 1662, previous decades of excessive hunting meant few red deer remained in Sherwood compared to earlier surveys.[83]

As wood-pasture commons, these timber woods were under silvopastoral management to protect multiple rights and usages. According to Parliamentary surveys in 1656, they were heavily grazed by at least eight townships, including seven forest villages inside Sherwood, (Edwinstowe, Warsop, Thoresby, Perlthorpe, Budby, Clipstone, and Ollerton), and one outside (Walesby). The villagers claimed *estovers*, or wood rights, including fencebote to repair gates and enclosures, firebote for fuel, and housebote to repair their houses. In summer, cattle sheltered in the woods away from heat and flies. In the late 1650s, these villages had defended their right to pannage, herbage, and estovers in Birkland and Bilhagh before the Parliamentary Committee of Appeals, which was considering an abortive disafforestation plan.[84] With the end of the Protectorate, the plan was shelved; however, memories of the dispute were recent when Thomas Corbin arrived in Sherwood in 1662. Numerous claims from these villages registered pannage rights in Corbin's 1662 survey, for which they paid two pence per pig sent into Birkland and Bilhagh.[85] Rights of pasture for cattle, not goats or geese, pannage for swine at pannage-time when acorns fell, and reasonable estovers were all ancient rights claimed by owners and tenants of Edwinstowe.[86] In 1663, Sherwood's eyre heard and confirmed their ancient commons in the woods.[87] While the village commoners, their animals, and the royal deer posed challenges for Thomas Corbin's plan to protect the oak timber woods after felling, Birkland and Bilhagh still supplied navy timber in 1686 and again in the 1770s.[88]

Conclusion

In Sherwood Forest, trees grew by natural regeneration and no one was allowed to dig or "grub" up tree roots unless the roots were dead. This was embedded into all early modern wood leases on the Newcastle estates. Only on rare occasions were dead roots sold,[89] because good practice meant keeping all roots alive for the next crop of timber or coppice. The importance of Sherwood's strong oak roots was at the heart of Thomas Corbin's report to the Crown in 1662. This practice continued in Sherwood as the most effective method of natural regeneration. In April 1725, Lord Harley, Earl of Oxford, the new owner of Welbeck Park and the Newcastle family's forest manors, described the mighty oaks in his park. The woods were not grown from acorns, "but the stump after a former fall of timber 2, 3, 4, 5 and sometimes 6 growing up together from the same common stock."[90] As a private wood-pasture without commons rights, Harley fenced Welbeck's woods after felling, unlike Sherwood's open wood-pasture in Birkland and Bilhagh. However, both Harley and Corbin found similar conditions in 1662 and 1725, where the living oak roots literally were the roots of conservation and natural regeneration in Sherwood.

The evidence from the royal forest of Sherwood during the Stuart century (1603–1714) indicates that the key element for protecting trees and woods for future use occurred with enclosure after felling, which allowed natural regeneration from stumps and roots. Forest laws supported and reinforced traditional woodland practices within Sherwood when the woods could be fenced—for example, the enclosure of woods after felling in Thorneywood Chase coppices, and in the private woods of the Newcastle family and in Clipstone Park. Here, forest laws, the courts, and their officers strengthened traditional practice and the early modern statutes for sustainable management of woodlands.

In open woods under the wood-pasture system, it was more problematic to conserve woods by natural regeneration. The royal woods of Birkland and Bilhagh were ancient timber woods, which supported hundreds of forest inhabitants with commons rights of grazing for cattle and pannage for pigs. These woods were under multiple uses and lay open for the royal deer to browse at will. In this case, forest laws protecting game and commons access were in opposition to customary practice and statutes to protect woods by natural regeneration. The main problem in these open woods was that forest laws protected both venison and vert. As forest law took precedence over customary law in royal forests, which were primarily hunting forests, the stewardship of the deer in Sherwood was often at the expense of natural regeneration in Birkland and Bilhagh's oak woods.

Sara Morrison is an Associate Professor at Brescia University College, Canada's only women's college, affiliated to Western University in London, Ontario, Canada. She is completing a book, *The Stuart Royal Forests: An Environmental History*. Recently, she has published articles on "Good Stewardship and the Challenges of Managing the Stuart Royal Forests" (2014) and on red deer in eighteenth-century Sherwood Forest entitled, "Bambi in Sherwood Forest and the Great Deer Escape, 1703–11" (2017).

Notes

1. The National Archives (TNA), PRO 30/24/31/6 f. 68.
2. TNA, PRO 30/24/31/6 f. 68.
3. John Evelyn, "To the Reader," address to the Royal Society, 15 October 1662, preface in John Evelyn, *Sylva, or A Discourse of Forest-Trees, and the Propagation of Timber in His Majesty's Dominions* (London: Martin and Allestry, 1664); Noel D. G. James, *A History of English Forestry* (Oxford: Basil Blackwell, 1981), 138–51.
4. Robert G. Albion, *Forests and Sea Power: The Timber Problem of the Royal Navy 1652–1862* (Cambridge, MA: Harvard University Press, 1926), 108–9, with a map of the royal forests and oak producing counties. Sara Morrison, "Forests of Masts and Seas of Trees: The Restoration Navy and the English Royal Forests," in *English Atlantics Revisited: Essays Honouring Professor Ian K. Steele*, ed. Nancy L. Rhoden (Montreal: McGill-Queens University Press, 2007), 136–73.
5. Aged oaks recorded in Sherwood surveys in 1598, 1608, 1656, and 1662. Nottinghamshire Archives (NA), DD4P 75/19, 19; TNA, SP 14/42, f. 35; E 317/Nottinghamshire 11, 23; NA, DDFJ 10/9/8.
6. "An Act for the Preservation of Woods," 1543, 35 Henry VIII, c. 17. This and all Acts of Parliament cited hereafter are edited in the *Statutes of the Realm, 1215–1713*, 9 vols., ed. Alexander Luders et al. (London: G. Eyre and A. Strahan, 1810–1828).
7. "An Act for the Increase and Preservation of Timber within the Forest of Dean," 1668, 20 Car II, c. 3; "An Act for the Increase and Preservation of Timber in the New Forest and the County of Southampton," 1698, 9 & 10 Will

III, c. 36. See: Sara Morrison, "Forests of Masts," 157; James, *English Forestry*, 138–51, 308; Cyril E. Hart, *The Forest of Dean: New History 1550–1818* (Stroud: Alan Sutton Publishing Ltd, 1995); Colin R. Tubbs, *New Forest: An Ecological History* (Newton Abbot: David & Charles, 1968).

8. In 1547, Henry VIII created two Masters and two Surveyors of Woods, each side of the River Trent, reporting to the Court of Augmentations, which in 1554 was absorbed by the Exchequer and overseen by Surveyors-General of Woods, Forests, Parks and Chases. This reduced the importance of earlier Tudor Chief Justices and forest eyres. See Raymond Grant, *The Royal Forests of England* (Wolfeboro Falls, NH: Alan Sutton, 1991), 184, 188–98.
9. British Library (BL), Add. Ch. 10,659; TNA, LRRO 5/46.
10. J. Charles Cox, *The Royal Forests of England* (London: Methuen & Company, 1905), 216–18, refers to the 1539 forest session and deer survey; NA, DD4P 75/19, 19.
11. Unlike most forests south of the Trent (Windsor, Waltham, Dean, and New Forest), Sherwood did not have a forest eyre in the 1630s. See George Hammersley, "The Revival of the Forest Laws under Charles I," *History* 45 (1960): 85–102.
12. Hart, *Forest of Dean*; Tubbs, *New Forest.*
13. John Manwood, *A Treatise and Discourse of the Laws of the Forrest*, 2nd ed. (London: Societe of Stationers, 1615), 18.
14. For definitions and terminology associated with forest laws, courts, and institutions (forest bounds, vert, venison, verderers, court of attachment, swanimote, and eyre), see the preface of any edition of Manwood, *Treatise and Discourse*.
15. On verderers, see *Manwood's Treatise on the Laws of the Forest*, 4th ed., ed. William Nelson (London: E. Nutt, 1717), 248–353. On Sherwood's verderers see: NA, DDFJ 10/9/6, 104–10, 116–18; Sara Morrison, "Fee Trees and Rogue Verderers in Early Eighteenth Century Sherwood Forest," *Transactions of the Thoroton Society* 117 (2013): 97–107.
16. Charles R. Young, "Conservation Policies in the Royal Forests of Medieval England," *Albion* 10, no. 2 (1978): 102–3. On medieval forests, see: Charles R. Young, *The Royal Forests of Medieval England* (Leicester: Leicester University Press, 1979); Cox, *Royal Forests of England*; Grant, *Royal Forests of England*, chapters 1–12; George J. Turner, *Select Pleas of the Forest*, Publications of the Selden Society, 13 (London: B. Quaritch, 1901); Margaret L. Bazeley, "The Extent of the English Forest in the Thirteenth Century," *Transactions of the Royal Historical Society* 4th Ser., 4 (1921): 140–72.
17. Young, "Conservation Policies," 102–3.
18. Grant, *Royal Forests of England*, 169–198.
19. Young, "Conservation Policies," 102–3; Grant, *Royal Forests of England*, 169–86.
20. "An Act for Inclosing of Woods," 1482, 22 Edward IV, c. 7.
21. "An Act for the Preservation of Woods," 1543, 35 Henry VIII, c. 17.

22. "An Act for Reviving and Continuance of Certain Statutes," 1570, 13 Elizabeth I, c. 25; sections 18 and 19 on "inclosure of woods after felling, as in 35 Henry VIII."
23. On Elizabeth's 1581 and 1585 acts protecting timber in the Weald and fuel supply near London, see James, *History of English Forestry*, 121–27, 307.
24. Andrew Yarranton, *England's Improvement by Sea and Land to Out-do the Dutch without Fighting, To Pay Debts without Moneys, To Set at Work all the Poor of England within the Growth of Our Own Lands* (London: R. Everingham, 1677). See Sara Morrison, "Good Stewardship and the Challenges of Managing the Stuart Royal Forests in England, 1603–1714," *Journal of Markets and Morality* 17, no. 2 (2014): 417.
25. Oliver Rackham, *Trees and Woodland in the British Landscape* (London: J. M. Dent, 1976), 91–95.
26. NA, DD4P 75/19. Elizabeth used Exchequer commissions extensively to deal with forest wood offences: TNA, E 178/1774, E 178/1786, E 178/7371, and E 178/1808.
27. "An Act for the Punishment of Unlawful Cutting or Stealing or Spoiling of Wood and Under-wood and Destroyers of Young Timber-trees," 1663, 15 Car II, c. 2. See Morrison "Forests of Masts," 140–41.
28. "An Act Concerning Tanners, Curriers, Shoemakers Cutting Leather," 1604, 1 James I, c. 22; sections 18–21 on bark shortage and restricting oak felling period from 1 April to 30 June. See James, *History of English Forestry*, 307.
29. Robert Plot, "A Discourse Concerning the Most Seasonable Time of Felling Timber," *Philosophical Transactions* 192 (1691): 455–61; "Commissioners . . . to Enquire into the . . . Woods, Forests and Land Revenues of the Crown," 11th Report: "General Naval Timber" (1792), Appendix 33.
30. "An Acte Against Burninge of Linge and Heath, and Other Moorburninge in the Counties of Yorke, Durham, Northumberland, Cumberland, Westmorland, Lancaster, Derbye, Nottingham, Leicester, at Unseasonable Tymes," 1610, 7 James I, c. 17.
31. Swainmote court offences: 14 Sept 1630, for pulling ling, caused great damage and disturbed deer during fence month (10 May–end June), NA, DDFJ10/9/6, 110; and 19 May 1638, punishing any man burning ling, cutting brakes, fern or heath in the forest, or trying to sell it; see Manuscripts and Special Collections, the University of Nottingham (UNMSC), Pw V5, 322.
32. "An Act for the Better Preservation of the Game," 1706, 6 Anne, c. 16, sect 10, Sherwood Forest; "An Acte for the Making of the Act of 5 Anne for the Better Preservation of the Game Perpetual and for Making the Same more Effectual," 1710, 9 Anne, c. 27. Fern ashes were used as fertilizer on sandy soils.
33. On Queen Anne's deer problems, see Sara Morrison, "Bambi in Sherwood Forest and the Great Deer Escape c. 1703–1711," in *Environmental History in the Making*, vol. 1: *Explaining*, ed. Estelita Vaz, Cristina Joanaz de Melo, and Lígia M. Costa Pinto (Basel: Springer Publishing, 2017), 341–57.
34. TNA, SP 12/243, f. 197; SP 14/28, ff. 180–196.

35. George Hammersley, "The Crown Woods and Their Exploitation in the Sixteenth and Seventeenth Centuries," *Bulletin of the Institute of Historical Research* 30 (1957): 136–61.
36. Peter R. Seddon, "The Application of Forest Law in Sherwood Forest, c. 1630–1680," *Transactions of the Thoroton Society* 82 (1978): 37–44.
37. NA, DDFJ/10/9/6.
38. Manwood, *Treatise and Discourse* (1598), preface. See also later editions of 1615, 1665, 1717, and 1741.
39. Hammersley, "The Crown Woods," 136–61; Hammersley, "Revival of the Forest Laws;" Philip A. J. Pettit, *The Royal Forests of Northamptonshire: A Study in their Economy, 1558–1714*, Publications of the Northamptonshire Record Society, 23 (Northampton: Northamptonshire Record Society, 1968); Grant, *Royal Forests of England*, 186–204. See also: Joan Thirsk, "The Crown as Projector on Its Own Estates, from Elizabeth I to Charles I," in *The Estates of the English Crown 1558*–1640, ed. Richard W. Hoyle (Cambridge, UK: Cambridge University Press, 1992), 297–352; Richard W. Hoyle, "Disafforestation and Drainage: The Crown as Entrepreneur?" in *Estates of the English Crown*, 353–88; and Peter Large, "From Swanimote to Disafforestation: Feckenham Forest in the Early Seventeenth Century," in *Estates of the English Crown*, 389–417.
40. Seddon, "Application of Forest Law"; Morrison, "Bambi in Sherwood Forest." On the revival of forest law in eighteenth-century Windsor Forest, see Edward P. Thompson, *Whigs and Hunters: The Origin of the Black Act* (New York: Pantheon Books, 1975).
41. NA, DDFJ 10/9/6. John Gosling records attachment court business, including correspondence between verderers and the Lord Warden of Sherwood.
42. TNA, LRRO 5/46 listed the vert offences since the last Justice Seat, including illegal felling, thefts, and waste offences in private and royal woods.
43. David Crook, "The Records of Forest Eyres in the Public Record Office, 1179 to 1670," *Journal of the Society of Archivists* 17, no. 2 (1996): 183–93.
44. Manwood, *Manwood's Treatise of the Forest Laws*, 4th ed. (1717), 145–46, 370, 381, 383–85.
45. Stephanos Mastoris and Sue Groves, eds., *Sherwood Forest in 1609: A Crown Survey by Richard Bankes*, Thoroton Society of Nottinghamshire Record Series 40 (Nottingham: Thoroton Society of Nottinghamshire, 1997); TNA, LR 2/194, f. 35.
46. TNA, LR 2/203, ff. 1–15.
47. "Commissioners . . . to Enquire into the . . . Woods," 14th Report: "Sherwood Forest" (1793); acreages cited from this report. The 1609 survey is often vague, especially concerning partly wooded closes or forest wastes; see David Crook, "Introduction," in *Sherwood Forest in 1609*, ed. Mastoris and Groves, xix–xxxix.
48. Dudley Fowkes, "The Breck System of Sherwood Forest," *Transactions of the Thoroton Society* 81 (1977): 55–61.

49. Mastoris and Groves, *Sherwood Forest in 1609*; TNA, LR 2/194, f. 21; SP 14/42, f. 35.
50. BL, Add. Ch. 10,659; Mastoris and Groves, *Sherwood Forest in 1609.*
51. NA, DD2P 27/11. Royal patent set out enclosure for nine years, cutting a third of the wood every three years.
52. The Crown sold several manors in Sherwood to raise money in 1601, 1603, and 1632; see "Commissioners . . . to Enquire into the . . . Woods," 14th Report, App 10, 32–33.
53. NA: DDP 15/55–56, 61; DD3P 7/1–3.
54. Philip Riden, "The Charcoal Industry in the East Midlands, 1580–1780," *Derbyshire Archaeological Journal* 111 (1991): 64–84.
55. R. Johnson, "17th Century Iron Works at Bulwell and Kirkby," *Transactions of the Thoroton Society* 64 (1960): 44–46; Philip Riden, ed., *George Sitwell's Letterbook, 1662–66*, Publications of the Derbyshire Record Society 10 (Chesterfield: Derbyshire Record Society, 1985); NA, DD4P 84/34.
56. Rackham, *Trees and Woodland*, 91–95.
57. Continuous court record 1703–23, see NA, DDP 27/3; SO.FR 6.
58. NA, DDFJ 10/9/6, 107–10.
59. NA, DD 382/2, 202–24.
60. Mastoris and Groves, *Sherwood Forest in 1609*. Coppices on Carlton, Burton Joyce, Lambley, Woodborough, and Nottingham maps.
61. Forest books contain attachment and swanimote cases in NA, DDFJ10/9/6, 107–10; DD 382/2; DDP27/3, SO FR 6.
62. NA, DDFJ10/9/6, 107–10.
63. On 23 March 1641, Lord Chaworth allowed to cut 1,300 "dotard and decayed trees, plus 80 ashes and maples on Ansley Hills, which left 1,000 oaks and other trees remaining," NA, DDFJ10/9/6, 109.
64. NUMASC, Pw V5, 340.
65. NUMASC, Pw V5, 334. Swainmotes held three times a year: 8 June, 14 Sept., and 17 Nov.
66. NA, DD382/2, 202–4.
67. NA, DD382/2, 213.
68. NA, DD382/2, 217–24.
69. NA, DDP27/3; Morrison, "Bambi in Sherwood Forest."
70. Cavendish purchased the park in 1630 and went into exile in the 1650s. The park was seven miles around and had the best timber trees on Cavendish's estates. David Crook, "Clipstone Park and Peel," *Transactions of the Thoroton Society* 80 (1976): 35–46.
71. NA, DDP 27/23g.
72. NA, DDFJ 10/9/6, ff. 116–18.
73. TNA, SP 18/96 ff. 89–93.
74. TNA, SP 18/96, ff. 89–93; NA, DDFJ 10/9/6, ff. 116–18.
75. NA, DD4P 84/34; Sir George Sitwell, "The Loyal Duke of Newcastle," *Derbyshire Archaeological and Natural History Society Journal* 13 (1891): 1–9.

See Riden, *George Sitwell's Letterbook*. On the damage committed at Clipstone during the interregnum, see Margaret Cavendish, Duchess of Newcastle, *The Life of William Cavendish, Duke of Newcastle*, ed. Ernest Rhys (London: J. M. Dent, 1915 [London: A. Maxwell, 1667]), 92–93.

76. Sherwood timber also used in 1609–12 and 1634–35; see Morrison, "Forests of Masts," 162, n. 13.
77. Morrison, "Forests of Masts," 140–41.
78. TNA, SP 14/42, f. 35; NA DDFJ 10/9/8, ff. 1–14; DDFJ 10/9/6, 131. 1608: 49,900 oaks; 1630: 44,000 oaks. Compares with Dean in 1661–62 survey of 30,233 trees. Morrison, "Forests of Masts," 162, n. 9.
79. TNA, SP 46/136 ff. 69, 70, 70A.
80. TNA, SP 46/136 f. 376a; SP 46/137 ff. 153, 155.
81. TNA, PRO 30/24/31/6, f. 68.
82. Oliver Rackham, *Ancient Woodland: Its History, Vegetation and Uses in England* (London: E. Arnold, 1980), 188–91.
83. Corbin found only eight red deer, NA, DDFJ/10/9/8, compared to 258 in 1648; see Seddon, "Application of Forest Law," 38.
84. TNA, E 317/Nottinghamshire 11, 23, 25; NA, DD2P 27/25–26. Surveys began in September 1656 and returned in 1658–59 record local protests. Walesby was offered one hundred acres in lieu of the town's commons in Birkland and Bilhagh. The scheme to sell four forests included Sherwood; see Sidney J. Madge, *The Domesday of Crown Lands* (London: Frank Cass and Co. Ltd., 1968); Silvanus Taylor, *Common-Good or the Improvement of Commons, Forrests, and Chases by Inclosure* (London: Francis Tyton, 1652), 32–33.
85. NA, DDFJ 10/9/8, ff. 15–18.
86. Grant 37 Edward III, see "Commissioners . . . to Enquire into the . . . Woods," 14th Report, 8; App 14, 38.
87. "Commissioners . . . to Enquire into the . . . Woods," 14th Report, App 15–16, 38–42.
88. NA, DDFJ 10/9/8, ff. 15–18 lists claimants who fed their swine in the woods. "Commissioners . . . to Enquire into the . . . Woods," 14th Report, App. 28, 61; App 32, 69. By 1662, zealous deer hunting in previous decades heavily reduced red deer herds, but Charles II's restocking allowed deer numbers to recover. See Morrison, "Bambi in Sherwood Forest."
89. NA, DD382/2, 215. John Foxe of Warsop petitioned the Chief Justice of Sherwood and was granted a license in November 1674 for permission to take "many old Oake roots, whereupon trees have formerly grown." These lay adjacent to Birkland wood, but were now useless. Verderers viewed the "stubby roots," which were not in Crown woods.
90. Historical Manuscripts Commission, *Duke of Portland Manuscripts* (London: HMSO, 1892–1931), vol. 5, 87.

Bibliography

Albion, Robert G. *Forests and Sea Power: The Timber Problem of the Royal Navy 1652–1862.* Cambridge, MA: Harvard University Press, 1926.

Bazeley, Margaret L. "The Extent of the English Forest in the Thirteenth Century." *Transactions of the Royal Historical Society* 4th Ser., 4 (1921): 140–72.

Cavendish, Margaret, Duchess of Newcastle. *The Life of William Cavendish, Duke of Newcastle*, edited by Ernest Rhys. London: J. M. Dent, 1915. [London: A. Maxwell, 1667.]

"Commissioners . . . to Enquire into the . . . Woods, Forests and Land Revenues of the Crown." Reports, 1787–93. Reprinted in *House of Commons Sessional Papers of the Eighteenth Century*. Vols. 76–78, edited by Sheila Lambert. Wilmington, DE: Scholarly Resources, 1975.

Cox, J. Charles. *The Royal Forests of England.* London: Methuen & Company, 1905.

Crook, David. "Clipstone Park and Peel." *Transactions of the Thoroton Society* 80 (1976): 35–46.

_____. "Introduction." In *Sherwood Forest in 1609: A Crown Survey by Richard Bankes*, edited by Stephanos Mastoris and Sue Groves. Thoroton Society Record Series 40. Nottingham: Thoroton Society of Nottinghamshire, 1997.

_____. "Lindhurst: No Man's Wood and the Manor of Mansfield." *Transactions of the Thoroton Society* 85 (1981): 78–89.

_____. "The Records of Forest Eyres in the Public Record Office, 1179 to 1670." *Journal of the Society of Archivists* 17, no. 2 (1996): 183–93.

Evelyn, John. *Sylva, or A Discourse of Forest-Trees and the Propagation of Timber in His Majesty's Dominions.* London: Martin and Allestry, 1664.

Fowkes, Dudley. "The Breck System of Sherwood Forest." *Transactions of the Thoroton Society* 81 (1977): 55–61.

Grant, Raymond. *The Royal Forests of England.* Wolfeboro Falls, NH: Alan Sutton, 1991.

Hammersley, George. "The Crown Woods and Their Exploitation in the Sixteenth and Seventeenth Centuries." *Bulletin of the Institute of Historical Research* 30 (1957): 136–61.

_____. "The Revival of the Forest Laws under Charles I." *History* 45 (1960): 85–102.

Hart, Cyril E. *The Forest of Dean: New History 1550–1818*. Stroud: Alan Sutton Publishing Ltd, 1995.

Historical Manuscripts Commission. *Duke of Portland Manuscripts*. London: HMSO, 1892–1931.

Hoyle, Richard W., ed. "Disafforestation and Drainage: The Crown as Entrepreneur?" In *The Estates of the English Crown 1558–1640*, edited by Richard W. Hoyle, 353–88. Cambridge, UK: Cambridge University Press, 1992.

_____. *The Estates of the English Crown 1558–1640.* Cambridge, UK: Cambridge University Press, 1992.

James, Noel D. G. *A History of English Forestry*. Oxford: Basil Blackwell, 1981.
Johnson, R. "17th Century Iron Works at Bulwell and Kirkby." *Transactions of the Thoroton Society* 64 (1960): 44–46.
Lambert, Sheila, ed. *House of Commons Sessional Papers of the Eighteenth Century*. 147 vols. Wilmington, DE: Scholarly Resources, 1975.
Large, Peter. "From Swanimote to Disafforestation: Feckenham Forest in the Early Seventeenth Century." In *The Estates of the English Crown 1558–1640*, edited by Richard W. Hoyle, 389–417. Cambridge, UK: Cambridge University Press, 1992.
Madge, Sidney J. *The Domesday of Crown Lands*. London: Frank Cass and Co. Ltd., 1968.
Manwood, John. *A Treatise and Discourse of the Lawes of the Forest*. London: Thomas Wight and Bonham Norton, 1598. See also: 2nd ed. London: Societe of Stationers, 1615. 3rd ed. London: Company of Stationers, 1665. Later ed., entitled *Manwood's Treatise of the Forest Laws*, including: 4th ed., edited by William Nelson. London: E. Nutt, 1717. 5th ed., edited by William Nelson. London: Henry Lintot, 1741.
Mastoris, Stephanos, and Sue Groves, eds. *Sherwood Forest in 1609: A Crown Survey by Richard Bankes*. Thoroton Society of Nottinghamshire Record Series 40. Nottingham: Thoroton Society of Nottinghamshire, 1997.
Morrison, Sara. "Bambi in Sherwood Forest and the Great Deer Escape c. 1703–1711." In *Environmental History in the Making*. Vol. 1: *Explaining*, edited by Estelita Vaz, Cristina Joanaz de Melo, and Lígia M. Costa Pinto, 341–57. Basel: Springer Publishing, 2017.
_____. "Forests of Masts and Seas of Trees: The Restoration Navy and the English Royal Forests." In *English Atlantics Revisited: Essays Honouring Professor Ian K. Steele*, edited by Nancy L. Rhoden, 136–73. Montreal: McGill-Queens University Press, 2007.
_____. "Fee Trees and Rogue Verderers in Early Eighteenth Century Sherwood Forest." *Transactions of the Thoroton Society* 117 (2013): 97–107.
_____. "Good Stewardship and the Challenges of Managing the Stuart Royal Forests in England, 1603–1714." *Journal of Markets and Morality* 17, no. 2 (2014): 405–27.
Pettit, Philip A. J. *The Royal Forests of Northamptonshire: A Study in their Economy, 1558–1714*. Publications of the Northamptonshire Record Society, 23. Northampton: Northamptonshire Record Society, 1968.
Plot, Robert. "A Discourse Concerning the Most Seasonable Time of Felling Timber." *Philosophical Transactions* 192 (1691): 455–61.
Rackham, Oliver. *Ancient Woodland: Its History, Vegetation and Uses in England*. London: E. Arnold, 1980.
_____. *Trees and Woodland in the British Landscape*. London: J. M. Dent, 1976.
Riden, Philip. "The Charcoal Industry in the East Midlands, 1580–1780." *Derbyshire Archaeological Journal* 111 (1991): 64–84.

_____, ed. *George Sitwell's Letterbook, 1662–66*. Publications of the Derbyshire Record Society, 10. Chesterfield: Derbyshire Record Society, 1985.

Seddon, Peter R. "The Application of Forest Law in Sherwood Forest, c. 1630–1680." *Transactions of the Thoroton Society* 82 (1978): 37–44.

Sitwell, Sir George. "The Loyal Duke of Newcastle." *Derbyshire Archaeological and Natural History Society Journal* 13 (1891): 1–9.

Statutes of the Realm, 1215–1713. 9 vols., edited by Alexander Luders et al. London: G. Eyre and A. Strahan, 1810–1828. Acts of Parliament are cited by title, year, regnal year, and chapter number.

Taylor, Silvanus. *Common-Good, or the Improvement of Commons, Forrests, and Chases by Inclosure.* London: Francis Tyton, 1652.

Thirsk, Joan. "The Crown as Projector on Its Own Estates, from Elizabeth I to Charles I." In *The Estates of the English Crown 1558–1640*, edited by Richard W. Hoyle, 297–352. Cambridge, UK: Cambridge University Press, 1992.

Thompson, Edward P. *Whigs and Hunters: The Origin of the Black Act.* New York: Pantheon Books, 1975.

Tubbs, Colin R. *New Forest: An Ecological History.* Newton Abbot: David & Charles, 1968.

Turner, George J. *Select Pleas of the Forest.* Publications of the Selden Society, 13. London: B. Quaritch, 1901.

Yarranton, Andrew. *England's Improvement by Sea and Land to Out-do the Dutch without Fighting, To Pay Debts without Moneys, To Set at Work all the Poor of England within the Growth of Our Own Lands.* London: R. Everingham, 1677.

Young, Charles R. "Conservation Policies in the Royal Forests of Medieval England." *Albion* 10, no. 2 (1978): 95–103.

_____. *The Royal Forests of Medieval England.* Leicester: Leicester University Press, 1979.

CHAPTER 11

Sustainability Prior to Carlowitz's *Sylvicultura*?

A Study Based on Cases from Schleswig-Holstein

Oliver Auge

This chapter focuses on the early modern region of Schleswig-Holstein, which includes today's German state of this name along with the adjoining Danish region of Southern Jutland. Today and for the last few centuries, this region has had a very open landscape, with less of its territory covered by woodlands than in many other parts of central and northern Europe. Yet various types of evidence suggest that in prehistorical times and perhaps as late as the Central Middle Ages (ca. 1000–1300 CE), Schleswig-Holstein was heavily wooded. Learned writers from as early as the eleventh century CE regularly referred to the region's dense forests.[1] Other indicators corroborate the hypothesis of a former abundance of woodland in the region, such as the numerous place names referring to woodland or clearances as well as the many finds of old rootstocks and tree trunks in moors in areas now sparsely wooded.[2] Even as late as the end of the Middle Ages (ca. 1500 CE), it is estimated that as much as 20 to 25 percent of the region was still wooded.[3]

The character of these woodlands varied, based chiefly on the region's three main landscapes, all of which were formed by the Ice Age. They are, from west to east, the Marsh on the North Sea coast, which is the geologically youngest of the three, followed by the oldest formation and median ridge, the so-called "Moorland" (*Geest*), and finally the Schleswig-Holstein Uplands in the east. The dominant form of vegetation in the Uplands is beechwood forest, while mixed oak forests are typical of the Moorland. Woodland comprising mainly ash, alder, and elm prevail in the Marsh.

By the late eighteenth century, after centuries of overexploitation, woodlands covered just 5 percent of the land.[4] In comparison, at the same time, woodlands may have covered as much as 30 percent of the whole of Germany.[5] Although of course such estimates are problematic due to poor data, they provide some sense of the context in which management of woodland resources took place. Later efforts of reforestation were able to expand the percentage of woodland in Schleswig-Holstein, which today stands at 11 percent of the total area. Still, this region remains by far the least forested in Germany. In contrast, the regions at the top of the list, including Hesse and Rhineland-Palatinate, are currently estimated to have as much as 42 percent woodland.[6]

These estimates suggest why it is worthwhile to focus on Schleswig-Holstein to study medieval and early modern laments of wood shortage and the degree to which such a shortage may have encouraged strategies of sustainable management. Debate about such complaints and the responses they may have provoked has in fact been referred to as the cradle of environmental history in Germany.[7] Joachim Radkau opened this debate in the 1980s by criticizing what he called the "myth" of wood shortage, noting that the myth was "a commonplace in forest and environmental history that for a long time remained widely unquestioned."[8] As Radkau showed, lament about wood shortage could sometimes be a mere rhetorical figure that actually served a variety of different interests and aims. Therefore, according to Radkau, one should not let oneself be deceived by the recurrence of the seemingly factual complaints about wood supply, which also appear in Schleswig-Holstein, but take them with a grain of salt.[9] The authorities who used the common motif of wood shortage and who at the same time profited from the actual shortage felt by many of their subjects were actually pursuing their own regulatory and political objectives.[10]

This is especially true for preindustrial circumstances. Undoubtedly, people then faced a scarcity of woodland; but at the same time, the economy and daily life of the people went on so smoothly that we must assume that premodern society was able to make a virtue of necessity and established a subsistence economy on the basis of other raw materials, wood imports, and wood-saving measures.[11] While this may not have worked indefinitely, and the utilization of alternative raw materials, such as peat and cow dung,[12] was not entirely unproblematic, it did work for an extended period of time, and this puts the political explosiveness of the wood shortage lamentations into perspective.[13]

In this chapter, I argue that forms of sustainable thinking were voiced prior to Hans Carl von Carlowitz's (1645–1714) famous coinage of the term "sustainable" (*nachhalthend*) in a silvicultural sense in

1713.[14] Sustainability existed as an economic principle for people in the Middle Ages and preindustrial periods without having been formulated as a theoretical concept. Yet the impact of efforts to conserve woodlands varied considerably in different regions; and overall, they did not halt the overexploitation and near disappearance of woodlands from Schleswig-Holstein.

Complaints about Wood Shortage and Measures to Conserve Woodland

Generally speaking, wood shortages were first lamented in Schleswig and Holstein in ordinances, and for this reason, decrees and regulations issued by various authorities figure prominently in this chapter. It is important to recognize, however, that in this region, these and other efforts to dictate the terms of woodland exploitation did not emanate from any single, overarching authority. Instead, early modern Schleswig-Holstein was characterized by a high degree of decentralization. In the sixteenth century, the region had been divided among several parties, including the kings of Denmark, their close relatives, the dukes of Schleswig-Holstein-Gottorp, and the so-called "partitioned-off dukes" (*Abgeteilte Herren*). The kings and dukes jointly governed the nobility of Schleswig and Holstein, who enjoyed numerous privileges, including exemption from most taxes. On their own estates, the nobility acted as small-scale territorial sovereigns; neither duke nor king had the power to intervene in their resource management. Along the Baltic coast, especially, the nobility's export-oriented estate economy was predominant, focusing first on grain, then on large-scale cattle and dairy farming.[15]

A typical example of rulers' concerns about depleted woodlands appears in a directive of 1681 issued jointly by King Christian V (1646–1699) and Duke Christian Albrecht (1641–1694/95), which expressed their fear that timber was shipped from their territories in such quantities that "there would soon be a serious shortage."[16] Numerous regulations issued in the following decades justified restrictions with reference to the alleged disastrous state of the forests.[17] When forestry began to emerge as a science in the eighteenth century, wood shortage was taken up as an issue in treatises on the subject: in 1798, for example, the forester Christian Peter Laurop wrote a monograph entitled, "Frank Thoughts on Wood Shortage, Especially Firewood Shortage, in the Duchies of Schleswig and Holstein and Measures to Remedy It."[18] These wood shortage lamentations in ordinances, administrative writing, and scientific or statistical works have been reiterated in treatises on environmental and economic history of Schleswig-Holstein largely uncritically, often by simply copying them verbatim from the sources and without questioning them. Clearly, a more careful approach that goes beyond simple monocausalities is required, as Uwe Erhard Schmidt recently suggested.[19]

Responses to or measures to overcome or ease wood shortages can be traced back to an early date. From 1378 onward, a "wood penny" (*Holzpfennig*) had to be paid by peasants to obtain the right of use for woods near the bishop's residence at Schwabstedt and in the surrounding territory.[20] In the Tønder district, all duties paid in wood had been replaced by payments in peat or coin prior to 1568.[21] Strict bans on unauthorized cutting of timber are documented for the administrative area of Segeberg and the island of Fehmarn in the sixteenth century. The offence of illegal tree felling appears to have been a widespread phenomenon.[22] Each of these examples may serve as an argument for an acute and continually spreading wood shortage. Still, the regulatory intentions actually connected to those measures must not be disregarded. Generally speaking, they revolved around one central question: to whom does the forest belong? Often, the authorities were not even interested in the trees as such but in other functions of the forests. One such function was the hunt, which among other interests fostered the kings' and dukes' will to sustain and expand their woodlands.[23]

Imports of black coal from England started slowly around 1600; occasional imports from Sunderland and Newcastle to Schleswig-Holstein on ships from Tønder and Ribe are well documented for 1616. Unlike the ubiquitous charcoal, which was used in private homes as well as commercially, this black coal was probably mainly intended for the proto-industrial sector.[24]

Toward the middle of the eighteenth century, timber and firewood were beginning to be imported from Norway, Pomerania, and other Baltic regions.[25] Most notably, however, peat as a fuel gained more and more importance for private households as well as for the economy in general when wood supplies grew increasingly scarce.[26] It cannot have been a coincidence that peat cutting and peat trading were addressed specifically in those village rules established between 1540 and 1652 that pertained to communities settled on the Moorland, an area that was especially sparsely wooded.[27] In some regions, for example Stapelholm or the district of Tønder, peat extraction and sale became one of the most important lines of business.[28] At the same time, the export of wood and timber became increasingly restricted. In the period between 1506 and 1721, bans on exporting timber, firewood, bark, or charcoal accumulated.[29] Moreover, the construction of wooden houses (i.e., houses built entirely with wooden planks) was forbidden between 1554 and 1557 in order to save timber. Instead, houses were to be made of bricks or at least have bases of stone.[30] Measures such as these enabled local communities to exist and operate economically for an extended period of time without depleting their own wood supplies.

Methods of dealing with wood scarcity were not limited to the intensified use of alternative fuels or the attempt to reduce consumption through bans and limitations. From early on, there were also efforts to protect remaining stretches of woodland and foster the regeneration of forests. These attempts were limited to certain areas in Schleswig-Holstein because the various ordinances enacted since the seventeenth century in cities and in royal and princely lands had no validity for the vast estates of the nobility, who held about one-third of the land.[31] In fact, since 1524, Schleswig's and Holstein's "Ancient Knights" (*Equites Originarii*—i.e., the long established noble families) had been granted independence from sovereign forest regulations by Duke Frederick I (1471–1533) as a way of thanking them for helping him attain the Danish throne; they were allowed to exploit their woodland at their own discretion.[32] The first forest regulation for the king's territories in the duchies of Schleswig and Holstein was enacted by Christian V of Denmark (1646–1699) on 1 September 1671 and renewed in 1680.[33] Another ordinance followed in 1737, while similar regulations for the ducal territories were issued in 1712, 1741, and 1770.[34]

Attempts to deal with wood scarcity were also limited to certain regions because the actual implementation of royal and princely prescriptions often was rather ineffective, as was also true in the case of hunting rights.[35] We cannot deduce from a decree that it was fully implemented, or that it was enforced over an extended period of time. A number of documented breaches of these regulations, which led to the offenders being prosecuted in court, underscore this.[36] The regulations themselves often address the inefficacy of preceding ordinances, as does, for instance, the king's ordinance from 1 September 1680.[37] The following analysis, however, is concerned more with the conception and underlying intentions of forest regulations than with their realization.

In the princely wood ordinance for Glücksburg from 1681, for example, Duke Christian of Schleswig-Holstein-Sonderburg-Glücksburg (1627–1698) decreed that "only useless, especially soft alder wood and old hollow oak trees, branches etc." were to be used for charcoal production. Also, charcoal kilns were to be built "only in places where it would not harm the forest in any way."[38] Fourteen years later, the princely wood ordinance for Schleswig-Holstein—issued by Duke Friedrich IV of Schleswig-Holstein-Gottorp (1671–1702) and applying to the ducal territories in Schleswig and Holstein—threatened with punishment all those who damaged oaks, beeches, or other fruit-bearing trees or who set heather fires in such a way that they spread to peatbogs or woodland. Moreover, the ordinance prohibited grazing goats in or near forests and ordered that "capital-trees [profitable trees] are to be spared and no fruit-bearing tree is to be designated for felling." The shrubs needed for fencing were to be cut in places where it would do the woodland the least harm.[39]

The intentions behind regulations such as these have already been uncovered for other regions by Peter Blickle, Winfried Freitag, and Paul Warde.[40] The authorities were certainly trying to protect the forests and their wood resources, although this protection must not be understood in terms of our modern ecological concerns but rather strictly economic ones. For example, a set of village rules from 1655 states: "daily experience shows that the forests in our territories of Fünen and Lolland are diminished in such a way that in the future we will be lacking in both firewood and timber."[41] Similarly, lords were also and probably even more interested in the assertion of their hunting prerogative and their political power.[42] Lisa Kragh was able to prove this for the second half of the eighteenth century by analyzing the ouster of peasants, and more specifically of those who owned a full farm (*Hufners*), from the sovereigns' enclosed forests in Schleswig and Holstein.[43] Still, further studies of the interrelations among peasants, dukes, princes, and the cities in Schleswig-Holstein are indispensable, especially regarding periods before the eighteenth century.

In practice, the regulations, even if they were only implemented here and there, must have helped to stabilize the forests, especially because since the middle of the sixteenth century, special personnel called "forest reeves" (*Waldmeister*) were employed to protect forests and supervise authorized logging.[44] The ordinances were even more effective—in principle at least—when they prescribed such measures as the strict demarcation of forest and farmland through the erection of fences or hedge banks, as did the wood ordinance issued in 1728 by prince-bishop Adolf Friedrich of Lübeck (1710–1771), or similar examples from Schleswig on the Angeln peninsula.[45]

Some decrees also advocated the principle of automatic forest regeneration. According to an ordinance from 1728, each landlord had to plant four trees a year and young men were only allowed to marry after having planted ten trees.[46] King Christian VI's (1699–1746) hunting and forest ordinance of 24 April 1737 seized this idea and ordered any bridal couple to plant ten oaks or fifteen beeches, which they had to "keep growing for three years after their planting, or get to the third leaf."[47] The resultant woodlots were called "bride" or "bridegroom paddocks" or "enclosures" (*Braut-* or *Bräutigamskoppel*), some of which still exist today.[48]

Large-Scale Reforestation Projects and the Idea of Sustainability

In the seventeenth century, authorities began mandating systematic reforestation, an approach that gradually became more common thereafter. An ordinance enacted in 1655 for Funen and Lolland in Denmark, where the building of war ships required huge amounts of lumber, prescribed that any tree felled had to be immediately replaced by the planting of a new one.[49] In 1805, this stipulation was extended to the whole of Denmark.[50] The king of Denmark's wood ordinance from 1671, too, prescribed the annual sowing and planting of oaks and beeches on empty land.[51] Such a principle of regeneration was first enacted in the constitution of Stapelholm, in Schleswig-Holstein, of 27 January 1623, which ordered the planting of the same amount of oaks and beeches that were felled.[52] To what extent and how successfully all of this was put into practice cannot be determined. The impact of these measures is relativized by current research with good reason.[53]

Still, some of the later regulations very clearly imply thinking in terms of sustainability. In the royal forest regulation from 24 April 1737, for instance, the reasoning goes as follows: "And we took the promotion of this beneficial deed into consideration out of paternal benevolence, even more so as it is of greatest importance to the community and to posterity."[54] An edict from 15 June 1785 states, "Experience taught that Our paternal intent, aimed at sustaining the proprietorial forests for posterity, is not achieved."[55]

The late eighteenth century also saw the first efforts of reforestation on noble estates and in the state forests that are known to have had some success. These projects increasingly aimed to plant fast-growing coniferous trees, such as fir, pine, and larch, where formerly deciduous trees, such as oak, alder, and birch, had grown. It was only from this period that extensive coniferous forests emerged in Schleswig-Holstein.[56] Today's "forest nursery" (*Forstbaumschule*) bears witness to these reforestation projects. Now a beer garden, the Royal Danish School of Forestry was initiated on 1 August 1785, with a tree nursery being added in 1788.[57] While the school's objective was to educate prospective foresters, the nursery was supposed to further "the promotion and dissemination of usable woody plants in the country's forests," and to provide saplings for ambitious reforestation plans.[58]

These measures were part of broader efforts to intensify and reform agriculture and economy in their entirety, which the Danish state undertook in the spirit of enlightenment, aiming at so-called "state prosperity."[59] It is not by chance, therefore, that Christian August Heinrich Niemann (1761–1832), a professor of "State Finance" (*Kameralistik*) and a financial councilor (*Etatsrat*) to the Danish state, became the School of Forestry's first director and initiator of its nursery. Not surprisingly, these initiatives and the more explicit ideas about sustainability that they entailed emerged under the aegis of rapidly developing state administrative structures. The Hunting and Forest Decree of 2 July 1784, devised by the head of the Danish forest administration, Daniel von Warnstedt (1729–1802), split up the formerly combined office of agriculture and forestry.[60] Even before, in 1778, hunting and forest offices had been separated to some extent within the Danish financial administration (*Rentekammer*), allowing for a distinct "bureau of forestry" (*Forstkontor*). Forestry and hunting, however, still remained administratively linked after the 1784 decree: the senior forester was still subordinate to the hunting official, and forest reeves could be responsible for matters concerning both woods and hunting. Finally, starting in 1792, the sovereign's forests were methodically surveyed in order to take stock of what was actually there. Only from then onward do we possess truly reliable data on the extent of forests in Schleswig-Holstein.

The attempts at reforestation toward the end of the eighteenth century were not, however, the first documented for Schleswig-Holstein. The King of Denmark's steward and renowned humanist Heinrich Rantzau (1526–1598) founded an arboretum close to Itzehoe toward the end of the sixteenth century. Ancient spruces grown there reportedly still existed in 1809.[61] Rantzau supposedly procured seeds of coniferous trees from the Elector of Brandenburg, and therefore is known as a pioneer of planting firs and spruces in Schleswig-Holstein, where these trees do not naturally occur.[62] It is noteworthy that he had this arboretum pictured in the so-called Rantzau Panel, which was created after 1591. Its center features a family tree of the Rantzaus drawn against the backdrop of a map of Schleswig-Holstein, Jutland, Funen, and Scania; it also illustrates the Rantzaus' great deeds and monuments, especially those of Heinrich.[63] The arboretum is clearly recognizable in the left-hand corner of the panel. It is accompanied by the following Latin text: "Heinrich Rantzau's arboretum." A stone pillar is depicted in front of the arboretum—probably identical to the original memorial stone commemorating Rantzau's plantation[64]—that also specifically mentions the founder: "Heinrich Rantzau." The text placed above the depiction of the arboretum reads, in Latin: "Anno 1580 Heinrich Rantzau planted these oaks, firs, and birches. He had the year of planting and of the foundation added here so that posterity would be able to determine the trees' age. This he recommends to the eternal God for all eternity."[65] It is an indication of the great importance that Rantzau ascribed to his arboretum, and maybe also of his not exclusively economic motivation, that he had it placed so prominently on this formidable monument of noble self-portrayal from around 1600.

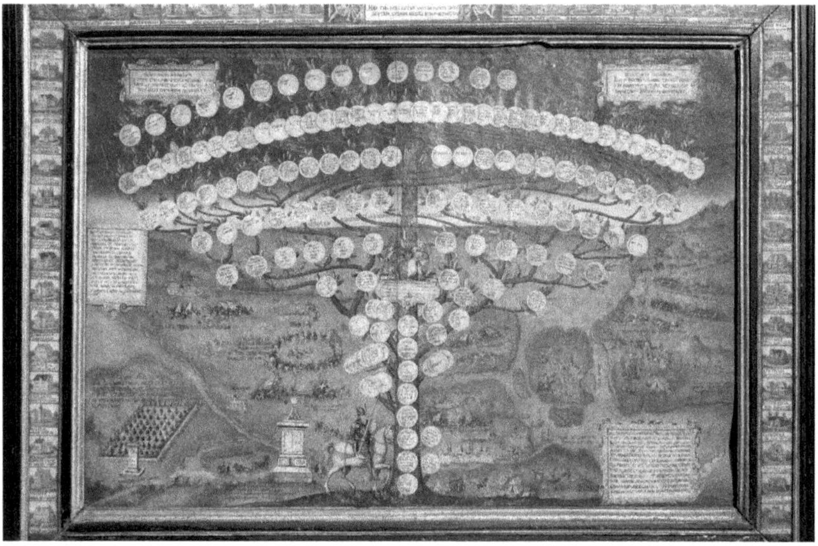

Figure 11.1. Rantzau Panel, ca. 1600. Photograph by Oliver Auge.

Figure 11.2. Rantzau Panel, detail: Heinrich Rantzau's Arboretum. Photograph by Oliver Auge.

This text reinforces the evidence presented above concerning woodland ordinances, which shows that forms of sustainable thinking had been voiced prior to Carlowitz's famous coinage of the term in 1713.[66] Even if it was not yet recognized as a formal, theoretical concept, a notion of sustainability existed as an economic principle for people in the Middle Ages and preindustrial periods. The environmental historian Bernd Hermann even deems such a praxis vital for life in premodern times.[67]

An Example of Sustainable—and for Profit—Coppice Silviculture

Hermann's hypothesis can be strengthened by another example, in this case from the late medieval period. In his archaeological dissertation on the use of firewood and charcoal in Schleswig-Holstein, Arne Paysen argues persuasively that the Carthusian monks in Ahrensbök, near Lübeck "managed a profit-oriented but sustainable coppice silviculture and were concerned with sustaining all possible usages of the forests, even if the felling area was in the hands of farmers."[68] His argument is based on both the anthracological analysis of charcoal kilns on the abbey's land and on detailed monastic financial registers dating from around 1500.[69] Both types of evidence reveal "a major, monastically administered enterprise of charcoal production,"[70] though it was the peasants, really, who were responsible for the actual charcoal burning. Written sources covering a period of approximately one hundred years (1455–1566) show that charcoal burning was a professionally organized craft set in a rural context, probably producing not only for local demand but also for the main market in Lübeck. Certain family names are encountered again and again in connection with charcoal burning; also, some farmers used several woodlots at the same time. Paysen calculated a contractual regeneration phase of eleven to twenty years until a leased woodlot was cut again for charcoal production. This corresponds to the sprouting times of alder and hornbeam in coppicewood. Also, permission to cut wood was only given in autumn, in order to fully exploit the crop cycle. Paysen therefore concludes: "the abbey itself created the necessary parameters for a maximally productive coppice silviculture. The farmers subservient to the monks did the work of charcoal burning. . . . The abbey controlled the sustainable use of the forests by carefully assigning rights of usage and enforcing the necessary phases of regeneration."[71]

Conclusion

Obviously, Paysen's results for the abbey in Ahrensbök and its sustainable woodland management cannot simply be generalized for all of Schleswig-Holstein.[72] However, further anthracological studies indicate that oak was generally not used as fuel, probably because it was too valuable as a building material and as a prerequisite for pannage. Hornbeam, beech, birch, alder, and hazel were preferred due to their lower use-value—a purposeful selection of wood that can be viewed as a step toward an adaptive or at least partially sustainable silviculture.[73]

At the same time, however, we encounter economic systems that were not considerate of sustainability at all. These systems increased in scope over time. As Günther Bock has remarked, people "mainly lived from hand to mouth, as it were."[74] This is especially true for protoindustries such as iron or glass works in Schleswig and Holstein, some of which had to close down their businesses due to wood shortage—though in the current state of research, whether this was due to industrial overexploitation or other factors is not known.[75] Among all the sectors responsible for the exploitation of wood resources, shipbuilding, in the seventeenth and eighteenth century especially of war vessels, played a prominent role. Danish woodlands were exhausted, at least for naval timber, by the beginning of the early modern period, so that from then onward, Schleswig-Holstein's woodlands were heavily drawn on. For instance, five hundred tall oaks were needed to build the large warship with several decks that was laid down near Itzehoe in 1609 by order of King Christian IV (1577–1648). But smaller vessels, which were built in Schleswig-Holstein in great numbers for coastal and domestic shipping, also swallowed up large quantities of wood. Shipbuilding was a thriving business. In 1845, eighty-three shipyards existed in the region. In Kiel alone, 157 ships were built between 1766 and 1807.[76]

The impact of woodland and forest use in late medieval and early modern Schleswig and Holstein varied considerably in different regions. At that time, careful treatment of wood resources is best documented in places where the owners' profits were dependent on drawing from the same woodlots again and again—as was the case for the monks from Ahrensbök.[77] The exploitation of wood resources and the near disappearance of woodlands from Schleswig-Holstein, however, were not ultimately prevented by these early indications of resource protection and regeneration, whether their actual impact can be verified or not.

Oliver Auge studied History and Latin philology in Tübingen, earning his doctorate in 2001. After several years as a research associate in Medieval History at the University of Greifswald, in 2008, he received his post-doctoral degree upon completion of his habilitation thesis. Since 2009, he has held the chair for Regional History at the University of Kiel. He has published several articles regarding sustainability in a historical perspective, and also organized a special session on this topic at the Meeting of German Historians (Verband der Historiker und Historikerinnen Deutschlands) in Mainz in 2012.

Notes

The author thanks Franz Steiner Verlag for the permission to reproduce here much of the substance of his earlier essay "Auf dem Weg zur Nachhaltigkeit? Ansätze zu Ressourcenschutz und Ressourcenregeneration im spätmittelalterlichen und frühneuzeitlichen Schleswig-Holstein," in *Wirtschaft und Umwelt vom Spätmittelalter bis zur Gegenwart: Auf dem Weg zu Nachhaltigkeit?*, ed. Günther Schulz and Reinhold Reith, 31–51. Stuttgart: Franz Steiner Verlag, 2015.

1. Bernhard Schmeidler, ed., *Magistri Adam Bremensis Gesta Hammaburgensis ecclesiae pontificum* (Hannover: Verlag Hahnsche Buchhandlung, 1917), 72.
2. Bo Fritzbøger, *Danske skove 1500–1800: En landskabshistorisk undersøgelse* (Odense: Landbohistorisk Selskab, 1992) 73f.; August Niemann, *Forststatistik der dänischen Staaten: Mit drei Schautafeln* (Altona: Hammerich, 1809), 196.
3. Concerning this and the following cf. Klaus-Joachim Lorenzen-Schmidt, "Waldverlust und Waldaufbau in Schleswig-Holstein vom Mittelalter bis 1914," in *Dünger und Dynamit: Beiträge zur Umweltgeschichte Schleswig-Holsteins und Dänemarks*, ed. Manfred Jakubowski-Tiessen and Klaus-Joachim Lorenzen-Schmidt (Neumünster: Wachholtz, 1999), 43–48.
4. Thomas Riis, *Wirtschafts- und Sozialgeschichte Schleswig-Holsteins: Leben und Arbeiten in Schleswig-Holstein vor 1800* (Kiel: Ludwig, 2009), 202, 204, 364; concerning Schleswig, cf. Friedrich Mager, *Entwicklungsgeschichte der Kulturlandschaft auf der Geest und im östlichen Hügelland des Herzogtums Schleswig bis zur Verkoppelungszeit* (Breslau: Hirt, 1930), 44–149, plus the annexed map of medieval forested areas in the duchy of Schleswig; Hubertus Neuschäffer, "Die Waldnutzung in der Geschichte Schleswig-Holsteins," *Schleswig-Holstein* 6 (1985): 8–12; Neuschäffer, *Geschichte von Wald und Forst in Schleswig-Holstein und der alte Rendsburger Wald mit dem Forstamt Barlohe* (Rendsburg: Möller, 1986), 16. Concerning the whole issue, cf. Oliver Auge, "Auf dem Weg zur Nachhaltigkeit? Ansätze zu Ressourcenschutz und Ressourcenregeneration im spätmittelalterlichen und frühneuzeitlichen Schleswig-Holstein," in *Wirtschaft und Umwelt vom Spätmittelalter bis zur Gegenwart: Auf dem Weg zu Nachhaltigkeit?*, ed. Günther Schulz and Reinhold Reith (Stuttgart: Franz Steiner Verlag, 2015), 31–51, especially 35–40; Lisa Kragh, "'Zweibeinigte Holzwürmer?' Die Bauern und der Waldrückgang in Schleswig-Holstein in Mittelalter und Früher Neuzeit," *Zeitschrift der Gesellschaft für Schleswig-Holsteinische Geschichte* 139 (2014): 30–44.
5. Hans-Rudolf Bork, ed., *Landschaftsentwicklung in Mitteleuropa: Wirkung des Menschen auf Landschaften* (Gotha: Klett-Perthes, 1998).
6. Figures from http://www.sdw.de/waldwissen/wald-in-deutschland/waldanteil (last accessed 10 May 2017). Other figures, which show a similar tendency, are cited by Lorenzen-Schmidt, "Waldverlust," 41.
7. Nils Freytag, "Deutsche Umweltgeschichte—Umweltgeschichte in Deutschland," *Historische Zeitschrift* 283 (2006): 388.
8. Quote, translated from German, taken from Bernd-Stefan Grewe, "'Man sollte sehen und weinen!' Holznotalarm und Waldzerstörung vor der Industrialisierung," in *Wird Kassandra heiser? Die Geschichte falscher Öko-Alarme*, ed. Frank Uekötter and Jan Hohensee (Stuttgart: Steiner, 2004), 34f. Joachim Radkau, "Zur angeblichen Energiekrise des 18. Jahrhunderts:

Revisionistische Betrachtungen über die Holznot," *Vierteljahrschrift für Sozial- und Wirtschaftsgeschichte* 73 (1986): 1–37; Radkau, "Holzverknappung und Krisenbewußtsein im 18. Jahrhundert," *Geschichte und Gesellschaft* 9 (1983): 513–43; Radkau and Ingrid Schäfer: Holz: Ein Naturstoff in der Technikgeschichte (Reinbek bei Hamburg: Rowohlt, 1987), 149–60.

9. For contemporary examples of laments of wood shortage, cf. Christian Peter Laurop, *Freimüthige Gedanken über den Holzmangel, vorzüglich über den Brennholzmangel in den Herzogthümern Schleswig und Holstein und die Mittel, ihm abzuhelfen* (Altona: Hammerich, 1798); Niemann, *Forststatistik*, 201 and especially 204f., including a reference to a persiflage, published in the *Altonaischer Naturalmanach*, on two-legged woodworms (= people as forest pests).
10. Kragh, "Zweibeinigte Holzwürmer," 45.
11. Oliver Auge, "Nachhaltigkeit als historisches Thema: Das Beispiel Schleswig-Holstein in Spätmittelalter und früher Neuzeit," *Natur- und Landeskunde. Zeitschrift für Schleswig-Holstein, Hamburg und Mecklenburg* 120 (2013): 142.
12. On deforested islands such as Fehmarn, the poor collected and dried cow dung to use it as heating fuel. Cf. Kragh, "Zweibeinigte Holzwürmer," 26f.
13. Lorenzen-Schmidt, "Waldverlust"; Riis, *Wirtschafts- und Sozialgeschichte*, passim.
14. Concerning Carlowitz and his work "Sylvicultura oeconomica" from 1713, cf., e.g., Reinhold Reith, "Nachhaltigkeit," in *Enzyklopädie der Neuzeit*, vol. 8, ed. Friedrich Jaeger (Stuttgart: Metzler, 2008), 1010; Reith, *Umweltgeschichte der frühen Neuzeit* (Munich: Oldenbourg, 2011), 135. Cf. also, Oliver Auge, "'Nachhaltigkeit' als historisches Thema—eine Hinführung," *Jahrbuch für Regionalgeschichte* 32 (2014): 45–54.
15. Oliver Auge and Joachim Schneider, "Der Niederadel: Schleswig-Holstein/Franken," in *Handbuch Landesgeschichte*, ed. Werner Freitag, Michael Kißener, Christine Reinle, and Sabine Ullmann (Berlin: De Gruyter Oldenbourg, 2018), 303–34.
16. Quote translated from German. The original reads as follows: "daß bald großer Mangel daran sein wird." Cited from Walter Hase, "Abriß der Wald- und Forstgeschichte Schleswig-Holsteins im letzten Jahrtausend," *Schriften des Naturwissenschaftlichen Vereins für Schleswig-Holstein* 53 (1983): 110.
17. Cf. Kragh, "Zweibeinigte Holzwürmer," 23f.
18. Laurop, *Freimüthige Gedanken.*
19. Uwe Eduard Schmidt, *Der Wald in Deutschland im 18. und 19. Jahrhundert: Das Problem der Ressourcenknappheit dargestellt am Beispiel der Waldressourcenknappheit in Deutschland im 18. und 19. Jahrhundert. Eine historisch-politische Analyse* (Saarbrücken: Conte-Verlag, 2002).
20. Werner Carstens, ed., *Schleswig-Holsteinische Regesten und Urkunden*, vol. 6 (Neumünster: Wachholtz, 1971), n.179, 117 (26 March 1378).
21. Mager, *Entwicklungsgeschichte*, 187; Riis, *Wirtschafts- und Sozialgeschichte*, 203.

22. Max Schönwandt, "Die Familie der Kirchspielvögte von Würtzen zu Nortorf," *Rendsburger Jahrbuch* 34 (1984): 66; Johann Eike Benesch, "Das Bordesholmische Brücheregister von 1619/20," *Jahrbuch Bordesholm* 6 (2004): 128; Hans Wilhelm Schwarz, "Aus dem Brücheregister der Hanerauer Amtsrechnung 1619," *Rendsburger Jahrbuch* 53 (2003): 209 (wood theft); Schwarz, "Aus dem Brücheregister der Hanerauer Amtsrechnung 1622," *Rendsburger Jahrbuch* 55 (2005): 87, 96 (offences against the authorities 1 and 2).
23. Kragh, "Zweibeinigte Holzwürmer," 34. Current research argues that historians have overestimated the importance of hunting in forest jurisdiction due to its prominence in early forest law. Cf. Dolly Jørgensen, "The Roots of the English Royal Forest," *Anglo-Norman Studies* 32 (2010): 114–28.
24. Ludwig Andresen, "Einfuhr von Steinkohle 1616," *Die Heimat* 15 (1930): 19.
25. Wilhelm Rust, *Das Flensburger Bauhandwerk von 1388 bis 1966: Ein Beitrag zur Baugeschichte der Stadt* (Flensburg: Innung d. Baugewerbes, 1967), 189f. Cf. also Riis, *Wirtschafts- und Sozialgeschichte*, 333.
26. Jan Carstensen, *Torfgewinnung: Geschichte und Bedeutung in Schleswig-Holstein* (Osnabrück: Scholl, 1985), 20–42; Riis, *Wirtschafts- und Sozialgeschichte*, 235.
27. Martin Rheinheimer, *Die Dorfordnungen im Herzogtum Schleswig*, vol. 1, 149 ff., vol. 2, n. 2 § 17 (Abild 1649), n. 3 § 6 (Achtrup 1633), n. 7 § 15 (Agerskov 1592), n. 111 § 10 (Großenwiehe 1623), n. 119 § 9 (Harreby 1601), n. 131 § 1 (Høgsbro 1568), n. 143 § 9 (Humptrup 1624), n. 145 §§ 2, 4, 10 (Immenstedt 1625), n. 171 § 4 (Leck 1540), n. 182 § 6 (Mildstedt 1571), n. 210 §§ 3, 6 (Nørre Vollum 1623), n. 224 § 6 (Ostenfeld 1625), n. 231 § 7 (Ottersbø 1619), n. 237 §§ 12, 14 (Ravstedt 1593), n. 270 § 2 (Sønder Vollum 1623), n. 297 § 4 (Terring 1619), n. 298 § 5 (Tielen 1593), n. 305 § 2 (Toghale 1652) (Stuttgart: Lucius & Lucius, 1999). Cf. also Martin Rheinheimer, "Umweltzerstörung und dörfliche Rechtssetzung im Herzogtum Schleswig (1500–1800)," in *Dünger und Dynamit: Beiträge zur Umweltgeschichte Schleswig-Holsteins und Dänemarks*, ed. Manfred Jakubowski-Tiessen and Klaus-Joachim Lorenzen-Schmidt (Neumünster: Wachholtz, 1999), 86f.
28. Riis, *Wirtschafts- und Sozialgeschichte*, 363.
29. Lorenzen-Schmidt, "Waldverlust," 60f.; Hase, "Abriß," 89, 108 (for the Kremper- and Wilstermarsch, 1550–1575).
30. Hase, "Abriß," 90, 109; concerning timber, cf. Mager, *Entwicklungsgeschichte*, 195f.
31. Lorenzen-Schmidt, "Waldverlust," 61; Hase, "Abriß," 97.
32. Friedrich Christoph Jensen, and Dietrich Hermann Hegewisch, eds., *Privilegien der Schleswig-Holsteinischen Ritterschaft von den in der Privilegienlade befindlichen Originalen genau abgeschrieben und mit denselben verglichen* (Kiel: Christian Friedrich Mohr, 1797), Nr. 23; Hase, "Abriß," 86.
33. LASH, Abt. 401 Königliche Verordnungen, n. 1680 1.9.
34. Lorenzen-Schmidt, "Waldverlust," 61.

35. Hubertus Hiller, *Untertanen und obrigkeitliche Jagd: Zu einem konfliktträchtigen Verhältnis in Schleswig-Holstein zwischen 1600 und 1848* (Neumünster: Wachholtz, 1992).
36. Cf., e.g., LASH, Abt. 65.1, n. 1201.
37. LASH, Abt. 401 Königliche Verordnungen, n. 1680 1.9.
38. Quotes translated from German. Originals from Mager, *Entwicklungsgeschichte*, 188, following LASH, Abt. 400.5, n. 606.
39. LASH, Abt. 7, n. 3026. Concerning this and the following, cf. Mager, *Entwicklungsgeschichte*, 250f.
40. Peter Blickle, "Wem gehörte der Wald? Konflikte zwischen Bauern und Obrigkeiten um Nutzungs- und Eigentumsansprüche," *Zeitschrift für Württembergische Landesgeschichte* 45 (1986): 167–78; Winfried Freitag, "Landbevölkerung, Forstpersonal und 'gute Waldordnung' in der 'Bayrischen Vorstordnung' von 1568," *Zeitschrift für Agrargeschichte und Agrarsoziologie* 55 (2007): 32–57; Paul Warde, *Ecology, Economy and State Formation in Early Modern Germany* (Cambridge, UK: Cambridge University Press, 2006).
41. Vilhelm Adolf Secher, ed., *Corpus Constitutionum Daniae*, vol. 6 (Copenhagen: Selskabet for Udgivelse af Kilder til Dansk Historie, 1918), 260, n. 205: "Eftersom den daglig erfarenhed udviser, at skoufvene i vore lande Fyn oh Laaland saaledis formindskis, at det er at befrygte, at hverken til ildebrand elle bygnyning i frembtiden fornødenhed skal finds."
42. Cf. also Karl Hasel, *Forstgeschichte: Ein Grundriß für Studium und Praxis* (Hamburg: Parey, 1985), 115f.; Kurt Mantel, *Forstgeschichte des 16. Jahrhunderts unter dem Einfluß der Forstordnungen und Noe Meurers* (Hamburg: Parey, 1980); Mantel, *Wald und Forst in der Geschichte. Ein Lehr- und Handbuch* (Hanover: Schaper, 1990), 61f., 84f.; Blickle, "Wem gehörte der Wald," 169f. Concerning the issue of hunting, cf. Werner Rösener, *Die Geschichte der Jagd: Kultur, Gesellschaft und Jagdwesen im Wandel der Zeit* (Darmstadt: Wissenschaftliche Buchgesellschaft, 2004), 215–24; Martin Knoll, *Umwelt—Herrschaft—Gesellschaft: Die landesherrliche Jagd Kurbayerns im 18. Jahrhundert* (Sankt Katharinen: Scripta-Mercaturae-Verlag, 2004), 2–4 and ch. 5; Knoll, "Dominanz als Postulat. Höfische Jagd: Natur und Gesellschaft im 'Absolutismus,'" in *Umwelt und Herrschaft in der Geschichte*, ed. François Duceppe-Lamarre and Jens Ivo Engels (Munich: De Gruyter, 2008), 73–91. Concerning Schleswig-Holstein, cf. the short remarks in Hase, "Abriß," 87, 94, or Riis, *Wirtschafts- und Sozialgeschichte*, 202.
43. Kragh, "Zweibeinigte Holzwürmer," 34.
44. Hase, "Abriß," 94; Mager, *Entwicklungsgeschichte*, 238–45.
45. Hans Jessen, "Die Bewaldung in Angeln im Jahre 1730," *Jahrbuch des Angler Heimatvereins* 52 (1988): 48–58.
46. LASH, Abt. 260, n. 611; Otto Jarchow, "Waldsterben im Eutinischen vor 250 Jahren," *Jahrbuch für Heimatkunde Eutin* 17 (1983): 34; Neuschäffer, "Die Waldnutzung," 9–12; Neuschäffer, *Geschichte von Wald und Forst*, 16, 24f.; Riis, *Wirtschafts- und Sozialgeschichte*, 364.

47. Quote translated from German. Friderich Detlef Carl von Cronhelm, ed., *Corpus Constitutionum Regio-Holsaticarum, oder allerhöchst-autorisirte Sammlung der in dem Herzogthum Holstein königl. Antheils, samt incorporirten Landen wie auch der Herrschaft Pinneberg, Stadt Altona und Graffschafft Rantzau, in Krafft eines beständigen Gesetzen ergangenen Constitutionen, Edicten, Mandaten, Decreten, Resolutionen, Privilegien, Concessionen, und anderen Verfügungen*, vol. 1 (Altona: Gebrüder Burmester, 1749), 1245–87; Hase, "Abriß," 97.
48. Hase, "Abriß," 97. A photo of a so-called "Bräutigamskoppel" (bridegroom paddock) can be found in Lorenzen-Schmidt, "Waldverlust," 62.
49. Secher, *Corpus*, 260; cf. Riis, *Wirtschafts- und Sozialgeschichte*, 204. Also concerning the following.
50. Concerning the wood shortage of 1805, cf. Thorkild Kjærgaard, *Den danske Revolution 1500–1800: En økohistorisk tolkning* (Copenhagen: C.A. Reitzels forlag, 1991), 122.
51. Hase, "Abriß," 110.
52. Cai Lorenz von Brockdorff, ed., *Corpus statutorum Slesvicensium, oder: Sammlung der in dem Herzogthum Schleswig geltenden Land- und Stadt-Rechte. Nebst den für diese Gegenden erlassenen neueren Verfügungen. Mit Anmerkungen begleitet*, vol. 1 (Schleswig: Serringhausen, 1794), 653f.; Dieter Stegmann, "Die Stapelholmer Konstitution von 1623: Zugleich ein Beitrag zur Rezeptionsgeschichte Schleswig-Holsteins" (PhD diss., Christian-Albrechts-Universität zu Kiel, 1967), 89f.; Niemann, *Forststatistik der dänischen Staaten*, 259; Hase, "Abriß," 97.
53. Lorenzen-Schmidt, "Waldverlust," 61f.
54. Quote translated from German. The original reads as follows: "Und Wir dann die Beförderung dieses heilsamen Wercks aus Landes-Väterlicher Vorsorge um destomehr in allergnädigste Erwegung gezogen, als daran zugleich dem allgemeinen Wesen und der Nachwelt höchsten gelegen." Von Cronhelm, *Corpus*, 1245.
55. Translated from German. LASH, Abt. 401 Königliche Verordnungen, n. 1785 15.6.
56. Hase, "Abriß," 94, 111; Neuschäffer, "Waldnutzung," 12; Riis, *Wirtschafts- und Sozialgeschichte*, 364. Concerning the sheep farm Halloh near Neumünster, cf. LASH, Abt. 105, n. 451–1.
57. Martin Nickol, "Forstbaumschule," in *Kiel Lexikon*, ed. Doris Tillmann and Johannes Rosenplänter (Kiel: Wachholtz, 2011), 101f.
58. Cf. also Neuschäffer, *Geschichte*, 17f., 24.
59. Franklin Kopitzsch, "Schleswig-Holstein im Gesamtstaat 1721–1830: Absolutismus, Aufklärung und Reform," in *Geschichte Schleswig-Holsteins: Von den Anfängen bis zur Gegenwart*, ed. Ulrich Lange (Neumünster: Wachholtz, 2003), 281–332, here 288–293 (mention of the tree nursery on p. 290).
60. Hase, "Abriß," 98f., 111; Neuschäffer, *Geschichte*, 24; Neuschäffer, "Waldnutzung," 8.

61. Mantel, *Forstgeschichte*, 688f.
62. Karl-Ernst Behre, *Landschaftsgeschichte Norddeutschlands* (Neumünster: Wachholtz, 2008), 210.
63. Oliver Auge, "Adlige Selbstdarstellung und Legitimation um 1600: Die sogenannte Rantzausche Tafel auf Krengerup (Fünen)," *Nordelbingen* 80 (2011): 35–55.
64. Mantel, *Forstgeschichte*, 689.
65. Auge, "Adelige Selbstdarstellung," 43: "Anno D(omi)ni 1580 Hen(ricus) Ran(zouius) has querqus, abietes et petulas plantauit et initiu(m) sationis adscribi jiussit, ut ear(um) aetatem exploraret posteritas. Qua in omnia orbis saecula aternae divinitati commendat."
66. Concerning Carlowitz and his work "Sylvicultura oeconomica" from 1713, cf., e.g., Reith, "Nachhaltigkeit," 1010; Jaeger, *Umweltgeschichte der frühen Neuzeit* (Munich: Oldenbourg, 2011), 135. Cf. also Oliver Auge, "'Nachhaltigkeit' als historisches Thema—eine Hinführung," 45–54.
67. Bernd Herrmann, "Kritisches Nachwort eines Umwelthistorikers," *Jahrbuch für Regionalgeschichte* 32 (2014): 116–19.
68. Quote translated from German. Arne Paysen, "Nachhaltige Energiewirtschaft? Brenn- und Kohlholznutzung in Schleswig-Holstein in Mittelalter und Früher Neuzeit" (PhD diss., Christian-Albrechts-Universität zu Kiel, 2009), 222. http://macau.uni-kiel.de/receive/dissertation_diss_00006895.
69. Paysen, "Nachhaltige Energiewirtschaft," 207–22, 313–32. A synopsis is provided in Paysen, "Die Waldwirtschaft des Klosters Ahrensbök als Beispiel für eine nachhaltige Ressourcennutzung," *Jahrbuch für Regionalgeschichte* 32 (2014): 73–86. The register of feudal payments including the register of charcoal burning has been published as an edition: *Schleswig-Holsteinische Regesten und Urkunden* X, ed. Wolfgang Prange (Neumünster: Wachholtz, 1989), 171–383, especially the register of charcoal burning on 368–77.
70. Quote translated from German. Paysen, "Nachhaltige Energiewirtschaft," 218.
71. Quotations from Paysen, "Nachhaltige Energiewirtschaft," 219, translated from German.
72. Günther Bock, "Stormarn im Mittelalter: Bevölkerung, Herrschaft, Siedlung und Ressourcennutzung," in *Nutzung gestaltet Raum: Regionalhistorische Perspektiven zwischen Stormarn und Dänemark*, ed. Oliver Auge and Norbert Fischer (Frankfurt am Main: Peter Lang Edition, 2017), 79–81.
73. Following Paysen, "Nachhaltige Energiewirtschaft," 262f.
74. Bock, "Stormarn im Mittelalter," 81 for Stormarn.
75. Paysen, "Nachhaltige Energiewirtschaft," 257–68; Riis, *Wirtschafts- und Sozialgeschichte*, 203.
76. Lorenzen-Schmidt, "Waldverlust," 56f.
77. Paysen, "Nachhaltige Energiewirtschaft," 267.

Bibliography

Andresen, Ludwig. "Einfuhr von Steinkohle 1616." *Die Heimat* 15 (1930): 19.

Auge, Oliver. "Adlige Selbstdarstellung und Legitimation um 1600: Die sogenannte Rantzausche Tafel auf Krengerup (Fünen)." *Nordelbingen* 80 (2011): 35–55.

_____. "Auf dem Weg zur Nachhaltigkeit? Ansätze zu Ressourcenschutz und Ressourcenregeneration im spätmittelalterlichen und frühneuzeitlichen Schleswig-Holstein." In *Wirtschaft und Umwelt vom Spätmittelalter bis zur Gegenwart: Auf dem Weg zu Nachhaltigkeit?*, edited by Günther Schulz and Reinhold Reith, 31–51. Stuttgart: Franz Steiner Verlag, 2015.

_____. "Nachhaltigkeit als historisches Thema: Das Beispiel Schleswig-Holstein in Spätmittelalter und früher Neuzeit." *Natur- und Landeskunde. Zeitschrift für Schleswig-Holstein, Hamburg und Mecklenburg* 120 (2013): 141–49.

_____. "'Nachhaltigkeit' als historisches Thema—eine Hinführung." *Jahrbuch für Regionalgeschichte* 32 (2014): 45–54.

Auge, Oliver, and Joachim Schneider. "Der Niederadel: Schleswig-Holstein/Franken." In *Handbuch Landesgeschichte*, edited by Werner Freitag, Michael Kißener, Christine Reinle, and Sabine Ullmann, 303–34. Berlin: De Gruyter Oldenbourg, 2018.

Behre, Karl-Ernst. *Landschaftsgeschichte Norddeutschlands*. Neumünster: Wachholtz, 2008.

Benesch, Johann Eike. "Das Bordesholmische Brücheregister von 1619/20." *Jahrbuch Bordesholm* 6 (2004): 117–33.

Blickle, Peter. "Wem gehörte der Wald? Konflikte zwischen Bauern und Obrigkeiten um Nutzungs- und Eigentumsansprüche." *Zeitschrift für Württembergische Landesgeschichte* 45 (1986): 167–78.

Bock, Günther. "Stormarn im Mittelalter: Bevölkerung, Herrschaft, Siedlung und Ressourcennutzung." In *Nutzung gestaltet Raum: Regionalhistorische Perspektiven zwischen Stormarn und Dänemark*, edited by Oliver Auge and Norbert Fischer, 45–82. Frankfurt am Main: Peter Lang Edition, 2017.

Bork, Hans-Rudolf, ed. *Landschaftsentwicklung in Mitteleuropa: Wirkung des Menschen auf Landschaften*. Gotha: Klett-Perthes, 1998.

Carstens, Werner, ed. *Schleswig-Holsteinische Regesten und Urkunden*. Vol. 6. Neumünster: Wachholtz, 1971.

Carstensen, Jan. *Torfgewinnung: Geschichte und Bedeutung in Schleswig-Holstein*. Osnabrück: Scholl, 1985.

Dahlmann, Friedrich Christoph. "Neocorus, der Dithmersche." *Kieler Blätter* 5 (1818): 181–252.

Freitag, Winfried. "Landbevölkerung, Forstpersonal und 'gute Waldordnung' in der 'Bayrischen Vorstordnung' von 1568." *Zeitschrift für Agrargeschichte und Agrarsoziologie* 55 (2007): 32–57.

Freytag, Nils. "Deutsche Umweltgeschichte—Umweltgeschichte in Deutschland." *Historische Zeitschrift* 283 (2006): 383–407.

Fritzbøger, Bo. *Danske skove 1500–1800: En landskabshistorisk undersøgelse.* Odense: Landbohistorisk Selskab, 1992.

Grewe, Bernd-Stefan. "'Man sollte sehen und weinen!' Holznotalarm und Waldzerstörung vor der Industrialisierung." In *Wird Kassandra heiser? Die Geschichte falscher Öko-Alarme*, edited by Frank Uekötter and Jan Hohensee, 24–40. Stuttgart: Steiner, 2004.

Hase, Walter. "Abriß der Wald- und Forstgeschichte Schleswig-Holsteins im letzten Jahrtausend." *Schriften des Naturwissenschaftlichen Vereins für Schleswig-Holstein* 53 (1983): 83–124.

Hasel, Karl. *Forstgeschichte: Ein Grundriß für Studium und Praxis.* Hamburg: Parey, 1985.

Herrmann, Bernd. "Kritisches Nachwort eines Umwelthistorikers." *Jahrbuch für Regionalgeschichte* 32 (2014): 109–20.

Hiller, Hubertus. *Untertanen und obrigkeitliche Jagd: Zu einem konfliktträchtigen Verhältnis in Schleswig-Holstein zwischen 1600 und 1848.* Neumünster: Wachholtz, 1992.

Jaeger, Friedrich. *Umweltgeschichte der frühen Neuzeit.* Munich: Oldenbourg, 2011.

Jarchow, Otto. "Waldsterben im Eutinischen vor 250 Jahren." *Jahrbuch für Heimatkunde Eutin* 17 (1983): 33–35.

Jensen, Friedrich Christoph, and Dietrich Hermann Hegewisch, eds. *Privilegien der Schleswig-Holsteinischen Ritterschaft von den in der Privilegienlade befindlichen Originalen genau abgeschrieben und mit denselben verglichen.* Kiel: Christian Friedrich Mohr, 1797.

Jessen, Hans. "Die Bewaldung in Angeln im Jahre 1730." *Jahrbuch des Angler Heimatvereins* 52 (1988): 48–58.

Jørgensen, Dolly. "The Roots of the English Royal Forest." *Anglo-Norman Studies* 32 (2010): 114–28.

Kjærgaard, Thorkild. *Den danske Revolution 1500–1800: En økohistorisk tolkning.* Copenhagen: C.A. Reitzels forlag, 1991.

Knoll, Martin. "Dominanz als Postulat. Höfische Jagd: Natur und Gesellschaft im 'Absolutismus.'" In *Umwelt und Herrschaft in der Geschichte*, edited by François Duceppe-Lamarre and Jens Ivo Engels, 73–91. Munich: De Gruyter, 2008.

_____. *Umwelt—Herrschaft—Gesellschaft: Die landesherrliche Jagd Kurbayerns im 18. Jahrhundert.* Sankt Katharinen: Scripta-Mercaturae-Verlag, 2004.

Kopitzsch, Franklin. "Schleswig-Holstein im Gesamtstaat 1721–1830: Absolutismus, Aufklärung und Reform." In *Geschichte Schleswig-Holsteins: Von den Anfängen bis zur Gegenwart*, edited by Ulrich Lange, 281–332. Neumünster: Wachholtz, 2003.

Kragh, Lisa. "'Zweibeinigte Holzwürmer?' Die Bauern und der Waldrückgang in Schleswig-Holstein in Mittelalter und Früher Neuzeit." *Zeitschrift der Gesellschaft für Schleswig-Holsteinische Geschichte* 139 (2014): 23–48.

Landesarchiv Schleswig-Holstein, Abt. 7, n. 3026, n. 4127. Abt. 260, n. 611. Abt. 400.5, n. 606. Abt. 401 Königliche Verordnungen, n. 1785.

Laurop, Christian Peter. *Freimüthige Gedanken über den Holzmangel, vorzüglich über den Brennholzmangel in den Herzogthümern Schleswig und Holstein und die Mittel, ihm abzuhelfen*. Altona: Hammerich, 1798.

Lorenzen-Schmidt, Klaus-Joachim. "Waldverlust und Waldaufbau in Schleswig-Holstein vom Mittelalter bis 1914." In *Dünger und Dynamit: Beiträge zur Umweltgeschichte Schleswig-Holsteins und Dänemarks*, edited by Manfred Jakubowski-Tiessen and Klaus-Joachim Lorenzen-Schmidt, 41–64. Neumünster: Wachholtz, 1999.

Mager, Friedrich. *Entwicklungsgeschichte der Kulturlandschaft auf der Geest und im östlichen Hügelland des Herzogtums Schleswig bis zur Verkoppelungszeit*. Breslau: Hirt, 1930.

Mantel, Kurt. *Forstgeschichte des 16. Jahrhunderts unter dem Einfluß der Forstordnungen und Noe Meurers*. Hamburg: Parey, 1980.

_____. *Wald und Forst in der Geschichte: Ein Lehr- und Handbuch*. Hanover: Schaper, 1990.

Neuschäffer, Hubertus. "Die Waldnutzung in der Geschichte Schleswig-Holsteins." *Schleswig-Holstein* 6 (1985): 8–12.

_____. *Geschichte von Wald und Forst in Schleswig-Holstein und der alte Rendsburger Wald mit dem Forstamt Barlohe*. Rendsburg: Möller, 1986.

Nickol, Martin. "Forstbaumschule." In *Kiel Lexikon*, edited by Doris Tillmann and Johannes Rosenplänter, 101–2. Kiel: Wachholtz, 2011.

Niemann, August. *Forststatistik der dänischen Staaten: Mit drei Schautafeln*. Altona: Hammerich, 1809.

Paysen, Arne. "Die Waldwirtschaft des Klosters Ahrensbök als Beispiel für eine nachhaltige Ressourcennutzung." *Jahrbuch für Regionalgeschichte* 32 (2014): 73–86.

_____. "Nachhaltige Energiewirtschaft? Brenn- und Kohlholznutzung in Schleswig-Holstein in Mittelalter und Früher Neuzeit." PhD diss., Christian-Albrechts-Universität zu Kiel, 2009. Retrieved 21 January 2020. http://macau.uni-kiel.de/receive/dissertation_diss_00006895.

Prange, Wolfgang, ed. *Schleswig-Holsteinische Regesten und Urkunden*. Vol. 10. Neumünster: Wachholtz, 1989.

Radkau, Joachim. "Holzverknappung und Krisenbewußtsein im 18. Jahrhundert." *Geschichte und Gesellschaft* 9 (1983): 513–43.

_____. "Zur angeblichen Energiekrise des 18. Jahrhunderts: Revisionistische Betrachtungen über die Holznot." *Vierteljahrschrift für Sozial- und Wirtschaftsgeschichte* 73 (1986): 1–37.

Radkau, Joachim, and Ingrid Schäfer. *Holz: Ein Naturstoff in der Technikgeschichte*. Reinbek bei Hamburg: Rowohlt, 1987.

Rantzau, Heinrich. *Statthalter in Schleswig und Holstein: Ein Humanist beschreibt sein Land*. Schleswig: Hamburg University Press, 1999.

Reith, Reinhold. "Nachhaltigkeit." In *Enzyklopädie der Neuzeit*. Vol. 8, edited by Friedrich Jaeger, 1009–12. Stuttgart: Metzler, 2008.

_____. *Umweltgeschichte der frühen Neuzeit*. Munich: Oldenbourg, 2011.

Rheinheimer, Martin. *Die Dorfordnungen im Herzogtum Schleswig*. Vols. 1 & 2. Stuttgart: Lucius & Lucius, 1999.

_____. "Umweltzerstörung und dörfliche Rechtssetzung im Herzogtum Schleswig (1500–1800)." In *Dünger und Dynamit: Beiträge zur Umweltgeschichte Schleswig-Holsteins und Dänemarks*, edited by Manfred Jakubowski-Tiessen and Klaus-Joachim Lorenzen-Schmidt, 81–92. Neumünster: Wachholtz, 1999.

Riis, Thomas. *Wirtschafts- und Sozialgeschichte Schleswig-Holsteins: Leben und Arbeiten in Schleswig-Holstein vor 1800*. Kiel: Ludwig, 2009.

Rösener, Werner. *Die Geschichte der Jagd: Kultur, Gesellschaft und Jagdwesen im Wandel der Zeit*. Darmstadt: Wissenschaftliche Buchgesellschaft, 2004.

Rust, Wilhelm. *Das Flensburger Bauhandwerk von 1388 bis 1966: Ein Beitrag zur Baugeschichte der Stadt*. Flensburg: Innung d. Baugewerbes, 1967.

Schmeidler, Bernhard, ed. *Magistri Adam Bremensis Gesta Hammaburgensis ecclesiae pontificum*. Hannover: Verlag Hahnsche Buchhandlung, 1917.

Schmidt, Uwe Eduard. *Der Wald in Deutschland im 18. und 19. Jahrhundert: Das Problem der Ressourcenknappheit dargestellt am Beispiel der Waldressourcenknappheit in Deutschland im 18. und 19. Jahrhundert. Eine historisch-politische Analyse*. Saarbrücken: Conte-Verlag, 2002.

Schönwandt, Max. "Die Familie der Kirchspielvögte von Würtzen zu Nortorf." *Rendsburger Jahrbuch* 34 (1984): 61–72.

Schwarz, Hans Wilhelm. "Aus dem Brücheregister der Hanerauer Amtsrechnung 1619." *Rendsburger Jahrbuch* 53 (2003): 209.

_____. "Aus dem Brücheregister der Hanerauer Amtsrechnung 1622." *Rendsburger Jahrbuch* 55 (2005): 87, 96.

Secher, Vilhelm Adolf, ed. *Corpus Constitutionum Daniae*. Vol. 6. Copenhagen: Selskabet for Udgivelse af Kilder til Dansk Historie, 1918.

Stegmann, Dieter. "Die Stapelholmer Konstitution von 1623: Zugleich ein Beitrag zur Rezeptionsgeschichte Schleswig-Holsteins." PhD diss., Christian-Albrechts-Universität zu Kiel, 1967.

von Brockdorff, Cai Lorenz, ed. *Corpus statutorum Slesvicensium, oder: Sammlung der in dem Herzogthum Schleswig geltenden Land- und Stadt-Rechte. Nebst den für diese Gegenden erlassenen neueren Verfügungen. Mit Anmerkungen begleitet*. Vol. 1. Schleswig: Serringhausen, 1794.

von Cronhelm, Friderich Detlef Carl, ed. *Corpus Constitutionum Regio-Holsaticarum, oder allerhöchst-autorisirte Sammlung der in dem Herzogthum Holstein königl: Antheils, samt incorporirten Landen wie auch der Herrschaft Pinneberg, Stadt Altona und Graffschafft Rantzau, in Krafft eines beständigen Gesetzen ergangenen Constitutionen, Edicten, Mandaten, Decreten, Resolutionen, Privilegien, Concessionen, und anderen Verfügungen*. Vol. 1. Altona: Gebrüder Burmester, 1749.

Warde, Paul. *Ecology, Economy and State Formation in Early Modern Germany*. Cambridge, UK: Cambridge University Press, 2006.

CHAPTER 12

Traditional Woodland Management, Forest Legislation, and Modern Nature Conservation in East-Central Europe

Péter Szabó

[There is a wood called] Scheyben, which was at the time [of the survey] six years old. When it reaches seven years, it is estimated to be worth twenty marks.

—Extract from the rent-roll (*urbarium*) of the Mikulov estate from 1414

These woods, since there is in them a good growth of underwood and standards for building, can be utilized by wood-sales every year without causing damage for c. 20 florins.

—Extract from the rent-roll of Maršovice from 1651

The origins of the coppices that belong to Dobrau may be exclusively attributed to the faulty management of broadleaved forests.

—Extract from the forest management plan of Dúbrava Wood from 1864

Planned cutting in forest stands younger than eighty years is forbidden.

—Paragraph 33.4 of the Forest Law of the Czech Republic, 1995

[A] change of the current policy of growing high forest towards re-establishment of active coppicing . . . is a solution supported by both economic and conservation considerations.

—Comment by Beneš et al. in a 2006 paper on butterfly conservation in Milovice Wood

Although they concern different locations and therefore caution is needed when connecting them, the five quotations above express the main message of this chapter. They describe a full circle from medieval and early modern practices in woodland management through their rejection in the nineteenth and twentieth centuries to rediscovery and rehabilitation in recent times.[1] However, I do not simply argue that there is nothing new under the sun. The motivations behind the actions of those who produced these sentences were obviously not the same. Nor can we create a unified story for all places and all woodland management types. In this chapter, I will describe in some detail three management types (coppicing, wood-pasturing, and litter raking) and their contested and entangled histories in the past seven centuries in east-central Europe.

This chapter mainly concerns the territory of what is now the Czech Republic. This area contains three geographical–political units, together loosely referred to as the "Czech Lands." Bohemia in the west and Moravia in the east formed, for most of their history, stable parts of the Czech kingdom and shared a ruler. The history of Silesia in the northeast was more checkered, and currently most of Silesia belongs to Poland. Most of my examples will come from Moravia.[2] Together with a handful of other areas in the region, in the sixteenth century, the Czech Lands became part of the Habsburg Monarchy and remained there until the formation of Czechoslovakia in the early twentieth century. The larger region where the Czech Lands are situated is sometimes referred to as east-central Europe. The term, which is political rather than spatial, refers to today's Poland, western Ukraine, Slovakia, Czech Republic, Hungary, and western Romania. "East-central" implies certain transitional characteristics in the historical development of the region between the Orthodox East and the (often idealized and simplified) West.[3] Allowing for some broad generalizations that mask much variety in local developments, the region was transformed based on the western model in the thirteenth century and subsequently somewhat diverted from that model in the Early Modern Period (sixteenth to eighteenth centuries).[4] I will use some examples from Hungary as instructive parallels to developments in the Czech Lands, whereas I will rely much less on Polish material.

Traditional Woodland Management

The general features of traditional woodland management have been described many times, including in this book.[5] The term itself was coined and is usually used to contrast with modern ("scientific") forestry. There are three important features that distinguish traditional woodland management from modern forestry, which became the dominant force in European woodland management in the nineteenth century. First, as opposed to the heavy use of plantations in modern forestry, traditional management relied almost exclusively on the self-renewing capacity of trees. Although references to sowing seeds of trees in forests can be found in premodern documents, these were negligible oddities in a system that used the generative and vegetative abilities of trees to grow again after felling. Peasants in medieval and early modern east-central Europe were well-aware of what modern ecologists call "succession": that in most circumstances in temperate Europe, a piece of land, if left alone, grows into a forest in time.[6] Perhaps the most succinct expression of this is the Hungarian word for "forest" (*erdő*), which, etymologically, can be loosely translated as "the thing that grows by itself."[7]

Second, traditional management was driven by the "smallest tree that will do the job" principle.[8] Lacking power tools to move and chop up large trees, people strove to use the smallest piece of wood that was suitable for a given purpose. Anything larger was considered to lead to a waste of human energy, which people sought to avoid.[9] The 1414 rent-roll of the Mikulov estate, quoted at the beginning of this chapter, referred to woods older than the normal seven-year cutting cycle as overgrown and difficult to sell. Firewood was made of small underwood, and timber for the building of houses or fortresses was unusually small to the modern eye. Trees larger than around 30 cm in diameter tended to be used only in exceptional situations and in major constructions.

Third, traditional management relied on multiple uses of the same piece of woodland. The best-known example of this is wood-pasture, which combined animal grazing with wood production. Much less appreciated is the fact that even intensive coppice systems often included other types of uses. For example, late seventeenth and early eighteenth century woodland management instructions (*Waldordnungen*) for the Mikulov estate prescribed that grazing animals could be allowed into coppices three years after cutting and that grass cutting was not allowed for four or five years.[10] With a coppice cycle of roughly thirteen years in this period,[11] any given part of the local coppice woods was under multiple-use for approximately two thirds of the time.

Practice

The methods of traditional woodland management are many. Some have received more attention, especially those involving tree cutting. Broadleaved trees were usually coppiced (i.e., cut on a short rotation—seven to twenty-five years) to produce firewood. Building timber was grown in the form of standards: straight trees from seed that were allowed to grow through several coppice cycles. Archaeological evidence shows that coppice underwood was not selective as to species.[12] By contrast, in many parts of Europe, oak was promoted as a standard tree because of its superior qualities in building.[13] Coniferous trees do not grow from stumps and are therefore unsuitable for coppice management. Their traditional management is much less known, and is usually described within the framework of selective cutting.[14] In any case, in east-central Europe, woods were usually more copious at higher elevations, and therefore their management was arguably less intensive than that of sparse broadleaved forests in the lowlands.

The other prominent method of traditional woodland management is wood-pasture. This has created or maintained many iconic European landscapes from Windsor Great Park to Mediterranean savannas, and therefore it drew much research attention.[15] Local versions include, for example, the combination of hay meadows and trees in Scandinavia.[16] Much less attention was paid to other uses, especially to those that did not involve tree cutting. This partly stems from the fact that modern forestry (and therefore forestry historians) relegated these other methods of woodland management to "minor" uses. However, significance is a question of viewpoint. Recent research shows that such "minor" uses could be ubiquitous with considerable consequences.[17] Frequent litter raking (the removal of leaves from broadleaved and coniferous forests to be used as bedding for farm animals and subsequently as fertilizer) was argued to remove as much biomass from forests as tree cutting itself.[18] Bürgi and Gimmi rightly pointed out that the ecologically most significant management forms may not always be those best visible in historical sources.[19]

As in other parts of Europe, people in medieval and early modern east-central Europe realized that they needed to manage their natural resources so that they would be "maintained in good condition." For example, woods around Ostřetín (eastern Bohemia) were described at the end of the fifteenth century as follows: "From these woods firewood can be sold each year for 600 groš without overcutting them. The pine stands that are here are banned so that they are not sold but preserved for [future] needs."[20] This was a utilitarian view that treated woods similarly to other natural resources, such as meadows or fishponds. Nonetheless, it was based on remarkable foresight. By looking to "[future] needs," the overlord of the Pardubice estate embraced a long-time perspective. It takes thirty to fifty years for a pine to reach timber-size—in other words, whoever created this woodland management regime was rather unlikely to see these pines cut down. That, however, was beside the point. Rather, plans were made so that the estate as a whole manages its woodland resources in a sustainable manner. It was easy to plan for a short coppice cycle, at least time-wise. Spatial organization was more of a challenge: the same document described the sequence in which individual woods that belonged to the castle of Pardubice were to be harvested on a nine-year rotation, showing that preference was given to larger areas to be cut at once rather than subdividing each wood into nine parts.[21] Parallel to this short-rotation and spatially sophisticated system, the estate (which included dozens of villages) needed to maintain a sufficient cohort of timber trees. Building activity was different from heating and cooking in the sense that it was intermittent and largely unpredictable. For example, after a village fire, many timber trees were needed at once. On the Mikulov estate, 616 standards were cut down in the winter of 1684–85 to help the inhabitants of Sedlec rebuild their settlement that had burnt down.[22] Because there were only a few thousand standards on the entire estate, this represented a significant proportion of the available resources. The aim of estate managers was to have the largest number of building timbers available at any given point without compromising coppice regrowth. Nonetheless, the "smallest tree" principle was applied to timber as well; therefore, standards that reached a certain age were harvested and sold. For example, some seven hundred standard trees were sold on the market on the Choustník estate (southern Bohemia) in 1447.[23]

Wood-pasture and other non-timber uses were organized along similarly sustainable lines. Care was taken not to overexploit forest resources, but this was understood not as a reference to some ideal state but keeping the needs of the local community in mind. For example, in Dúbrava Wood, two detailed rent-rolls described wood-pasture management in the seventeenth century.[24] The system was rather sophisticated. It regulated the exact areas where each neighboring settlement had access rights. Smaller sections of these wooded areas were prescribed for specific animals (cows, cattle, goats, pigs, etc.) as well as for hay cutting. A clear distinction was made between all-year grazing and pannage (letting pigs in the wood to devour the occasional good crop of oak acorns).[25] At the same time, local woodland managers also had to make sure grazing animals did not get into freshly cut coppice areas. Lastly, the documents also contained information on whether or not payment was due for each use. The resulting landscape was a combination of coppices and wood-pastures with large and spreading trees in grassland.[26] A similar management regime was noted down in the oak woods around Litovel in central Moravia: wood cutting (coppice and standards), hay cutting (on woodland meadows and among trees), pannage, and pasture (of horses and cows) were all combined in a single forest.[27] In another example from the region, Samojlik and colleagues mapped traditional uses from the first to the eighteenth centuries in Białowieża forest (Poland), where the system was similar but apparently (partly because of the sheer size of the forest) less intensive.[28]

Normative sources described woodland management along similar lines. Apparently there was very little effort to enforce top-down regulation different from what was already happening on the ground. Normative sources differed in their spatial scale but were similar in that each one concerned the forests in the possession of a single overlord. The first efforts at centrally regulating all forests in the region originated in the eighteenth century.[29] Information on forests either was included in general instructions on the management of particular estates or formed separate "forest instructions." The latter were compiled in larger numbers from the sixteenth century onward.[30] Royal regulations, among which the *Majestas Carolina* (ca. 1350) and sixteenth-century instructions to the "highest hunter of the Czech kingdom" (i.e., the highest forest official) are best known, did the same but pertained to woodland on royal estates.[31] The Habsburg rulers issued similar documents in this period in neighboring Hungary as well.[32] The contents of all these documents were rather similar, although attention to detail greatly varied. They almost always contained three elements. First, they emphasized the need for controlled cutting that had to be evidenced; the woodlot to be harvested had to be determined by local foresters and paid for in advance; all other wood-cutting activities were banned; and foresters were obliged to record the transaction in an account book. There are plenty of examples for such detailed account books, especially from the seventeenth century onwards. While all this did not present any novelty as far as tree cutting was concerned, it represented an aim to introduce complete control over what was cut down. Second, the wood harvested had to be removed from forests within a limited amount of time, by the beginning of spring at the latest.[33] This was intended to facilitate regrowth. Third, and with a similar purpose, animal grazing was banned in freshly cut areas for a number of years.[34] Two more points were frequently, but not always, included in such instructions: the warning not to cause forests fires and the right of locals to gather dry and windthrown wood for their own use.

Some documents contain remarkable insights into management techniques. For example, the general instructions of the Duchy of Frýdlant (1628) include a passage ordering foresters to always cut at the edges of forests because cutting in the middle of a stand made it more susceptible to windthrow.[35] *Why* woods should be managed in this way was rarely deemed worthy of explanation. Similarly to economic documents, normative sources sometimes mention a wish to "avoid damage," and occasional references of "favor to the people" can also be found.[36] A prominent document in which such a utilitarian perspective was not the only one is the above-mentioned *Majestas Carolina*, a failed attempt to codify Czech law by Emperor Charles IV (d. 1378). This included several sentences about the beauty of Czech forests, which was a source of admiration, and connected this to their preservation.[37]

Rejection

The eighteenth and, especially, the nineteenth centuries brought significant changes not so much in immediate management as in attitudes. Mainly governed by central authorities representing the modern state (in east-central Europe enlightened absolutism), the traditional system of woodland management was no longer deemed adequate and was suppressed by various means. The story is well-known, and it is not my aim to retell it here.[38] However, I would like to point out that realities on the ground could and often did differ from textbook examples. Coppicing was accepted as a normal management practice throughout the eighteenth century. In contrast, nineteenth-century foresters often snubbed it. However, where firewood was needed, and timber, especially coniferous timber, was unlikely to grow well, foresters were happy to continue with coppicing until the mid-twentieth century. The managers of the Mikulov estate woodlands noted in 1905: "the market for firewood is satisfactory, although it decreased in the past years with increasing coal consumption. However, it will still take a long while to bring all the necessary quantity [of coal]."[39] As a result, coppice management continued well into the twentieth century until it was abandoned in the 1920s–1940s.[40] This was typical for large parts of northwestern Europe. In Dúbrava Wood, the process was more peculiar. The system described above in connection with wood-pasture went through major changes in the late eighteenth century.[41] Woodland and pasture were separated, and coppicing, which had played only a lesser role in the previous regime, was enforced as the exclusive management of wooded areas. In effect, most former wood-pastures were transformed into coppices on thousands of hectares. Arguably because of site conditions, the attempt failed to bring the required results, or so those foresters that oversaw Dúbrava Wood in the mid-nineteenth century argued and whose dry remark on mismanagement I quoted at the beginning of this chapter. However, it took another century for the amount of high forests to overtake that of coppices. Notwithstanding the opinion of local forest managers, the area of coppices showed no decline until the 1930s. This was reflected in legislation as well: the 1928 Czechoslovak forest law still included twenty-year coppice cycles, and the 1960 forest law prescribed the transformation of coppices into high forests rather than banning coppice management as such.

Wood-pasture has had a worse press than coppicing in forestry history; but in fact, it was not more readily or generally rejected by most forest managers. As examples above showed, there was much awareness of the detrimental effects of grazing on young shoots in coppices already in early periods. Quite logically, when greater efforts were made to maintain only one use for each piece of land and also with greater emphasis on wood production, the presence of animals was seen as increasingly harmful. There was a seamless transition from the prohibition of grazing animals in freshly cut areas into encouraging their complete removal from forests: the argumentation did not have to change. For example, the forest code of 1754 for the Czech kingdom, issued by Empress Maria Theresa (d. 1780), acknowledged that a complete grazing ban in forests would be desirable but too onerous, especially for lesser woodland owners, and suggested letting animals into one half of every forest while leaving the other half free of grazing in a spatially dynamic system.[42] The 1769 forest code for Hungary already aimed to introduce a blanket ban and suggested that woodland owners plant willows and poplars outside forests to be used as wood-pasture.[43] Nonetheless, the 1852 forest law of the Habsburg Monarchy took woodland grazing for granted and only aimed at keeping it within reasonable limits. In Czechoslovakia, it was only the 1960 forest law (the first comprehensive legislation since 1852) that completely banned grazing in all forests. Basically the same story can be told in Hungary, where a blanket ban came into effect with the 1961 forest law.[44]

Litter raking featured much less in twentieth-century forestry history than woodland grazing. This is in sharp contrast to the attention that foresters in the nineteenth century devoted to it. For example, in the 1850s and 1860s there was a lively debate in the journal of Bohemian foresters (*Vereinsschrift für Forst-, Jagd- und Naturkunde*) about what was called the "forest-litter question" (*Waldstreu-Frage*). The opening piece readily acknowledged that litter raking was a widespread practice and tried to point out its negative effects on the soil.[45] Some fifteen pieces by various authors followed. Their main focus was calculating the economic value of litter raking and juxtaposing this with economic loss through reduced tree growth. One Franz Kreß even conducted a twelve-year experiment in which he recorded the amount of litter raked and tree increment in a pinewood.[46] Other pieces dealt with the effects of raking on the chemical properties of the soil. Similar articles appeared in the journal of Moravian foresters as well.[47] The issue made it to the highest circles of forestry education. Sometime between 1864 and 1888, Anton Urbanek wrote his essay for the state forest examination (the examination necessary to be employed as a forester) on litter raking.[48] Basically all authors agreed that litter raking was bad for tree-growth and proposed to limit it. However, similarly to grazing, litter raking was still widely practiced in the first half of the twentieth century. A summary paper on grazing and litter raking in Czechoslovakia in 1948 basically repeated the claims and arguments that had been heard for a century adding only that both uses were now mainly practiced in smaller woodlands outside state ownership.[49] Once again it was the 1960 forest law that codified the complete ban of litter raking in forests.

The combination of pressure from official forestry, the substitution of forest products with alternative sources of energy, and restrictive legislation finally pushed traditional woodland management into sharp decline in the mid-twentieth century. In 1930, there were still more than 130,000 hectares of coppices in the Czech parts of Czechoslovakia. This went down to around 103,000 hectares in the 1940s, 78,000 hectares in 1950, and 30,000 hectares in 1980.[50] Similar stories can be told for wood-pasture and litter raking in the Czech Lands as well as in neighboring countries. By the end of the twentieth century, traditional management practices all but disappeared from east-central European forests, lingering on only in more remote areas.[51]

Rediscovery

In the 1960s, coppicing was rediscovered by a small group of British woodland ecologists. Their motivation was quite clearly rooted in nature conservation (in the modern sense). Coppice woods came to be valued because they harbored a large diversity of often rare plants. Because many of these plants are light-demanding, it quickly became clear that their continued presence was dependent on management. Since then, countless studies in inactive coppices all over Europe demonstrated that the closing in of the tree canopy through succession after the abandonment of management was detrimental to these plants. Logically, then, ecologists proposed the reintroduction of coppicing simply to save populations of rare plants. One of the first such sites was Hayley Wood in Essex, England.[52] In a reversal of the logical relationship between history and ecology, in many European regions, ecologists started to compile lists of "ancient woodland indicator" plants (i.e., those plants that tend to grow in woodlands with a habitat continuity on a centennial scale).[53] Together with Hooper's hedge dating rule,[54] this was arguably the first time ever that purely ecological data (species distribution) became historical data (indicator of land-cover continuity). In the following decades, more types of organisms were added to the list. Some woodlands (such as Wolves Wood in Suffolk, England) are managed in the interest of bird protection. Several types of insects, notably butterflies, were shown to be strongly dependent on active coppicing.[55] Keeping in mind the initial strong resistance of foresters to coppice reintroduction, in a somewhat unexpected turn of events, foresters have recently embraced coppicing and have put considerable effort into mapping its economic potential. One of their motivations was the growing need for renewable energy sources. It now appears that coppicing can make economic as well as conservation sense.[56]

Wood-pasture had a somewhat different story. While grazing in woodlands was, as we have seen above, very much against the ideals of enlightened land managers, the resulting park-like landscapes were among the first to be appreciated for aesthetic values in aristocratic circles. Freestanding trees (usually but not exclusively of wood-pasture origin) were often incorporated into landscaped parks. A further aristocratic tradition that could create and maintain wood-pasture was hunting, in which semiwild rather than domestic animals kept the stands open.[57] That said, existing research suggests that the value of wood-pastures for modern nature conservation was realized later than that of coppice woods: in the 1990s in England and even later in the rest of Europe.[58] Early and much of later research centered on invertebrates: wood-pastures harbor the vast majority of very old trees in Europe, which are crucial for many of these creatures.[59] Perhaps because of this relatively late start, wood-pasture is perhaps the most prolific topic currently among traditional management types in the context of nature protection. As opposed to other types, the social aspects of wood-pasture are also intensively researched.[60] Because, especially in east-central Europe, wood-pasture is still a living practice, anthropological studies have tried to gather and analyze this knowledge for the purposes of nature conservation.[61] As far as legislation is concerned, wood-pasture is in a difficult position. Being a transitory category between grassland and forest, European legislative systems—which to a large extent take their land management concepts from the Enlightenment—find it difficult to deal with and therefore legally protect wood-pasture.

Litter raking has lagged behind coppicing and wood-pasture in being recognized as a possibly valuable contributor to nature conservation. The first review describing previous experiments with litter raking in forest stands came out in 2006, and a year later, Bürgi and Gimmi offered litter collecting as a "new tool for nature conservation" with a question mark at the end.[62] Ongoing experiments are not quite unanimous about the effects of litter raking. Clearly, there is mechanical disturbance and nutrient removal, but whether this is advantageous for rare plants that prefer poorer soils remains to be seen.[63] Further types of traditional nontimber forest uses (such as hay cutting or traditional bee keeping) are very little researched from the perspective of contemporary nature conservation. These examples show that the rediscovery of traditional forest uses is still in progress. One possible conclusion drawn from this process is to interpret traditional uses as interconnected parts of a coherent whole, which brings us back to the third distinguishing feature of traditional woodland management I described at the beginning of this chapter.

Originally, as I argued above, the motivation of ecologists in proposing the reintroduction of traditional woodland management practices was mainly biological. They were little concerned with traditional ways of living or the cultural value of the accumulated ecological knowledge.[64] This has changed, especially since the 2000s and with research on wooded pastures. Among the aims of historical ecology, one of the disciplines dealing with the historical connections between people and nature is to map and preserve "traditional ecological knowledge," which is seen as a value in itself.[65] Clearly the motivations of medieval and early modern woodland managers were different, even though similarities can be contextualized within a common framework, as the Introduction to the present volume exemplifies. It is, nonetheless, worth pointing out that choosing not to explain the rationale, as it were, of motivations was not specific to the pre-1700 period. In Latham's 2015 excellent survey of the development of European forest conservation, "conservation" is contrasted with "protection" and "nature" with practical, aesthetic, cultural, and political reasons for protection. These latter are elaborated on in some detail, but what "nature" is and why it needs to be conserved seems too obvious to be explained.[66] Intriguingly and with direct relevance to the subject matter of the present chapter, Latham explicitly connected what he saw as the current "utilitarian" turn in nature conservation to medieval and early modern attitudes. Indeed, "ecosystem goods and services"—one of the buzz phrases of conservation ecology since the 2000s—have gained considerable momentum and are now central to much conservation research and legislation. The ecosystem goods and services concept is based on the idea that "healthy" ecosystems will provide the most services. Even though the definitions of this "health" would arguably differ between the Middle Ages and the twenty-first century, one can agree with Latham that a parallel can certainly be drawn. It kindles the imagination that the Millennium Ecosystem Assessment should talk about "damage" to ecosystems in a rather similar manner to the Czech rent-roll from 1651 I quoted at the beginning of this chapter.[67]

Unintended Consequences

The last aspect of the connections between medieval/early modern "conservation" and modern nature conservation is what I call unintended consequences. The degradationist narrative of the Enlightenment and scientific forestry may be gone, but one of its legacies is still very much with us: human societies are not seen as part of nature (the ecological jargon for human action is "anthropogenic disturbance"). The claim that this should not be so is nothing new. Löwenthal famously called nature and culture "Siamese twins, separated only at high risk of the demise of both."[68] There is, however, more to this claim than the fact that *Homo sapiens* is just another species. There is growing evidence to show that humans and nature have coevolved with often surprising consequences. "Anthropogenic disturbance" is sometimes the *cause* of biological diversity rather than a nuisance in its way. In some types of ecosystems, this is understood: meadows, for example, have long been recognized as human-made and at the same time valued for their high biodiversity.[69] Woodland, however, perhaps because it is seen as the natural (i.e., human-free) vegetation of most of temperate Europe, was rarely interpreted in this framework. Coppices are obviously a good example of the coevolution, but the phenomenon goes deeper than the presence of light-demanding plants that disappear in lack of management. Entire ecosystems may be driven by human action and result in forest landscapes that are seen as "natural" in modern nature conservation. Perhaps the most sweeping example is the extinction of late Pleistocene megafauna. This obviously caused major changes in subsequent vegetation development in Europe (and elsewhere), and human hunting is one of the possible explanations.[70]

Less grandiose examples can be found in east-central Europe as well. With my colleagues, we conducted several studies in Dúbrava Wood to find out whether the protected oak woods have been, as is usually presumed, dominant at the site since the mid-Holocene. Medieval and early modern documents confirmed this view, but palynological results show a more complicated picture. While open vegetation with considerable amounts of oak has been present at the site for more than seven millennia, the current absolute dominance of oak has a much shorter history. In the fourteenth century, active management was introduced to protect and promote oak. This coincided with a significant increase in oak pollen. Gradually, even the name of the Wood changed. Earlier, it was called *Klečka* ("a place with shrubs"), but during the fourteenth through sixteenth centuries, this name gradually changed into Dúbrava ("oak wood"), arguably as a reflection of vegetation change.[71] Does this make Dúbrava Wood less valuable for nature conservation? The answer is no, but it does turn its interpretative framework around.

Oak seems to be particularly suspect when it comes to coevolution with humans.

My final example comes from Děvín—one of the woods on the Mikulov estate that has featured extensively in this chapter. Here, we found that the establishment of new oak standards simply stopped after management was abandoned in the late 1930s.[72] This probably has to do with light availability, but no straightforward explanation is available. If oaks in the wood are so much dependent on management, how did they get there in the first place, and how did they survive periods without humans? Or did they get into the wood together with people in prehistory, and were there no periods of neglect long enough to make them disappear? Or did they in fact periodically disappear only to reappear when people returned?

As these examples show, some of the consequences of historical human action in forests are very much intended. The medieval managers of Dúbrava Wood wanted to have more oaks in their forests and achieved this aim. However, they would not have dreamt of the vegetation association *Carici fritschii-Quercetum roboris* or Natura 2000 Sites of Community Interest. Through their activities, they still contributed to the emergence of these modern categories. In some senses, they actually created them. Areas of objectivized natural value at many sites did not survive in spite of anthropogenic disturbance; they were rather cocreated by past human societies. At least in long-settled lowland areas, much of what modern nature conservation values according to its own standards can be seen as partially the unintended consequences of past human action.

Péter Szabó is an environmental historian and historical ecologist. He works at the Institute of Botany of the Czech Academy of Sciences in Brno, Czech Republic. He completed his PhD in medieval studies at the Central European University in Budapest, Hungary. His research interests lie in the long-term interactions between human societies and wooded environments, with a special focus on what historical knowledge can contribute to today's nature conservation. Between 2017 and 2019, he served as the President of the European Society for Environmental History.

Notes

The writing of this chapter was supported by long-term research development project no. RVO 67985939 and by grant GA17-09283S from the Czech Science Foundation.

1. Sources for the citations in order of appearance: Bertold Bretholz, ed., *Das Urbar der Liechtensteinischen Herrschaften Nikolsburg, Dürnholz, Lundenburg, Falkenstein, Feldsberg, Rabensburg, Mistelbach, Hagenberg und Gnadendorf* (Reichenberg: Verlag der Anstalt für Sudetendeutsche Heimatforschung, 1930), 109; Moravský zemský archiv (hereafter MZA) F 177 kniha 3; MZA F 5 kniha 233; Jiří Benes, Oldřich Čížek, Jozef Dovala, and Martin Konvička, "Intensive Game Keeping, Coppicing and Butterflies: The Story of Milovicky Wood, Czech Republic," *Forest Ecology and Management* 237 (2006): 353–65. All translations from the original Czech or German are by the author. All sites are in the southeastern Czech Republic.
2. The best general introduction to the history of Moravia in the study period is Josef Válka, *Dějiny Moravy*, 2 vols. (Brno: Muzejní a vlastivědná společnost, 1991–1996).
3. For a summary of the historiography on the topic, see Nora Berend, Premysław Urbańczyk, and Premysław Wiszewski, *Central Europe in the High Middle Ages: Bohemia, Hungary and Poland c. 900–c. 1300* (Cambridge, UK: Cambridge University Press, 2013), 1–39.
4. Jan Klápště, *The Czech Lands in Medieval Transformation* (Leiden: Brill, 2011); Adrienne Körmendy, *Melioratio Terrae: Vergleichende Untersuchungen über die Siedlungsbewegung im östlichen Mitteleuropa im 13.-14. Jahrhundert* (Poznan: Publishing House of the Poznan Society for the Advancement of the Arts and Sciences, 1995).
5. See Richard Keyser's contribution in this volume.
6. However, such a forest need not be a closed forest of tall trees.
7. Péter Szabó, *Woodland and Forests in Medieval Hungary* (Oxford: Archaeopress, 2005), 57–60.
8. Oliver Rackham, *Woodlands* (London: Collins, 2006), 280.
9. This strongly resonates with traditional frugality as described in the Introduction to this book.
10. MZA F 18 inv. č. 7679.
11. Péter Szabó, "Driving Forces of Stability and Change in Woodland Structure: A Case-Study from the Czech Lowlands," *Forest Ecology and Management* 259 (2010): 650–56.
12. E.g., Alexa Dufraisse, "Firewood Management and Woodland Exploitation during the Late Neolithic at Lac de Chalain (Jura, France)," *Vegetation History and Archaeobotany* 17 (2008): 199–210; Welmoed A. Out, "Firewood

Collection Strategies at Dutch Wetland Sites in the Process of Neolithisation," *Holocene* 20 (2010): 191–204.

13. Kristof Haneca, Katarina Čufar, and Hans Beeckman, "Oaks, Tree-Rings and Wooden Cultural Heritage: A Review of the Main Characteristics and Applications of Oak Dendrochronology in Europe," *Journal of Archaeological Science* 36 (2009): 1–11.
14. Elisabeth Johann, "Traditional Forest Management under the Influence of Science and Industry: The Story of the Alpine Cultural Landscapes," *Forest Ecology and Management* 249 (2007): 54–62.
15. Dolly Jørgensen and Peter Quelch, "The Origins and History of Medieval Wood-Pastures," in *European Wood-pastures in Transition: A Social-Ecological Approach*, ed. Tibor Hartel and Tobias Plieninger (Abingdon: Routledge, 2014), 55–69; Tibor Hartel, Tobias Plieninger, and Anna Varga, "Wood-Pastures in Europe," in *Europe's Changing Woods and Forests: From Wildwood to Managed Landscapes*, ed. Keith J. Kirby and Charles Watkins (Wallingford: CABI, 2015), 61–76.
16. Matts Lindbladh and Richard Bradshaw, "The Development and Demise of a Medieval Forest-Meadow System at Linnaeus' Birthplace in Southern Sweden: Implications for Conservation and Forest History," *Vegetation History and Archaeobotany* 4 (1995): 153–60.
17. Matthias Bürgi, Urs Gimmi, and Martin Stuber, "Assessing Traditional Knowledge on Forest Uses to Understand Forest Ecosystem Dynamics," *Forest Ecology and Management* 289 (2013): 115–22.
18. Walter Kilian, "Forest Site Degradation—Temporary Deviation from the Natural Site Potential," *Ecological Engineering* 10 (1998): 5–18l.
19. Matthias Bürgi and Urs Gimmi, "Three Objectives of Historical Ecology: The Case of Litter Collecting in Central European Forests," *Landscape Ecology* 22 (2007): 77–87.
20. *Archiv český, čili staré písemné památky české i moravské sebrané z archivů domácích i cizích*, vol. 17 (Praha, 1899), 446. [author's translation from original Czech]
21. Johann, "Traditional Forest Management," mentioned cutting areas as large as 400 ha all at once.
22. MZA F18 inv. č. 10362.
23. Robert Šimůnek, "Lesní správa na panství Choustník v polovině 15. století (s edicí rejstříku prodeje dřeva z roku 1447)," *Táborský archiv* 12 (2004): 87–150.
24. MZA F 5 kniha 1 and kniha 3a.
25. Péter Szabó, "Rethinking Pannage: Historical Interactions between Oak and Swine," in *Trees, Forested Landscapes and Grazing Animals: A European Perspective on Woodlands and Grazed Treescapes*, ed. Ian D. Rotherham (Abingdon: Routledge, 2013), 51–61.
26. This was described in detail in the mid-nineteenth century: MZA F 5 kniha 232.
27. MZA G 160 inv. č. 3630.

28. Tomasz Samojlik, Ian D. Rotherham, and Bogumiła Jędrzejewska, "Quantifying Historic Human Impacts on Forest Environments: A Case Study in Białowieża Forest, Poland," *Environmental History* 18 (2013): 576–602.
29. Josef Nožička, *Přehled vývoje našich lesů* (Praha: Státní zemědělské nakladatelství, 1957), 189–99; Károly Tagányi, *Magyar erdészeti oklevéltár*, vol. 1 (Budapest: Országos Erdészeti Egyesület, 1896), i–xl.
30. Nožička, *Přehled vývoje*, 49–51.
31. *Archiv český*, vol. 22 (1905), 241–51. On royal hunters in Moravia, see Libor Jan, *Vznik zemského soudu a správa středověké Moravy* (Brno: Masarykova univerzita, 2000), 192–208.
32. Eszter Magyar, *A feudalizmus kori erdőgazdálkodás az Alsó-Magyarországi bányavárosokban (1255–1747)* (Budapest: Akadémiai Kiadó, 1983).
33. Within four weeks: *Archiv český*, vol. 23 (1906), 271; until St. George's day: *Archiv český*, vol. 23 (1906), 52.
34. Six-year ban: *Archiv český*, vol. 23 (1906), 164.
35. *Archiv český*, vol. 23 (1906), 52.
36. *Archiv český*, vol. 22 (1905), 306.
37. *Archiv český*, vol. 3 (1844), 65–180.
38. James C. Scott, *Seeing like a State: How Certain Schemes to Improve the Human Condition Have Failed* (New Haven: Yale University Press, 1998), 11–24; Henry E. Lowood, "The Calculating Forester: Quantification, Cameral Science, and the Emergence of Scientific Forestry Management in Germany," in *The Quantifying Spirit in the Eighteenth Century*, ed. Tore Frängsmyr, J. L. Heilbron, and Robin E. Rider (Berkeley: University of California Press, 1990), 315–42; Richard Hölzl, *Umkämpfte Wälder: die Geschichte einer ökologischen Reform in Deutschland 1760–1860* (Frankfurt: Campus Verlag, 2010).
39. MZA F 72 kniha 106.
40. Jana Müllerová, Péter Szabó, and Radim Hédl, "The Rise and Fall of Traditional Forest Management in Southern Moravia: A History of the Past 700 Years," *Forest Ecology and Management* 331 (2014): 104–15.
41. Péter Szabó, "The End of Common Uses and Traditional Management in a Central European Wood," in *Cultural Severance and the Environment: The Ending of Traditional and Customary Practice on Commons and Landscapes Managed in Common*, ed. Ian D. Rotherham (Dordrecht: Springer, 2013), 205–13.
42. Nožička, *Přehled vývoje*, 189–99.
43. Tagányi, *Magyar erdészeti oklevéltár*, vol. 2, 304.
44. Dénes Saláta, Soma Horváth, and Anna Varga, "Az erdei legeltetésre, a fás legelők és legelőerdők használatára vonatkozó 1791 és 1961 közötti törvények," *Tájökológiai Lapok* 7 (2009): 387–401.
45. *Vereinschrift für Forst-, Jagd- und Naturkunde* 15 (Neue Folge 1) (1853): 34–41. Anonymous contribution signed "W."
46. Franz Kreß, "Ueber die Schädlichkeit der Streunutzung," *Vereinschrift für Forst-, Jagd- und Naturkunde* 55 (1866): 3–10.

47. *Mitteilungen der k. k. Mährisch-Schlesischen Gesellschaft für Ackerbau, Natur- und Landeskunde.*
48. MZA B 41 fascikl 12.
49. Antonín Pfeffer, "Hrabání steliva v lesích a lesní pastva v normálních dobách," *Sborník Masarykovy akademie práce* 22, no. 124 (1948): 1–31.
50. Jan Malík, ed., *Statistická ročenka republiky Československé* (Praha: Státní úřad statistický, 1936), 47; Petr Maděra, Martin Machala, Tomáš Slach, Michal Friedl, Linda Cernušáková, Daniel Volarík, and Antonín Buček, "Predicted Occurrence of Ancient Coppice Woodlands in the Czech Republic," *i Forest-Biogeosciences and Forestry* 10 (2017): 788–95; *Report on the State of Forests and Forestry in the Czech Republic* (Praha: Ministry of Agriculture of the Czech Republic, 2000), 58.
51. Tibor Hartel, Ine Dorresteijn, Catherine Klein, Orsolya Máthé, Cosmin I. Moga, Kinga Öllerer, Marlene Roellig, Henrik von Wehrden, and Joern Fischer, "Wood-Pastures in a Traditional Rural Region of Eastern Europe: Characteristics, Management and Status," *Biological Conservation* 166 (2013): 267–75.
52. Oliver Rackham, *Hayley Wood: Its History and Ecology* (Cambridge, UK: Cambridgeshire & Isle of Ely Naturalists' Trust, 1975).
53. Martin Hermy, Olivier Honnay, Les Firbank, Carla Grashof-Bokdam, and Jonas E. Lawesson, "An Ecological Comparison between Ancient and Other Forest Plant Species of Europe, and the Implications for Forest Conservation," *Biological Conservation* 91 (1999): 9–22.
54. Ernest Pollard, Max D. Hooper, and Norman Winfrid Moore, *Hedges* (London: Collins, 1974).
55. Thomas Fartmann, Cornelia Müller, and Dominik Poniatowski, "Effects of Coppicing on Butterfly Communities of Woodlands," *Biological Conservation* 159 (2013): 396–404.
56. Jan Kadavý, ed., *Nízky les jako plnohodnotná alternativa hospodaření malých a středních vlastníků lesa* (Kostelec nad Černými lesy: Lesnická práce, 2011).
57. John Fletcher, "The Impact of Hunting on European Woodland from Medieval to Modern Times," in *Europe's Changing Woods and Forests: From Wildwood to Managed Landscapes*, ed. Keith J. Kirby and Charles Watkins (Wallingford: CABI, 2015), 116–26.
58. Keith J. Kirby and Ambroise Baker, "The Dynamics of Pre-Neolithic European Landscapes and Their Relevance to Modern Conservation," in *Trees, Forested Landscapes and Grazing Animals: A European perspective on Woodlands and Grazed Treescapes*, ed. Ian D. Rotherham (London: Routledge, 2013), 91–92; Hartel et al., "Wood-Pastures in Europe"; Keiko Matteson, *Forests in Revolutionary France: Conservation, Community, and Conflict, 1669–1848* (New York: Cambridge University Press, 2015), 262–64 describes the rediscovery of traditional practices, including wood-pasture in France.
59. Keith Alexander, "Ancient Trees, Grazing Landscapes and the Conservation of Deadwood and Wood Decay Invertebrates," in *Trees, Forested Landscapes*

and Grazing Animals: A European Perspective on Woodlands and Grazed Treescapes, ed. Ian D. Rotherham (Abingdon: Routledge, 2013), 330–38 with references to earlier research.
60. Tibor Hartel and Tobias Plieninger, eds., *European Wood-pastures in Transition: A Social-ecological Approach* (New York: Routledge, 2014).
61. Hartel et al., "Wood-Pastures in a Traditional Rural Region."
62. Emma J. Sayer, "Using Experimental Manipulation to Assess the Roles of Leaf Litter in the Functioning of Forest Ecosystems," *Biological Reviews* 81 (2006): 1–31; Bürgi and Gimmi, "Three Objectives."
63. See, for example, Ondřej Vild, Jesse M. Kalwij, and Radim Hédl, "Effects of Simulated Historical Tree Litter Raking on the Understorey Vegetation in a Central European Forest," *Applied Vegetation Science* 18: 569–78.
64. This generalization does not do justice to all. For example, Oliver Rackham was clearly aware of these aspects and commented on them several times.
65. Bürgi and Gimmi, "Three Objectives."
66. James Latham, "The Development of Forest Conservation in Europe," in *Europe's Changing Woods and Forests: From Wildwood to Managed Landscapes*, ed. Keith J. Kirby and Charles Watkins (Wallingford: CABI, 2015), 309–25.
67. *Millennium Ecosystem Assessment: Ecosystems and Human Well-being: Synthesis* (Washington, DC: Island Press, 2005).
68. David Lowenthal, "Natural and Cultural Heritage," *International Journal of Heritage Studies* 11 (2005): 81–92.
69. Tsipe Aavik, Ülle Jõgar, Jaan Liira, Ingmar Tulva, and Martin Zobel, "Plant Diversity in a Calcareous Wooded Meadow—The Significance of Management Continuity," *Journal of Vegetation Science* 19 (2008): 475–84; Peter Poschlod, Andre Baumann, and Petr Karlík, "Origin and Development of Grasslands in Central Europe," in *Grasslands in Europe of High Nature Value*, ed. Peter Veen, Richard Jefferson, Jacques de Smidt, and Jan van der Straaten (Zeist: KNNV Publishing, 2009), 15–25.
70. Paul L. Koch and Anthony D. Barnosky, "Late Quaternary Extinctions: State of the Debate," *Annual Review of Ecology, Evolution, and Systematics* 37 (2006): 215–50.
71. Eva Jamrichová, Péter Szabó, Radim Hédl, Petr Kuneš, Přemysl Bobek, and Barbora Pelánková, "Continuity and Change in the Vegetation of a Central European Oakwood," *Holocene* 23 (2013): 46–56.
72. Jan Altman, Radim Hédl, Péter Szabó, Petr Mazůrek, Vladan Riedl, Jana Müllerová, Martin Kopecký, and Jiří Doležal, "Tree-Rings Mirror Management Legacy: Dramatic Response of Standard Oaks to Past Coppicing in Central Europe," *PLoS ONE* 8, no. 2: e55770.

Bibliography

Aavik, Tsipe, Ülle Jõgar, Jaan Liira, Ingmar Tulva, and Martin Zobel. "Plant Diversity in a Calcareous Wooded Meadow—the Significance of Management Continuity." *Journal of Vegetation Science* 19 (2008): 475–84.

Alexander, Keith. "Ancient Trees, Grazing Landscapes and the Conservation of Deadwood and Wood Decay Invertebrates." In *Trees, Forested Landscapes and Grazing Animals: A European Perspective on Woodlands and Grazed Treescapes*, edited by Ian D. Rotherham, 330–38. Abingdon: Routledge, 2013.

Altman, Jan, Radim Hédl, Péter Szabó, Petr Mazůrek, Vladan Riedl, Jana Müllerová, Martin Kopecký, and Jiří Doležal. "Tree-Rings Mirror Management Legacy: Dramatic Response of Standard Oaks to Past Coppicing in Central Europe." *PLoS ONE* 8, no. 2: e55770.

Archiv český, čili staré písemné památky české i moravské sebrané z archivů domácích i cizích. 39 vols. Praha, 1840–2004.

Benes, Jiří, Oldřich Čížek, Jozef Dovala, and Martin Konvička. "Intensive Game Keeping, Coppicing and Butterflies: The Story of Milovicky Wood, Czech Republic." *Forest Ecology and Management* 237 (2006): 353–65.

Berend, Nora, Premysław Urbańczyk, and Premysław Wiszewski. *Central Europe in the High Middle Ages: Bohemia, Hungary and Poland c. 900–c. 1300*. Cambridge, UK: Cambridge University Press, 2013.

Bretholz, Bertold, ed. *Das Urbar der Liechtensteinischen Herrschaften Nikolsburg, Dürnholz, Lundenburg, Falkenstein, Feldsberg, Rabensburg, Mistelbach, Hagenberg und Gnadendorf*. Reichenberg: Verlag der Anstalt für Sudetendeutsche Heimatforschung, 1930.

Bürgi, Matthias, and Urs Gimmi. "Three Objectives of Historical Ecology: The Case of Litter Collecting in Central European Forests." *Landscape Ecology* 22 (2007): 77–87.

Bürgi, Matthias, Urs Gimmi, and Martin Stuber. "Assessing Traditional Knowledge on Forest Uses to Understand Forest Ecosystem Dynamics." *Forest Ecology and Management* 289 (2013): 115–22.

Dufraisse, Alexa. "Firewood Management and Woodland Exploitation during the Late Neolithic at Lac de Chalain (Jura, France)." *Vegetation History and Archaeobotany* 17 (2008): 199–210.

Fartmann, Thomas, Cornelia Müller, and Dominik Poniatowski. "Effects of Coppicing on Butterfly Communities of Woodlands." *Biological Conservation* 159 (2013): 396–404.

Fletcher, John. "The Impact of Hunting on European Woodland from Medieval to Modern Times." In *Europe's Changing Woods and Forests: From Wildwood to Managed Landscapes*, edited by Keith J. Kirby and Charles Watkins, 116–26. Wallingford: CABI, 2015.

Haneca, Kristof, Katarina Čufar, and Hans Beeckman. "Oaks, Tree-Rings and Wooden Cultural Heritage: A Review of the Main Characteristics and

Applications of Oak Dendrochronology in Europe." *Journal of Archaeological Science* 36 (2009): 1–11.

Hartel, Tibor, Ine Dorresteijn, Catherine Klein, Orsolya Máthé, Cosmin I. Moga, Kinga Öllerer, Marlene Roellig, Henrik von Wehrden, and Joern Fischer. "Wood-Pastures in a Traditional Rural Region of Eastern Europe: Characteristics, Management and Status." *Biological Conservation* 166 (2013): 267–75.

Hartel, Tibor, and Tobias Plieninger, eds. *European Wood-pastures in Transition: A Social-ecological Approach*. New York: Routledge, 2014.

Hartel, Tibor, Tobias Plieninger, and Anna Varga. "Wood-Pastures in Europe." In *Europe's Changing Woods and Forests: From Wildwood to Managed Landscapes*, edited by Keith J. Kirby and Charles Watkins, 61–76. Wallingford: CABI, 2015.

Hermy, Martin, Olivier Honnay, Les Firbank, Carla Grashof-Bokdam, and Jonas E. Lawesson. "An Ecological Comparison between Ancient and Other Forest Plant Species of Europe, and the Implications for Forest Conservation." *Biological Conservation* 91 (1999): 9–22.

Hölzl, Richard. *Umkämpfte Wälder: die Geschichte einer ökologischen Reform in Deutschland 1760–1860*. Frankfurt: Campus Verlag, 2010.

Jamrichová, Eva, Péter Szabó, Radim Hédl, Petr Kuneš, Přemysl Bobek, and Barbora Pelánková. "Continuity and Change in the Vegetation of a Central European Oakwood." *Holocene* 23 (2013): 46–56.

Jan, Libor. *Vznik zemského soudu a správa středověké Moravy*. Brno: Masarykova univerzita, 2000.

Johann, Elisabeth. "Traditional Forest Management under the Influence of Science and Industry: The Story of the Alpine Cultural Landscapes." *Forest Ecology and Management* 249 (2007): 54–62.

Jørgensen, Dolly, and Peter Quelch. "The Origins and History of Medieval Wood-Pastures." In *European Wood-pastures in Transition: A Social–Ecological Approach*, edited by Tibor Hartel and Tobias Plieninger, 55–69. Abingdon: Routledge, 2014.

Kadavý, Jan, ed. *Nízky les jako plnohodnotná alternativa hospodaření malých a středních vlastníků lesa*. Kostelec nad Černými lesy: Lesnická práce, 2011.

Kilian, Walter. "Forest Site Degradation—Temporary Deviation from the Natural Site Potential." *Ecological Engineering* 10 (1998): 5–18l.

Kirby, Keith J., and Ambroise Baker. "The Dynamics of Pre-Neolithic European Landscapes and Their Relevance to Modern Conservation." In *Trees, Forested Landscapes and Grazing Animals: A European perspective on Woodlands and Grazed Treescapes*, edited by Ian D. Rotherham, 87–98. London: Routledge, 2013.

Klápště, Jan. *The Czech Lands in Medieval Transformation*. Leiden: Brill, 2011.

Koch, Paul L., and Anthony D. Barnosky. "Late Quaternary Extinctions: State of the Debate." *Annual Review of Ecology, Evolution, and Systematics* 37 (2006): 215–50.

Körmendy, Adrienne. *Melioratio Terrae: Vergleichende Untersuchungen über die Siedlungsbewegung im östlichen Mitteleuropa im 13.-14. Jahrhundert*. Poznan: Publishing House of the Poznan Society for the Advancement of the Arts and Sciences, 1995.

Kreß, Franz. "Ueber die Schädlichkeit der Streunutzung." *Vereinsschrift für Forst-, Jagd- und Naturkunde* 55 (1866): 3–10.

Latham, James. "The Development of Forest Conservation in Europe." In *Europe's Changing Woods and Forests: From Wildwood to Managed Landscapes*, edited by Keith J. Kirby and Charles Watkins, 309–25. Wallingford: CABI, 2015.

Lindbladh, Matts, and Richard Bradshaw. "The Development and Demise of a Medieval Forest-Meadow System at Linnaeus' Birthplace in Southern Sweden: Implications for Conservation and Forest History." *Vegetation History and Archaeobotany* 4 (1995): 153–60.

Lowenthal, David. "Natural and Cultural Heritage." *International Journal of Heritage Studies* 11 (2005): 81–92.

Lowood, Henry E. "The Calculating Forester: Quantification, Cameral Science, and the Emergence of Scientific Forestry Management in Germany." In *The Quantifying Spirit in the Eighteenth Century*, edited by Tore Frängsmyr, J. L. Heilbron, and Robin E. Rider, 315–42. Berkeley: University of California Press, 1990.

Maděra, Petr, Martin Machala, Tomáš Slach, Michal Friedl, Linda Cernušáková, Daniel Volarík, and Antonín Buček. "Predicted Occurrence of Ancient Coppice Woodlands in the Czech Republic." *i Forest-Biogeosciences and Forestry* 10 (2017): 788–95.

Magyar, Eszter. *A feudalizmus kori erdőgazdálkodás az Alsó-Magyarországi bányavárosokban (1255–1747)*. Budapest: Akadémiai Kiadó, 1983.

Malík, Jan, ed. *Statistická ročenka republiky Československé*. Praha: Státní úřad statistický, 1936.

Matteson, Keiko. *Forests in Revolutionary France: Conservation, Community, and Conflict, 1669–1848*. New York: Cambridge University Press, 2015.

Millennium Ecosystem Assessment: Ecosystems and Human Well-being: Synthesis. Washington, DC: Island Press, 2005.

Müllerová, Jana, Péter Szabó, and Radim Hédl. "The Rise and Fall of Traditional Forest Management in Southern Moravia: A History of the Past 700 Years." *Forest Ecology and Management* 331 (2014): 104–15.

Nožička, Josef. *Přehled vývoje našich lesů*. Praha: Státní zemědělské nakladatelství, 1957.

Out, Welmoed A. "Firewood Collection Strategies at Dutch Wetland Sites in the Process of Neolithisation." *Holocene* 20 (2010): 191–204.

Pfeffer, Antonín. "Hrabání steliva v lesích a lesní pastva v normálních dobách." *Sborník Masarykovy akademie práce* 22, no. 124 (1948): 1–31.

Pollard, Ernest, Max D. Hooper, and Norman Winfrid Moore. *Hedges*. London: Collins, 1974.

Poschlod, Peter, Andre Baumann, and Petr Karlík. "Origin and Development of Grasslands in Central Europe." In *Grasslands in Europe of High Nature Value*, edited by Peter Veen, Richard Jefferson, Jacques de Smidt, and Jan van der Straaten, 15–25. Zeist: KNNV Publishing, 2009.

Rackham, Oliver. *Hayley Wood: Its History and Ecology*. Cambridge, UK: Cambridgeshire & Isle of Ely Naturalists' Trust, 1975.

_____. *Woodlands*. London: Collins, 2006.

Report on the State of Forests and Forestry in the Czech Republic. Praha: Ministry of Agriculture of the Czech Republic, 2000.

Saláta, Dénes, Soma Horváth, and Anna Varga. "Az erdei legeltetésre, a fás legelők és legelőerdők használatára vonatkozó 1791 és 1961 közötti törvények." *Tájökológiai Lapok* 7 (2009): 387–401.

Samojlik, Tomasz, Ian D. Rotherham, and Bogumiła Jędrzejewska. "Quantifying Historic Human Impacts on Forest Environments: A Case Study in Białowieża Forest, Poland." *Environmental History* 18 (2013): 576–602.

Sayer, Emma J. "Using Experimental Manipulation to Assess the Roles of Leaf Litter in the Functioning of Forest Ecosystems." *Biological Reviews* 81 (2006): 1–31.

Scott, James C. *Seeing like a State: How Certain Schemes to Improve the Human Condition Have Failed*. New Haven: Yale University Press, 1998.

Šimůnek, Robert. "Lesní správa na panství Choustník v polovině 15. století (s edicí rejstříku prodeje dřeva z roku 1447)." *Táborský archiv* 12 (2004): 87–150.

Szabó, Péter. "Driving Forces of Stability and Change in Woodland Structure: A Case-Study from the Czech Lowlands." *Forest Ecology and Management* 259 (2010): 650–56.

_____. "The End of Common Uses and Traditional Management in a Central European Wood." In *Cultural Severance and the Environment: The Ending of Traditional and Customary Practice on Commons and Landscapes Managed in Common*, edited by Ian D. Rotherham, 205–13. Dordrecht: Springer, 2013.

_____. "Rethinking Pannage: Historical Interactions between Oak and Swine." In *Trees, Forested Landscapes and Grazing Animals: A European Perspective on Woodlands and Grazed Treescapes*, edited by Ian D. Rotherham, 51–61. Abingdon: Routledge, 2013.

_____. *Woodland and Forests in Medieval Hungary*. Oxford: Archaeopress, 2005.

Tagányi, Károly. *Magyar erdészeti oklevéltár*. 3 vols. Budapest: Országos Erdészeti Egyesület, 1896.

Válka, Josef. *Dějiny Moravy*. 2 vols. Brno: Muzejní a vlastivědná společnost, 1991–1996.

Vereinsschrift für Forst-, Jagd- und Naturkunde 15 (Neue Folge 1) (1853): 34–41.

Vild, Ondřej, Jesse M. Kalwij, and Radim Hédl. "Effects of Simulated Historical Tree Litter Raking on the Understorey Vegetation in a Central European Forest." *Applied Vegetation Science* 18: 569–78.

Afterword

Paul Warde

One could argue that historical writing on premodern history has long displayed a concern for issues of conservation and sustainability. Not, as the Introduction to this volume shows, *avant les mots,* but before they became explicit themes in environmental history. Especially the second generation of the *Annales* school, with its deep connection to historical geography, was concerned with environments and the land. Much of the economic history of the postwar period sketched a Malthusian framework for understanding deep currents in European history, from authors such as M. M. Postan, Wilhelm Abel, and the "motionless history" (*histoire immobile*) of Emmanuel Le Roy Ladurie.[1] This work was intimately concerned with the relationship between land, resources, and population, often working on a grand scale but also inspiring numerous case studies to test its propositions. While many later scholars came to reject their arguments about the centrality of Malthusian mortality crises, nevertheless, this basic structure of concerns lives on in influential works, now also engaged with a discussion of climate, as in twenty-first century works from Bruce Campbell and Paolo Malanima.[2]

At the heart of this tradition of socioeconomic history was something also highlighted in the Introduction to this book: the centrality of scarcity in the premodern agrarian world, not simply in the very general sense of scarcity meaning all actions have an opportunity cost as a founding notion in economics, but also in the sense that scarcity was felt intimately by those living close to subsistence level; however, the earlier and *Annaliste* literature largely dealt with this as a question of the dynamics of whole societies, even if articulated at a micro level through demographic variables, practices of inheritance, price trends, and productivity. There was much less interest in dealing with scarcity as a question of everyday management. This is one thing that the cases in this book bring to the table: medieval and early modern people in knowing engagement with their environments. This move itself reflects a move in our contemporary understanding of scarcity from being a primarily *economic* question to one that wants to understand ways of living-in-the-world.

But what to call this? Those two words spring to mind: "conservation" and "sustainability." These are words both of their times, and our own time, and bring all the discomforts that historians traditionally feel in applying concepts developed, it would seem, after the period they study. Class has had the most spectacular career of these words, bloating out like a red giant star to swallow all around it before collapsing back into a dwarven shell of doubt as to whether "class" is even the right term to apply even to the periods that liberally applied it to themselves. Conservation, as Dowling and Keyser show, actually *was* familiar in the early modern period; but not in the sense of the historiography of environmental history that is associated with the so-named movements that appeared in the United States, Europe, and their empires in the latter part of the nineteenth century. The term was at that time often associated with the application of technical expertise, and thus distanced from the supposedly traditional and more ignorant peasant societies of the past. Indeed, a champion of American thinking, George Perkins Marsh, explicitly blamed feudalism and the Catholic church for bringing woe to the European landscape after the fall of Rome, swatting away centuries of medieval endeavor in one single maneuver.[3]

The chapters of this book demonstrate over and over that the managing of the landscape to maintain desired characteristics over time is no discovery of the moderns, but extends far back in time—and indeed embraces all farming and many earlier forms of hunter–gathering and herding (and here we must acknowledge much work in history, historical geography, anthropology, and archaeology that has addressed this). Conservation can justifiably be applied and extended, and if we labor still with a little fear of anachronism and pluralities of meaning, well, it is not as if we agree about what conservation is in the twentieth and twenty-first centuries either. Indeed, as well as opening up new terrain for investigation, the juxtaposition of modern movements with past practices is paradoxically a useful comparative exercise in its own right, helping us guard against flattening history and treating explanations as if they hold for all times and place. If in some later forms Christian beliefs could inspire a kind of cornucopian enthusiasm for ingenuity and a confidence in people's capacity to deliver limitless growth (becoming an explanation for a certain style of environmental politics), there is very little sign of such thinking in the manner that the inhabitants of a very religious—Christian, Jewish, Muslim—medieval Europe harbored their resources.[4] A little willful anachronism can demonstrate the weakness of generalizations.

The studies presented here show that efforts toward conservation were ubiquitous. So was change, and managing that was as negotiated and fraught as contestations in other areas of medieval and early modern life. The chapters do not attempt to provide laudatory case histories of success and prior environmental wisdom (although some cases could also be that). In any case, one can hardly take self-conscious ecological knowledge as the essential grounding for sustainable management. It is sadly far from true that modern environmental awareness can be accredited with providing a guarantee of successful environmental management. A visitor from another planet, or a historian of the future (who hopefully does not reside on Planet B) might rather correlate the enormous rise in environmentally conscious talk in modern times with an equally spectacular rise in destructive behavior. A lazy correlation might imply all this talk made things worse. We have not demonstrated—yet—that an explicitly ecological consciousness is a prerequisite for ecologically wise management.

I would like to conclude this book in what I take to be the spirit of the volume, taking some steps beyond the threshold it establishes into an old world with a new framing. I will do this by examining the histories presented through the lens of the second term raised above: "sustainability"—the history of which I am most familiar. What might a dialogue between our present understandings and preindustrial resource management highlight as themes and questions for further research?

The Uses of Sustainability

"Sustainability" is a very modern term, having taken off from around 1972.[5] As the Introduction has set out, the term *sustenare* is much older, and the verb "to sustain" has been present in English for centuries. In the nineteenth century, it became attached to the notion of "sustained yield" theory in forestry, and later fisheries; but cognates can be found, for example, in the idea of "sustained growth" discussed by postwar economists. The German term *Nachhaltigkeit* is much older and dates from the mid-eighteenth century, with cognates around *nachhaltende* (which could also be translated "lasting") being older still.[6] Thus, the fact that *Nachhaltigkeit* was then chosen as the translation of the English neologism "sustainability" gave German forestry a certain retrospective eminence in environmental thought, but this might be seen as a rather coincidental linkage by which "sustainability" in English drew upon a word that happened to be well-established in technical theories of resource management due to being chosen by earlier translators of German forestry. There is no evidence that the word in English arose from a direct engagement with the German *Nachhaltigkeit*, and the rather crude tool of a Google n-gram search suggests that the takeoff of its use in German from the mid-1980s is actually a consequence of the prestige attached to the English word.[7]

Why tell this convoluted modern history? It shows that the history of a word itself can only take us so far, even if that word has power and importance. We have to examine wider discourses and styles of thought. Indeed, as soon as it appeared, "sustainability," and the related "sustainable development," became a contested terrain, picked up and used by radically different political groupings and interests. This led to accusations of "greenwashing," but who really owns a word or its meaning? Sustainability became a concept everybody could agree was a good thing, but they then populate it with wildly different notions of the good life and the policies that would be required. Its power arguably lay more as a means—sometimes very consciously so—to generate consensus and steer debates according to certain precepts, rather than provide a more clear-cut guide to living. It is not an answer, but led to a style of questioning. It is with these thoughts in mind that I offer three different ways in which sustainability could be usefully applied to the past. None is the "right" one. And while it is always sensible to guard against anachronism, we may also remember that we are not, in such historical inquiry, diverting the idea of sustainability from its "true" meaning, because that would be a profound misunderstanding of how we experience discourses of sustainability today.

The three uses I am suggesting are as follows:

1) Sustainability as the maintenance of a flow of desired resources, which often requires the maintenance of a stock to provide that flow

2) Sustainability as a certain ethic of decision-making, orientated toward the future
3) Sustainability as a historical narrative, an existential question of destiny

Through these, I will seek to examine both common themes and particularities of the historical periods and practice. There will not be equal space devoted to each, because the chapters of this book, rightly, have given far more attention to the first, for reasons that I will lay out below.

Resource Sustainability

Many of the cases in this book are histories of the management of a particular resource or specific areas of land. This can be imagined as a very material, even quantitative problem. Often by necessity, the histories told are histories of consequences, because we do not have access to the reasoning or motivation behind people's actions. Especially in the case of "traditional" or peasant practices, contemporary descriptions are only provided by their critics, often late in the seventeenth and eighteenth centuries—those critics having little interest in delving into the motivations of ordinary people, or the specificities of localities.

An extensive literature already exists on the management of wood as a resource in many European countries, although as Richard Keyser points out in his chapter, this has a strong bias toward sources produced by the state and attempts by state authorities to extend their influence. This has also meant that the seventeenth and eighteenth centuries are much better studied, especially in anglophone environmental history literature. Oliver Auge's chapter makes available another case to an anglophone audience, and also importantly focuses on a relatively *wood-poor* area. Much of the existing literature has been drawn to heavily forested regions, where the nature of scarcities and contestation over resources, often involving industrial users, was rather different to that found in less wooded regions.[8] In contrast to writing on woodlands, fisheries have been under-researched both in their maritime and inland versions, and much of the discussion of the management of game and parks has taken place in a literature divorced from environmental history. It is an important service to bring work on all these resources, with their different ecological bases, into dialogue, also mixing different economies, regions, and languages.

This extension of an environmental history view to "new" resources, and resources in their multiplicity, raises a further question. Should we include *any* concern with *any* resource in the future as a dimension of sustainable practice? If that is the case, what is the relationship *between* resources? How might we think about the desire for children to inherit in this context, for example? What expectation was there that particular resources or areas of land would actually stay in a family's hands across the generations? Would your children be dependent on them or not? Were resources fungible and substitutable (in modern terms, should we think of "hard" or "soft" sustainability)? How might property rights and forms of tenancy affect behavior? Did those institutions such as monasteries or urban and village communes who could operate on different time-horizons and with different degrees of flexibility than industries or families behave differently as a consequence, as has been argued for early modern Germany by Winfried Schenk?[9] Much of the discussion of agrarian resource management has in fact focused on *collective* rather than privately held resources: various kinds of commons and shared spaces. Could a history of environmental management also embrace privately held assets? This is easiest, of course, in the kind of lordship studied by Abigail Dowling, where the challenge to the historian can become the embarrassment of sources rather than their paucity (and the problem of scarcity becomes one of research time!). There is plenty of work to do.

Resources themselves are not fixed. New resources can emerge. While the need for fuel or pasture was continuous, needs and diets changed. The premodern economy was not ossified in perpetuity, waiting for the shock of modernity. Péter Szabó's contribution, for example, shows how the practice of litter raking appeared in the nineteenth century as a response to the new need for stall bedding as fodder crops replaced open pasture. By the mid-twentieth century, this "traditional" practice, too, would be disappearing with the development of more modern animal housing.[10] The sources we draw upon, however, are often defined by jurisdictions that shape, but are not synonymous with, the location of resources or types of land. This is clear in the case of the legal entity of "forest," but applies to other categories too. Such boundaries may not fit well with the significant boundaries in the ecological functioning of landscapes, and a central question for the history of management must be how well the law and jurisdictions could adapt to change, while remembering that they—and thus we—have only a partial view of the actual historic environment. The important emphasis in Part I on multiple-resource-use helps guard against a tendency of modern authors to see certain environments as "transitional," as in-between worlds, not belonging to categories we expect. Wood-pasture is a very widespread and historically important case of this, which also had considerable potential to draw users into conflict. The challenge of attaching legal categories to nature's entangled bank equally shines through many of the chapters. It must surely be a rich area for future research, possibly including interdisciplinary analyses, including how landscapes functioned in cooperation with natural scientists.

The fact of multiple-use and the exploitation of resources that were seasonal, fluid (sometimes literally), and of different value to different people highlights that the older "Malthusian" approach of studying the population–resource balance misses much of what was important and delicate in environmental management. The question is never just "were there shortages" as an indicator of sustainability (most notably in the "Wood shortage" literature), but shortages of what and for whom? As well as the struggles over woodland space detailed in Part III, watery space represented a notable challenge, with all the varied possibilities for use from washing and drinking to motive power, as low friction means of transportation, or as a home for desirable prey. Richard Hoffmann, Eva Jakobsson, Michael Zeheter, and Cristina Arrigoni Martelli give us a rich view into those varied potential uses and associated competing interests, and the legal solutions adopted. The ubiquity of regulation of the water and woods (those areas supposedly outside the close attention of the propertied to their cultivated land) indicates the great value of the range of ways that people exploited them. Regulatory enforcement has a cost, and Richard Hoffmann carefully adumbrates the choices of whether to implement them and to what extent. Indeed, even not enforcing regulations has a cost in terms of lost prestige and habitual disregard for authority. In turn, as soon as there is a cost in obtaining a resource, especially in the context of multiple uses, someone else can make it more expensive than others would like. It is clear that many debates and fears over shortages should be understood in this light, which helps explain why so much environmental regulation was concerned with *allocation* and *access*. It is likely that we cannot—or indeed should not—disentangle "ecological" concerns from distributional ones in at least some examples. Measures to limit the rights to fishing and fowling in Tuscany described by Arrigoni Martelli, for example, may both have reduced the overall harvest by restricting the number of harvesters, and equally limited the capacity of those harvesters to compete for the best spots, thus creating barriers to entry in this market, keeping incomes higher for those who purchased rights, while also reducing the potential for explosive arguments (a modern analogy might be the way a city regulates taxi drivers).

When people argue or discuss or negotiate the allocation of resources, the question must arise as to what they believe to be the *right* or *just* amount of supply to themselves and others. By the central Middle Ages, Europe contained very little land (or water) that was not viewed as a resource by someone with accompanying claims upon it. Sustainable access to a resource for one person might mean exclusion and risk for another. The desired outcomes of management—and what might be considered "sustainable"—thus depended on the choices and opportunities open to a population, which itself will be of divergent interests. The complex patterns of management and access developed as a consequence are richly portrayed in Dolly Jørgensen's work on pigs. Yet this history of contestation and rubbing along together also raises a very significant point. It seems likely that for all the study of sustainable woodland management, very few woods disappeared because they were managed *unsustainably*, precisely because of all the managerial approaches outlined in this book. The results might be higher or lower yielding, or provide more of one resource than another, but outright failure is possibly a comparatively rare case. Yet woodland area did decline. This was not because management failed but because people (or at least the people able to assert their property claims) found alternatives, and in their eyes better uses, for the ground. Most woods in the British Isles disappeared, for example, because the increasing demand for food meant a better return to putting land down to crops and grass than growing trees, leading to the woods being grubbed up. Nevertheless, communities were also capable of coming to a consensus that certain landscapes needed a range of interlocking measures that maintained them as a whole, very probably in a state quite unlike that which would prevail without human intervention. This may have been more likely where the economy developed a dependence on a set of mutually interacting resources, a case presented to us by Maïka de Keyzer's study of the Campine region on the Belgian-Dutch border. One did not have to manage this landscape as a whole, like a modern planner, but rather understand that failure in one area had ramifying consequences that could not easily be compensated for by others. This in turn might presuppose a certain equality of exposure to risks, although not necessarily of wealth, across the community.

The fact that the presence of natural resources was a matter of choice (within certain environmental constraints) is just as clear with those that seem, to the modern eye, more wild than coppices or mill-runs (although in practice it is not clear how significant boundaries between the cultivated land and the *saltus* really were to medieval people). Keeping a population of huntable game such as deer and boar is a matter of management, coming with the opportunity cost of slower woodland regeneration and possibly trampled and gorged crops. As in later centuries, but with less effective weapons, predators were persecuted (as were poachers). And what might seem "success" to a modern conservationist ethos that values "wildlife" was certainly to the cost of other and generally poorer users of the landscape. In the areas most favored for the chase by monarchs and nobles, stocking rates of game were high enough to require winter feeding; animals were transported between different reserves to balance populations and feed, and the "right" amount of game was that which guaranteed satisfying sport.[11] This too was sustainability of sorts, but it forces us to recognize the relative nature of what counts as success. Whose nature is it? Such questions are not just the preserve of modern conservationist and environmental struggles, and the question of game could become a lightning rod for all sorts of struggle over sovereignty, status, and class. This brings me to my second type of sustainability: ethics.

An Ethic of Sustainability

Discussing sustainability in terms of ethics might seem surprising. Are the ethics of sustainability not driven by what is required to preserve the resources? Are we not, as Francis Bacon said, to command nature by "obeying her"?[12] But as we have seen, the value of resources in a varied community is a matter of choice (albeit a far from unlimited one). Sustainability is thus also a question of wants and desires, and in turn of the question: What does it take to live the good life? Indeed, a little anachronism can help us understand the issue here, because even as post 1970s ideas of sustainability can be presented as related to all sorts of technical and scientific knowledge, it is clear that much of the politics that swirls around the idea are to do with competing notions of the good life.

While there is plenty of medieval and early modern discussion of the good life, we have seen that it is much harder to discern precise motivations behind many kinds of regulation, the more so the earlier in time we go (although this book admirably demonstrates there is rather more material than some people might have thought). Thus, even when we discover practices to have positive effects in fostering the durability of desired environments, it is still hard to perceive what is driving this in the minds of actors at the time.

Nevertheless, there is still much to be learned from what is available. The ethics of sustainability do not have to be thought of in terms of the application of high-falutin' moral principles at every step. In fact, what might be rather more visible are everyday influences—what Krisitn Asdal in a rather different context called the "little tools" which sustainable lives are made.[13] De Keyzer's chapter thus shows us the importance of socialization into habits, the role of peer group pressure, and principles of neighborly conduct and respect that might be required even for quite rule-bound systems of management to work. Such "little tools" of sustainability were mental, social, and indeed physical, such as the windbreaks recognized an essential to protecting the sandy Campine soils.

The experience of managing a lordly estate described by Abigail P. Dowling was of course rather different from the Campine peasants, but the landscape was equally the product of a certain ethos. This might extend to perceptions of public duty (such as providing timber for wartime exigencies); but above all, it was about display and the maintenance of certain pleasures. One cannot reduce such management to either utilitarian or merely political concerns, because it also required the development of a kind of connoisseurship and engagement with nature. Nevertheless, such an estate should not be mistaken for an ecology in and of itself. The resources required to manage the estate were dependent on the payments of all sorts of feudal dues and rents, and, of course, the systematic exclusion of others from the park. Yet there is evidence, if scattered and across a great period of time, of an idea that crossed social scales: that there was an aesthetic value to natural things, and nature was of course very widely understood in art and literature as a rich source of metaphor and symbolism. Engaging with nature—that was not physically so very distant—was good; indeed, its proximity was desirable and enjoyable. This can be found among the noble hunters of Flanders and Tuscany, but also in the foresters and peasants of the Czech lands discussed by Péter Szabó—thus even when the direct motives for being up close to nature greatly differed. A history of emotion and sensibility might outline how very different those things were understood to be when linked with the experience of the outdoors in the nineteenth century after romanticism, or in the contented inquisitions of his parish's fauna by the eighteenth-century parson Gilbert White;[14] but it seems unreasonable to suppose that earlier landscapes were not also landscapes of feeling.

There was, however, an ethics to more straightforwardly economic activity too. Indeed, to my knowledge, historians are yet to bring together environmental history with the very extensive writings on the ethics of exchange, of just prices, of usury and interest that populate medieval writings. That might be a rich and enlightening exchange itself. In the case of the fisheries of Lake Constance, Michael Zeheter shows how different kinds of fishing could represent a different ethos orientated toward auto-consumption or commerce, with appropriately varied regulatory regimes. Yet we should take care not to assume these were hermetically sealed worlds, or that sustainable resource management was more clearly related to the necessary modesty and thrift of subsistence agriculture and pastoralism. We see cases throughout the book of commercial interests driving regulatory balances to ensure the market survived. In fact, most European households had a foot in the market, and it was often the ambiguity generated in a language that allowed legal and political claims on resources due to "needs" (*necessita*) but in which households partly provisioned through truck and barter that led to protracted disputes about resource use. Negotiating a desirable path and securing one's future within these parameters of need and opportunity was a challenge common to state, aristocrat, and peasant.

By the seventeenth century, the result of these processes was a visibly moralized landscape, a signifier of the virtues of its inhabitants. In this context, the enclosure of "waste" and common fields (which was by no means novel but had been ongoing for centuries) became represented as a struggle between progress and backwardness. The heaths and pastures and scraggy pollards became evidence of immoral and knavish nature of the inhabitants, but equally itself a kind of nursery that bred up new generations of idlers. Improvers would change all that—sweep away what they now read as a *histoire immobile.* Equally, their grand gardens, ordered estates, and libraries of husbandry manuals and wittily constructed wildernesses were demonstrations of character too, just as were the smart and sober lines of a Mennonite farm.

Yet the outer life, and the life outdoors, was not just a product of the inner life. It was still shaped by material and ecological possibilities, institutions, and incentives. The chapters of this book make these most visible to us through regulatory sources; but regulation should not merely be seen as a source of restraint. It could foster attentiveness and knowledge of measures that worked over time. Verderers described in Sara Morrison's work on Sherwood Forest, so often by-words for corruption and where historians have seen the forest courts perpetually in decline, were in fact called upon to make regular judgments about the appropriate measures for the lands they oversaw. Regulation in practice expected both discretion and the attendant capacities to judge. Fiscal policies affected behavior too. Cristina Arrigoni Martelli examines the system of licensing for fishing that limited harvesting and generated revenue for cities, but the act of auctioning the licenses must itself have given a commercial imperative to resource use that now required a return (a pattern repeated in monetary valuations of nature to the present). Similarly, commons and "wastes" survived into the nineteenth century and beyond not only because of their various functions in peasant economies, but frequently because they were tax-free and hence a "cheap" asset for all who had common rights.[15] Liberals saw this as an inequity and skewing the market: a different ethos. Imposing taxation on commons in the nineteenth century frequently led to their privatization.

Sustainability as History and Destiny

Today, when people talk about sustainability in regard to even the most trivial things—the make of their toothbrush, the local authorities' tree-planting scheme—it is understood by most to connect them to a much wider set of debates. These debates embrace themes no less than the future of our whole society and even the viability of life on our planet. The question of the good life is an existential concern. This seems to be a very distinct aspect to the modern world's conceptualization of resource management, which has been connected to the word "sustainability" since it emerged in the early 1970s—the moment of *Limits to Growth* and the United Nations' Stockholm conference on the Human Environment. Of course, earlier periods had their existential concerns. But this seems different.

Sustainability today is a story about destiny, a story we are narrating to ourselves as we live it. But it is also necessarily a story about trajectory and history. This is one reason why we need our history writing, for all periods, to engage with questions of sustainability; but criteria of sustainability we have developed today are also questions we can level at historical societies themselves. This connection of history and resource use is not, however, something that required the word "sustainability" to appear. I have argued elsewhere that it took on a distinctive and powerful form in public debate during the nineteenth century as anxieties about biological processes and dangers unleashed by population growth, urbanization, and imperial development led to explanations of the collapse of past societies in terms of environmental destruction.[16]

Medieval people also found important but different linkages between history and resource use. In customary systems of management, historical longevity of practices is often taken as evidence of their essential goodness. Of course, the idea of a practice being from "time immemorial" was itself generally fictive, and one of the strengths of custom was its pragmatic elasticity in adapting to changing circumstances, as historians of custom have frequently pointed out. More generally, how resource management fitted into conceptions of historical time might be worthy of further research. This is not least because today, a return to older practices of land management, as detailed by Péter Szabó in the case of coppicing, is itself embedded in narratives of heritage and the value of respect for the past. Long-enduring landscapes are seen as great reservoirs of biodiversity, and their management a means to connect people with nature. Yet the medieval and early modern world of intimate connection, as Dowling and Keyser note in the Introduction, was also one of poverty and vulnerability. It was one where, to couch it in terms of economics, the opportunity cost of labor as very low. This was what in part underpinned the management techniques themselves. How will respect and care for these landscapes be managed in a world where we value our time very much more highly, and where the products of traditional crafts are ill-fitted to the standardized and just-in-time supply chains of modern consumerism?

At a very simple but profound level, asking the question of sustainability is a way of making visible the things on which our society really depends; what so often is hiding in plain sight. This volume does some of that work, too, especially for medieval society. Medievalists have known that the sources existed to study resource management and even some of the motivations behind it. Yet somehow, no wider literature devoted to the question has emerged. That should change, is changing. Yet the literature that does exist should change too. It is no longer enough to demonstrate that what we now consider to be a question of sustainability and environment did exist, in its own way, long before the nineteenth century—that there is a phenomenon to study. The discipline now needs to move toward the more standard historical approaches of establishing chronologies, significance, the diffusion of ideas and techniques, and treating the interior of this very long period in all its differentiated particularity—issues that Richard Keyser points to in his reconstruction of French forestry regulation in the central Middle Ages. It is equally clear that this history is a European one—at the very least, possibly extending toward the Islamic world, Africa and Asia.

That scarcity of records may even prove a conceptual advantage to scholars working on earlier centuries. They are more aware than most how their histories are shaped by the limited availability of sources (even if some sources themselves are voluminous). Historians of modern times may be tempted, through the variety of things they touch upon, to think that their histories are more all-embracing than they really are—and where their view is still often steered by ease of access to material. It is the very lack of uniformity in governmental functions in medieval times that makes it difficult to "see like a state" or from the state's perspective, even if one wanted to. Ironically, this can helpfully draw historians into different ways of seeing and understanding actions at different scales—an issue that is really desirable, of course, for the study of all periods. And one might hypothesize—as I tentatively step out from the firm foundations that the authors in this volume have helped build—that it is the change in the scales of resource management, moving both up and down over time, that accounts for much of the difference we observe between periods in how questions of sustainability are addressed.

Paul Warde is Professor of Environmental History at the Faculty of History, University of Cambridge, UK. He has written extensively on the economic and environmental history of Europe. His books include *Power to the People: Energy in Europe over the Last Five Centuries* (2013, with Astrid Kander and Paolo Malanima), *The Invention of Sustainability* (2018), and *The Environment: A History of the Idea* (2018, with Sverker Sörlin and Libby Robin).

Notes

1. Michael M. Postan, *The Medieval Economy and Society: An Economic History of Britain in the Middle Ages* (London: Weidenfeld & Nicholson, 1972); Wilhelm Abel, *Geschichte der deutschen Landwirtschaft*, 2nd ed. (Stuttgart: E. Ulmer, 1967); Emmanuel Le Roy Ladurie, "History That Stands Still," in *The Mind and Method of the Historian* (Brighton: Harvester Press, 1981); Emmanuel Le Roy Ladurie, *Les paysans de Languedoc* (Paris: Flammarion, 1969).
2. Bruce M. S. Campbell, *The Great Transformation: Climate, Disease and Society in the Late Medieval World* (New York: Cambridge University Press, 2016); Paolo Malanima, *Premodern European Economy: One Thousand Years (10th-19th centuries)* (Leiden: Brill, 2009).
3. George Perkins Marsh, *Man and Nature: Or, Physical Geography as Modified by Human Action* (New York: Charles Scribner, 1864), 27, 48.
4. Lynn T. White Jr., "The Historical Roots of Our Ecologic Crisis," *Science* 155 (1967): 1203–7.
5. The historical development of the term has received very little detailed attention; however, for a collection of usages in modern times, see Charles V. Kidd, "The Evolution of Sustainability," *Journal of Agricultural and Environmental Ethics* 5 (1992): 1–26.
6. See Paul Warde, *The Invention of Sustainability: Nature and Destiny c. 1500–1870* (New York: Cambridge University Press, 2018), chapters 4–5.
7. Google n-gram consulted on 25 September 2019.
8. For a selection of excellent studies, see Elisabeth Johann, *Geschichte der Waldnutzung in Kärnten unter dem Einfluß der Berg-, Hütten- und Hammerwerke,* (Klagenfurt: Verlag des Geschichtsvereines für Kärnten 1968); Ingrid Schäfer, *"Ein Gespenst geht um": Politik mit der Holznot in Lippe 1750–1850. Eine Regionalstudie zur Wald und Technikgeschichte* (Detmold: Naturwissenschaftlicher u. Historischer Verein f. d. Land Lippe, 1992); Peter M. Steinsiek, *Nachhaltigkeit auf Zeit: Waldschutz im Westharz vor 1800* (Münster: Waxmann, 1999); Christoph Ernst, *Den Wald entwickeln: Ein Politik- und Konfliktfled in Hunsrück und Eifel im 18. Jahrhundert* (Munich: De Gruyter Oldenbourg, 2000); Uwe E. Schmidt, *Der Wald in Deutschland im 18. und 19. Jahrhundert: Das Problem der Ressourcenknappheit dargestellt am Beispiel der Waldressourcenknappheit in Deutschland im 18. und 19. Jahrhundert—eine historisch-politische Analyse* (Saarbrücken: Conte, 2002); Richard Hölzl, *Umkämpfte Wälder. Die Gescichte einer ökologischen Reform in Deutschland 1760–1860* (Frankfurt a.M: Klostermann, 2010).
9. Winfried Schenk, *Waldnutzung, Waldzustand und regionale Entwicklung in vorindustrieller Zeit im mittleren Deutschland* (Stuttgart: Steiner, 1996)
10. This is discussed in depth in Reinhold Schaal, *Waldzustände als Spiegel gesellschaftlicher Ansprüche: Die Entwicklung von Wäldern auf der mittleren*

Schwäbischen Alb und im nördlichen Oberschwaben seit dem 16. Jahrhundert (Ostfildern: Jan Thorbecke Verlag, 2011).
11. For example, see Martin Knoll, *Umwelt—Herrschaft—Gesellschaft: Die landesherrliche Jagd Kurbayerns im 18. Jahrhundert* (St. Katharinen: Scripta Mercaturae, 2004).
12. Warde, *Invention of Sustainability*, 108.
13. See many essays by Kristin Asdal, especially her essay, "On Politics and the Little Tools of Democracy," *Distinktion* 9 (2008): 11–26, and her subsequent "Little Tools Project" on measurement and the bio-economy: https://www.sv.uio.no/tik/english/research/projects/little-tools/index.html (last accessed 21 January 2020).
14. Gilbert White, *The Natural History of Selborne* (London: Penguin, 1977 [1789]).
15. See examples in Niels Grüne, Jonas Hübner and Gerhard Siegl, eds., *Ländliche Gemeingüter: Kollektive Ressourcennutzung in der europäischen Agrarwirtschaft / Rural Commons: Collective Use of Resources in the European Agrarian Economy* (Innsbruck: Studien Verlag, 2016).
16. Warde, *Invention of Sustainability.*

Bibliography

Abel, Wilhem. *Geschichte der deutschen Landwirtschaft*, 2nd ed. Stuttgart: E. Ulmer, 1967.

Asdal, Kristin. "On Politics and the Little Tools of Democracy." *Distinktion* 9 (2008): 11–26.

Campbell, Bruce M. S. *The Great Transformation: Climate, Disease and Society in the Late Medieval World.* New York: Cambridge University Press, 2016.

Ernst, Christoph. *Den Wald entwickeln: Ein Politik- und Konfliktfled in Hunsrück und Eifel im 18. Jahrhundert.* Munich: De Gruyter Oldenbourg, 2000.

Grüne, Niels, Jonas Hübner, and Gerhard Siegl, eds. *Ländliche Gemeingüter: Kollektive Ressourcennutzung in der europäischen Agrarwirtschaft / Rural Commons: Collective Use of Resources in the European Agrarian Economy.* Innsbruck: Studien Verlag, 2016.

Hölzl, Richard. *Umkämpfte Wälder: Die Gescichte einer ökologischen Reform in Deutschland 1760–1860.* Frankfurt a.M: Klostermann, 2010.

Johann, Elisabeth. *Geschichte der Waldnutzung in Kärnten unter dem Einfluß der Berg-, Hütten- und Hammerwerke.* Klagenfurt: Verlag des Geschichtsvereines für Kärnten, 1968.

Kidd, Charles V. "The Evolution of Sustainability." *Journal of Agricultural and Environmental Ethics* 5 (1992): 1–26.

Knoll, Martin. *Umwelt—Herrschaft—Gesellschaft: Die landesherrliche Jagd Kurbayerns im 18. Jahrhundert.* St. Katharinen: Scripta Mercaturae, 2004.

Ladurie, Emmanuel Le Roy. "History That Stands Still." In *The Mind and Method of the Historian*. Brighton: Harvester Press, 1981.

_____. *Les paysans de Languedoc*. Paris: Flammarion, 1969.

Malanima, Paolo. *Premodern European Economy: One Thousand Years (10th-19th centuries)*. Leiden: Brill, 2009.

Marsh, George Perkins. *Man and Nature: Or, Physical Geography as Modified by Human Action*. New York: Charles Scribner, 1864.

Postan, Michael M. *The Medieval Economy and Society: An Economic History of Britain in the Middle Ages*. London: Weidenfeld & Nicholson, 1972.

Schaal, Reinhold. *Waldzustände als Spiegel gesellschaftlicher Ansprüche: Die Entwicklung von Wäldern auf der mittleren Schwäbischen Alb und im nördlichen Oberschwaben seit dem 16. Jahrhundert*. Ostfildern: Jan Thorbecke Verlag, 2011.

Schäfer, Ingrid. *"Ein Gespenst geht um": Politik mit der Holznot in Lippe 1750–1850. Eine Regionalstudie zur Wald und Technikgeschichte*. Detmold: Naturwissenschaftlicher u. Historischer Verein f. d. Land Lippe, 1992.

Schenk, Winfried. *Waldnutzung, Waldzustand und regionale Entwicklung in vorindustrieller Zeit im mittleren Deutschland*. Stuttgart: Steiner, 1996.

Schmidt, Uwe E. *Der Wald in Deutschland im 18. und 19. Jahrhundert: Das Problem der Ressourcenknappheit dargestellt am Beispiel der Waldressourcenknappheit in Deutschland im 18. und 19. Jahrhundert—eine historisch-politische Analyse*. Saarbrücken: Conte, 2002.

Steinsiek, Peter M. *Nachhaltigkeit auf Zeit: Waldschutz im Westharz vor 1800*. Münster: Waxmann, 1999.

Warde, Paul. *The Invention of Sustainability: Nature and Destiny c. 1500–1870*. New York: Cambridge University Press, 2018.

White, Gilbert. *The Natural History of Selborne*. London: Penguin, 1977 [1789].

White, Lynn T., Jr. "The Historical Roots of Our Ecologic Crisis." *Science* 155 (1967): 1203–7.

Index

www.ingramcontent.com/pod-product-compliance
Ingram Content Group UK Ltd.
Pitfield, Milton Keynes, MK11 3LW, UK
UKHW021836100926
8556IPUK00004B/365

9 781789 206920